SOCIETY FOR EXPERIMENTAL BIOLOGY

SEMINAR SERIES · 18

PLANT BIOTECHNOLOGY

PLANT BIOTECHNOLOGY

Edited by

S. H. MANTELL

Research Associate in Botany, University of Leicester

and

H. SMITH

Professor of Botany, University of Leicester

CAMBRIDGE UNIVERSITY PRESS

Cambridge

London New York New Rochelle

Melbourne Sydney

Published the Press Syndicate of the University of Cambridge
The Pitt Building, Trumpington Street, Cambridge CB2 1RP
32 East 57th Street, New York, NY 10022, USA
296 Beaconsfield Parade, Middle Park, Melbourne 3206, Australia

First published 1983

Printed in Great Britain by the University Press, Cambridge

Library of Congress catalogue card number: 82-23534

British Library cataloguing in publication data

Plant biotechnology. – (Society for Experimental
Biology seminar series; 18)

1. Botany, Economic 2. Biochemical engineering
I. Mantell, S. H. II. Smith, H.
III. Series
581.6′4 SB107

ISBN 0 521 24550 8

UP

CONTENTS

Contributors vii *Preface* xi

Part I: Production of commercially useful compounds by plant cell cultures

Commercial applications and economic aspects of mass plant cell culture
M. W. Fowler 3

Novel experimental systems for studying the production of secondary metabolites by plant tissue cultures
K. Lindsey and M. M. Yeoman 39

Biotransformation of β-methyldigitoxin to β-methyldigoxin by cell cultures of *Digitalis lanata*
A. W. Alfermann, W. Bergmann, C. Figur, U. Helmbold, D. Schwantag, I. Schuller and E. Reinhard 67

Cultural factors that influence secondary metabolite accumulations in plant cell and tissue cultures
S. H. Mantell and H. Smith 75

Part II: Plant propagation by tissue culture

In vitro propagation of horticultural and agricultural crops
G. Hussey 111

In vitro propagation of tree crops
O. P. Jones 139

Part III: Germplasm maintenance and storage

Low-temperature preservation of living cells
E. James 163

Germplasm storage in plant biotechnology
L. A. Withers 187

In vitro approaches to the conservation and utilisation of global plant genetic resources
G. G. Henshaw and J. F. O'Hara 219

Part IV: In vitro approaches to the genetic manipulation of plants

Genetic transformation through somatic hybridisation
E. C. Cocking 241

Modification of agronomic traits using in vitro technology
S. Bright, V. Jarrett, R. Nelson, G. Creissen, A. Karp, J. Franklin, P. Norbury, J. Kueh, S. Rognes and B. Miflin 251

Part V: Genetic engineering of higher plants

The expression of eukaryotic genes in bacteria and its application to plant genes
A. A. Gatenby 269

The current status of plant viruses as potential DNA/RNA vector systems
R. Hull 299

Ti plasmid-mediated gene transfer to higher plant cells
J. Schröder, H. De Greve, J. P. Hernalsteens, J. Leemans, M. Van Montagu, L. Otten, G. Schröder, L. Willmitzer and J. Schell 313

Index 327

CONTRIBUTORS

Alfermann, A. W.
Pharmazeutisches Institut, Universität Tübingen, D-7400, Tübingen, Federal Republic of Germany.

Bergmann, W.
Pharmazeutisches Institut, Universität Tübingen, D-7400, Tübingen, Federal Republic of Germany.

Bright, S.
Department of Biochemistry, Rothamsted Experimental Station, Harpenden, Herts, AL5 2JQ, UK.

Cocking, E. C.
Department of Botany, School of Biological Sciences, University Park, Nottingham, NG7 2RD, UK.

Creissen, G.
Department of Biochemistry, Rothamsted Experimental Station, Harpenden, Herts. AL5 2JQ, UK.

De Greve, H.
Max-Planck-Institut für Züchtungsforschung, D-5000, Köln 30, Federal Republic of Germany.

Figur, C.
Pharmazeutisches Institut, Universität Tübingen, D-7400, Tübingen, Federal Republic of Germany.

Fowler, M. W.
Wolfson Institute of Biotechnology, University of Sheffield, Sheffield, S10 2TN, UK.

Franklin, J.
Department of Biochemistry, Rothamsted Experimental Station, Harpenden, Herts. AL5 2JQ, UK.

Gatenby, A. A.
Plant Breeding Institute, Maris Lane, Trumpington, Cambridge, CB2 2LQ, UK.

Helmbold, U.
Pharmazeutisches Institut, Universität Tübingen, D-7400, Tübingen, Federal Republic of Germany.

Henshaw, G. G.
School of Biological Sciences, University of Bath, Claverton Down, Bath, BA2 7AY, UK.

Hernalsteens, J. P.
Max-Planck-Institut für Züchtungsforschung, D-5000, Köln 30, Federal Republic of Germany.

Hull, R.
John Innes Institute, Colney Lane, Norwich, NR4 7UH, UK.

Hussey, G.
John Innes Institute, Colney Lane, Norwich, NR4 7UH, UK.

James, E.
London School of Hygiene and Tropical Medicine, Winches Farm Field Station, 395 Hatfield Road, St Albans, Herts, UK.

Jarrett, V.
Department of Biochemistry, Rothamsted Experimental Station, Harpenden, Herts, AL5 2JQ, UK.

Jones, O. P.
East Malling Research Station, East Malling, Maidstone, Kent, ME19 6BJ, UK.

Karp, A.
Department of Biochemistry, Rothamsted Experimental Station, Harpenden, Herts, AL5 2JQ, UK.

Kueh, J.
Department of Biochemistry, Rothamsted Experimental Station, Harpenden, Herts, AL5 2JQ, UK.

Leemans, J.
Max-Planck-Institut für Züchtungsforschung, D-5000, Köln 30, Federal Republic of Germany.

Lindsey, K.
Department of Botany, The King's Buildings, Mayfield Road, Edinburgh, EH9 3JH, UK.

Mantell, S. H.
Department of Botany, University of Leicester, University Road, Leicester, LE1 7RH, UK.

Miflin, B.
Department of Biochemistry, Rothamsted Experimental Station, Harpenden, Herts, AL5 2JQ, UK.

Nelson, R.
Department of Biochemistry, Rothamsted Experimental Station, Harpenden, Herts, AL5 2JQ, UK.

Norbury, P.
Department of Biochemistry, Rothamsted Experimental Station, Harpenden, Herts, AL5 2JQ, UK.

O'Hara, J. F.
School of Biological Sciences, University of Bath, Claverton Down, Bath, BA2 7AY, UK.

Otten, L.
Max-Planck-Institut für Züchtungsforschung, D-5000, Köln 30, Federal Republic of Germany.

Reinhard, E.
Pharmazeutisches Institut, Universität Tübingen, D-7400, Tübingen, Federal Republic of Germany.

Rognes, S.
Botanical Laboratory, University of Oslo, Norway.

Schell, J.
Max-Planck-Institut für Züchtungsforschung, D-5000, Köln 30, Federal Republic of Germany.

Schröder, G.
Max-Planck-Institut für Züchtungsforschung, D-5000, Köln 30, Federal Republic of Germany.

Schröder, J.
Max-Planck-Institut für Züchtungsforschung, D-5000, Köln 30, Federal Republic of Germany.

Schuller, I.
Pharmazeutisches Institut, Universität Tübingen, D-7400, Tübingen, Federal Republic of Germany.

Schwantag, D.
Pharmazeutisches Institut, Universität Tübingen, D-7400, Tübingen, Federal Republic of Germany.

Smith, H.
Department of Botany, University of Leicester, University Road, Leicester, LE1 7RH, UK.

Van Montagu, M.
Max-Planck-Institut für Züchtungsforschung, D-5000, Köln 30, Federal Republic of Germany.

Willmitzer, L.
Max-Planck-Institut für Züchtungsforschung, D-5000, Köln 30, Federal Republic of Germany.

Withers, L. A.
Department of Agriculture and Horticulture, School of Agriculture, Sutton Bonington, Loughborough, Leicestershire, LE12 5RD, UK.

Yeoman, M. M.
Department of Botany, The King's Buildings, Mayfield Road, Edinburgh, EH9 3JH, UK.

PREFACE

The last decade has seen dramatic advances in the biosciences, particularly in recombinant DNA technology, in vitro enzymology and the in vitro culture of plant and animal cells. The arrival of the microchip too has permitted a more widespread use of sophisticated instrumentation, opening up new horizons in cell and molecular biology as a result. The knowledge gained from these developments has established a foundation from which new areas of biotechnology can now spring.

This volume presents some of these developments in the context of plant biotechnology, which is the study of methods by which plant resources can be tailored to generate industrial processes and novel plant materials for use in agriculture, forestry and horticulture. As the different chapters show, the relative advancement of each branch of the field of plant biotechnology varies widely. For example, a great deal of what constitutes current plant biotechnology is based on the techniques of plant cell and protoplast culture that have evolved over a period of 20 years or more whereas the manipulation and engineering of specific nuclear genes and plastogenes of plants are some of the most recent and exciting developments in the field. We are particularly grateful to the respective groups at Nottingham University and the Max Planck, John Innes and Plant Breeding Institutes for making their contributions to this book on the latter areas of plant biotechnology (see the chapters by Cocking, Schröder *et al.*, Hull and Gatenby). There are no less important developments, however, in other areas, particularly in plant cell immobilisation (Lindsey & Yeoman) and in the use of these systems for industrial-scale biotransformations (Alfermann *et al.*). Cryopreservation (James), and the maintenance of plant germplasm on the laboratory scale (Withers) and on the global scale (Henshaw & O'Hara) are also essential ingredients to the new phase of plant biotechnology. Even the dilemma of recalcitrance in vitro of crop species is becoming less of a problem as research continues on agricultural and horticultural species (Hussey) and on trees (Jones). Furthermore, the application of plant cell fermentations to the production of certain valuable plant-derived compounds on an industrial scale appears to be closer at hand than ever before (Fowler).

We hope that one outcome of this volume will be an increased awareness of the fact that there is still a great need for strategic research in applied sciences like plant biotechnology. Before this new technology can realise its true potential to society, there is clearly a need for more fundamental knowledge. Current research sponsorship in this field should therefore be prepared to support both basic and applied spheres of study and not solely those forms of research which are assured of yielding money-spinning processes after short-term projects. The contents of this book also provide an indication of some of the directions in which plant biotechnology is likely to go in the coming years. At the very least, we feel that it should provide a source of background information and references to both students and researchers alike who wish to initiate or broaden their interests in the field. Unfortunately, the scope of this book does not allow for an in-depth treatment of plant enzymes and other proteins as these relate to plant biotechnology. It should not be forgotten that this is an area of significant potential.

It is noticeable that many of the techniques described in the five sections of this book tend to be complementary to each other, particularly where a combination of techniques may be required to achieve a desired objective. This strengthens a view held by us that plant biotechnology can rightly be considered a discrete area of research involving both the applied and fundamental aspects of plant cell biology.

We are grateful to those who contributed to this volume through their papers and poster presentations given at a seminar held during the 204th Meeting of the Society at the University of Leicester on 5–7 January 1982. We would especially like to thank the officers of the Society for their help in arranging the seminar, particularly Drs Bill Cockburn and Peter Shelton, who acted as local secretaries. We appreciate a donation of £100 which was made by Boots Co. Ltd (Industrial Division) to help to defray some of the costs of hosting the seminar.

May 1982

S. H. Mantell
H. Smith
Editors for the Society for Experimental Biology

PART I

Production of commercially useful compounds by plant cell cultures

M. W. FOWLER

Commercial applications and economic aspects of mass plant cell culture

Perspective

It is now some 80 years since Haberlandt first began experiments which could truly be described as cell culture in that he attempted to grow single cells isolated from a variety of plant tissues in simple nutrient solutions. Although Haberlandt was able to maintain his cells in these nutrient solutions, cell division was not observed until much later (Haberlandt, 1902). Since these early observations of Haberlandt, plant cell, tissue and organ culture have progressed a long way. Rapid developments in the 1950s and 1960s (Murashige, 1978) converted tissue and organ culture from an area primarily of academic interest into a major tool for the horticultural industry through the mass propagation of house plants and the development of unique lines with desirable characteristics. The horizon rapidly widened from horticulture to agriculture to cover important food and cash crops such as citrus fruits and jojoba. Following behind these developments has been a growing interest in other applications, not the least of which has been the possibility of using cell culture systems for the synthesis of a variety of natural products, which, while perhaps not having the same 'volume' impact as tissue culture in horticulture and agriculture, could have a significant impact on various sectors of the wider chemical industry.

For centuries the plant kingdom has been a major chemicals resource, particularly for pharmaceuticals and food additives. Recently there has been increased interest in plants as a source of chemicals for a wide variety of uses. Parallel advances made in cell culture technology, particularly since the early 1970s, have brought us to a point where we are perhaps able to provide answers to some, but by no means all, of the specific questions raised under the general question 'is there a potential in plant cell culture as an enabling technology for the further exploitation of the plant kingdom as a chemicals resource?' An attempt has been made in this review to provide a backcloth against which such judgements can perhaps be made, and also to pin-point both areas of constraint against which progress needs to be made and areas of potential that could be developed, if the objective of plant cell culture technology as a form of 'living factory' is to be achieved.

Before entering into details we should perhaps summarise the main reasons often quoted for looking upon plant cell culture as an alternative route to natural product synthesis. These are:

(a) independence from various environmental factors, including climate, pests, geographical and seasonal constraints;
(b) defined production systems, with production as and when required, and at the amount required, hence giving a close control over market supply;
(c) more consistent product quality and yield;
(d) reduction in land use for 'cash crops';
(e) freedom from political interference.

There may also be advantages in the downstream processing and recovery of the desired product as compared with conventional processes. At the end of the day, however, the key criterion has to be one of price or cost advantage as compared with conventional technology. Strategic and social considerations, except in certain exceptional cases, rank lower in importance. Even when considering price, caution needs to be exercised. Industry will only move to a cell culture route for synthesis if a cost advantage as opposed to a cost comparability may be perceived. Mass cell culture technology is initially capital intensive; to persuade industry to go down this road and to abandon, albeit gradually, their present investment in plantations and process equipment will not be easy. Such commercial considerations do not mean to say, however, that high-quality science of a fundamental nature is not a requisite for the development of cell culture technology. Two points should be made; firstly there are already indications that the development of cell culture technology is being held up because of a lack of fundamental knowledge; secondly, the rigorous criteria we need to apply to process development should lead us to apply more objectivity and quantitation to our experimental work. Good fundamental research properly targeted is the key to successful application in this and related areas of biotechnology.

The plant kingdom as a commercial resource

The properties and potentials of the plant kingdom are fundamental to the well-being of mankind. Plants contribute to a wide range of industries, including those of agriculture, food, construction, chemicals, fabrics and paper, and of course energy. Some of these industries are vital to our needs, others are peripheral. The chemical industry is a microcosm of the wider industrial plant-supported scene. It embraces pharmaceuticals, for which certain plant products are presently irreplaceable because of their therapeutic actions, alongside others like foods, and cosmetics, for which plant products are used merely to titillate the palate and enhance the appearance respectively.

Table 1. *Natural products from plants and their associated industries*

Industry	Plant product	Plant species	Industrial uses
Pharmaceuticals	Codeine (alkaloid)	*Papaver somniferum*	Analgesic
	Diosgenin (steroid)	*Dioscorea deltoidea*	Anti-fertility agents
	Quinine (alkaloid)	*Cinchona ledgeriana*	Antimalarial
	Digoxin (cardiac glycoside)	*Digitalis lanata*	Cardiatonic
	Scopolamine (alkaloid)	*Datura stramonium*	Antihypertensive
	Vincristine (alkaloid)	*Catharanthus roseus*	Antileukaemic
Agrochemicals	Pyrethrin	*Chrysanthemum cinerariaefolium*	Insecticide
Food and drink	Quinine (alkaloid)	*Cinchona ledgeriana*	Bittering agent
	Thaumatin (chalcone)	*Thaumatococcus danielli*	Non-nutritive sweetener
Cosmetics	Jasmine	*Jasminum* sp.	Perfume

The potentials for chemical synthesis within the plant kingdom are vast; something of the order of 2×10^4 structures from plants are known. These are currently being added to at a rate of about 1600 each year. It is interesting to note, however, that different substances are often confined in their synthesis to specific plant groups or families. Some indication of the industries involved, their key products and the range of chemical structures utilised may be gained from Table 1. Products range from substances of low molecular weight, e.g. sugar alcohols, to large polymers. Market volumes and costs also vary tremendously; for instance from some perfumes with a value of US \$6000 kg^{-1} and a market volume of a few tens of kilograms a year, to products like tobacco with a value of US \$6.00 kg^{-1}, and a market volume in excess of 40000 tonnes.

In general plant chemical products come under the heading of secondary metabolites, a term which has come in for much discussion over the last few years. The relatively loose definition of secondary metabolites as 'constituents of cells not essential for their survival' will be followed in this review. This excludes substances such as simple sugars and their derivatives as well as the protein amino-acids.

Plant chemical products range from highly pure compounds, such as many pharmaceuticals, via complex mixtures and blends, e.g. some food additives and cosmetics, to complex bulk materials such as cocoa butter fat and tobacco. The chemical complexity of many plant products has long been a source of concern to the chemical industry. In the case of single constituent products, attempts have been made to produce many of them through chemical synthesis, with varying degrees of success. Such an approach has often been constrained by low yields, high cost, difficult chemical conversions or the need for a high degree of purity of a particular isomer from a complex mixture. In the majority of cases the plant has proved to be the most effective means of synthesis. Let us briefly survey the key industrial chemical sectors to which plants contribute.

Medicinals

Medicinals constitute a major area of plant products, but in terms of high monetary value rather than volume. At the turn of the nineteenth century plant extracts and potions constituted the major part of the armoury of the general practitioner against disease and sickness. The discovery and development of the synthetic drugs and microbial agents during the period 1920–50 led to a marked reduction in the use of plant medicinals. Nonetheless, medicinal plants still make a major contribution to the pharmaceutical industry, comprising some 25% of prescribed drugs, and providing a market yield of billions of dollars worldwide. Table 2 lists the top 10 plant medicinals with their origin and clinical action. Note that these structures cover not only a great variety of chemical forms but also have a wide range of therapeutic or related activity, ranging from antihypertensives to analgesics and anti-fertility agents.

At the periphery of medicinals come the narcotics and stimulants. The narcotics are to a degree midway between the poisons and medicinal agents. At low and closely monitored concentrations they are often extremely efficacious, e.g. morphine. At high concentrations, or in continual application, they can become addictive, or even potential killers. In this guise they have come to be a major problem to law enforcement agencies. They represent a substantial, if illicit, group of natural products on the world economic scene. The most familiar ones are of course marijuana (or hash) from *Cannabis*, opium and heroin from *Papaver somniferum* and cocaine from *Erythroxylon*. Another plant which contributes to this group to a certain degree is tobacco, through the presence of nicotine. Nicotine is also regarded in some cases as a stimulant and leads us to the next section.

Stimulants represent a somewhat different area to narcotics and are not generally harmful. Most are based on the use of caffeine or the related base theobromine and are consumed in the form of beverages. Caffeine is found in a surprising range of plants worldwide, the two familiar ones being *Camellia sinensis* (tea) and *Coffea arabica* (coffee).

Poisons – perhaps the ultimate medicinal (!) – have a long history of association between the plant kingdom and man. Many have gained a certain notoriety; Socrates died of hemlock poisoning, and the Borgia family of Italy achieved positions of power and influence through their malevolent use of a whole range of plant poisons. At a less negative level, neurotoxic plant poisons are still used today by many tribes in Africa and South America for

Table 2. *The ten most prescribed medicinals from plant sources*

Medicinal agent	Activity	Plant source
Steroids from diosgenin	Anti-fertility agents	*Dioscorea deltoidea*
Codeine	Analgesic	*Papaver somniferum*
Atropine	Anticholinergic	*Atropa belladonna* L.
Reserpine	Antihypertensive	*Rauwolfia serpentina* L.
Hyoscyamine	Anticholinergic	*Hyoscyamus niger* L.
Digoxin	Cardiatonic	*Digitalis lanata* L.
Scopolamine	Anticholinergic	*Datura metel* L.
Digitoxin	Cardiovascular	*Digitalis purpurea* L.
Pilocarpine	Cholinergic	*Pilocarpus jabonandi*
Quinidine	Antimalarial	*Cinchona ledgeriana*

hunting animals for food, curare being a good example of such a poison. Many plant poisons are potent neurotoxins, e.g. ricin from castor bean. All plant poisons are extremely active at very low concentrations. Undoubtedly there is potential here for the development of biocides.

Agrochemicals

Apart from plant growth regulators, the most dramatic discovery in the agrochemicals sector in recent years has been that of the pyrethroids. Extracted from the flowers of *Chrysanthemum cinerariaefolium*, the pyrethroids are extremely potent insecticides. The natural pyrethroids are under competition from synthetic pyrethroids, but certain questions are beginning to be raised about the latter, particularly because of signs of insect resistance and cumulative toxicity.

Fine chemicals

'Fine chemicals' is an all-embracing title covering chemicals such as perfumes, flavours, aromas, colourants and food materials. Fine chemicals range from products of very high cost and low market volume to bulk products of low value. For example the perfume jasmine costs in the region of US \$6000 kg^{-1} with a market volume of perhaps only 20–30 kg p.a.; cocoa butter fat, a basic constituent of chocolate, costs about US \$4000 $tonne^{-1}$ with a market volume of some 20000 tonnes. Products range from single components such as quinine to complex mixtures such as the essential oils. These latter are typically monoterpenic, often volatile compounds which form the basis of the very extensivc aroma industry, an industry of very high added-value products but very susceptible to the whim of the consumer.

Against this background, it is worth noting the increased interest of industry in the plant kingdom as a chemicals resource. Given that it now costs between US \$50 and \$100m to develop a new synthetic drug, over a period of perhaps 10–15 years and having had to screen at least 10000 different chemicals, it is not difficult to understand the renewed interest in plants as synthetic factories.

Cell culture as a source of plant products

Cell cultures may contribute in at least four major ways to the production of plant natural products. These are:

(a) as a new route of synthesis to established products, e.g. codeine, quinine, pyrethroids;

(b) as a route of synthesis to a novel product from plants difficult to grow or establish, e.g. thebaine from *Papaver bracteatum*;

(c) as a source of novel chemicals in their own right, e.g. rutacultin, from cultures of *Ruta*;
(d) as biotransformation systems, either on their own, or as part of a larger chemical process, e.g. digoxin synthesis.

Products from cell cultures

The earliest detailed reference to plant cell culture as an industrial route to natural product synthesis is probably the patent application of Routier & Nickell in 1956. Despite the promise of this and related work (Nickell, 1980) progress in the following decade was slow. In many cases the desired products were either not produced in culture, or where they were, the yields were extremely low. Additionally it became almost a dogma that secondary product synthesis was unlikely unless there was advanced tissue or even organ formation in the culture. The regeneration from cultures of plants which produced both qualitatively and quantitatively the same spectrum of products as the plant from which the culture was initiated did, however, serve to show that the ability for product synthesis was not lost during culture but merely, for some unknown reason, not expressed (see Dougall 1979*a*). This period in the development of plant cell and tissue culture with its various problems has been more than adequately reviewed by Dougall (1972), Constabel, Gamborg, Kurz & Steck (1974), Butcher (1977), Staba (1977) and Street (1977) and will not be further gone into here. In spite of the generally low level of product yield observed during the period 1950 to the mid-1970s, the odd exception did occur; for example, diosgenin (Kaul, Stohs & Staba, 1969; Khanna & Mohan, 1973), ginseng saponins (Furuya & Ishii, 1972), harmin (Reinhard, Corduan & Volks, 1968) and visnagin (Teuscher, 1973), were all demonstrated in cell cultures at levels approaching or even exceeding those of the parent plant. There is one point, however, recently discussed by Dougall (1979*b*), which should be noted here. Comparisons between the yields from the higher plant and those from cell cultures, while being convenient, are artificial; the values quoted for the plant are typically from those parts which accumulate the highest amount of the desired product and from which it is usually harvested, e.g. those for the morphine alkaloids from the capsule of the opium poppy. If this yield is averaged across the whole plant, then quite obviously greatly reduced yields result. Comparisons with cell cultures must therefore be made with this in mind.

For reasons largely unknown, though probably reflecting our steadily improving knowledge of the biochemistry and physiology of cell cultures, the years 1973 and 1974 seem to have marked something of a turning-point in cell culture technology. Around that time a major increase occurred in the number of reports of cell cultures which gave reasonable yields of specific and desired secondary metabolites. These reports have been steadily increasing over the last five to six years (Table 3). In addition the range of substances observed in cultures of all types (i.e. callus, organ and tissue) has widened dramatically (Zenk, 1978; Nickell, 1980; Fowler, 1981; Table 4). Two points should be made regarding the information contained in Table 4. Firstly, the range of chemical structures involved again demonstrates the tremendous synthetic potential and versatility of the plant kingdom. Secondly, it should be stressed that only a relatively small number of the products listed are suitable targets in economic terms for commercialisation (see Economic considerations, p. 30).

Of particular note and encouragement has been the increasing number of reports of cell suspension cultures that are capable of synthesising specific products to levels equivalent to, or higher than, the plant from which they were derived (see Table 3). This is particularly important in terms of process development and scale-up (see Mass growth and production systems, p. 19). Cell suspensions are more amenable to scale-up from a biochemical engineering standpoint, requiring relatively simple bioreactors as compared with more organised tissue systems. This does not mean to say, of course, that large-scale

Table 3. *Natural product yields from cell cultures and whole plants.* (*For a more complete tabulation, see Dougall 1979a*)

		Yield	
Natural product	Species	Cell culture	Whole plant
Anthraquinones	*Morinda citrifolia*	900 nmol g^{-1} dry wt	Root, 110 nmol g^{-1} dry wt
Anthraquinones	*Cassia tora*	0.334% fresh wt	0.209% seed, dry wt
Ajmalicine and serpentine	*Catharanthus roseus*	1.3% dry wt	0.26% dry wt
Diosgenin	*Dioscorea deltoidea*	26 mg g^{-1} dry wt	20 mg g^{-1} dry wt tuber
Ginseng saponins	*Panax ginseng*	0.38% fresh wt	0.3–3.3% fresh wt
Nicotine	*Nicotiana tabacum*	3.4% dry wt	2–5% dry wt
Thebaine	*Papaver bracteatum*	130 mg g^{-1} dry wt	1400 mg g^{-1} dry wt leaf and 3000 mg g^{-1} dry wt root
Ubiquinone	*Nicotiana tabacum*	0.5 mg g^{-1} dry wt	16 mg g^{-1} dry wt leaf

operations cannot be designed for tissue systems. Indeed the greater the variety of process technology available for large-scale plant cell and tissue culture, the more chance there is of developing an industrial process.

Although there are signs that in an increasing number of cases it is possible to uncouple secondary metabolite synthesis from morphological development, this does not mean that the productive cells in the cell suspension do not show some degree of differentiation. Indeed, work in our own Institute with cell suspensions of *P. somniferum* (Morris & Fowler, 1980) indicates that the cultures are most productive in opiate alkaloids when a high number of specialised lactiferous cells are present (Fig. 1). Similar observations havc been made by other workers. In contrast, however, high alkaloid-yielding, rapidly growing cell suspensions of *Catharanthus roseus* show relatively little differentiation (Fig. 2), a feature observed with some other culture systems, including high nicotine-yielding tobacco cells (S. H. Mantell; personal communication, 1981) and carrot cells containing anthocyanin (Kinnersley & Dougall, 1980).

Table 4. *Substances reported from plant cell cultures*

Alkaloids	Latex
Allergens	Lipids
Anthraquinones	
Antileukaemic agents	Naphthoquinones
Antitumour agents	Nucleic acids
Antiviral agents	Nucleotides
Aromas	
	Oils
Benzoquinones	Opiates
	Organic acids
Carbohydrates (including polysaccharides)	
Cardiac glycosides	Peptides
Chalcones	Perfumes
	Phenols
Dianthrones	Pigments
	Plant growth regulators
Enzymes	Proteins
Enzyme inhibitors	
	Steroids and derivatives
Flavanoids, flavones	Sugars
Flavours (including sweeteners)	
Furanocoumarins	Tannins
Hormones	Terpenes and terpenoids
Insecticides	Vitamins

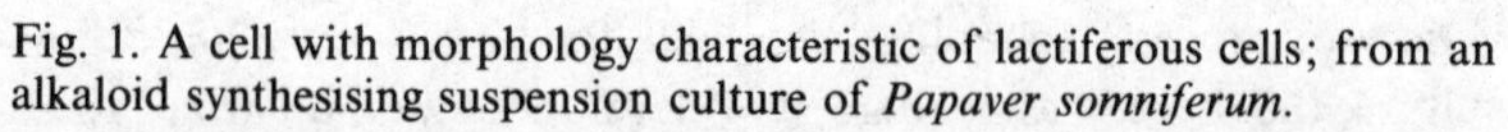

Fig. 1. A cell with morphology characteristic of lactiferous cells; from an alkaloid synthesising suspension culture of *Papaver somniferum*.

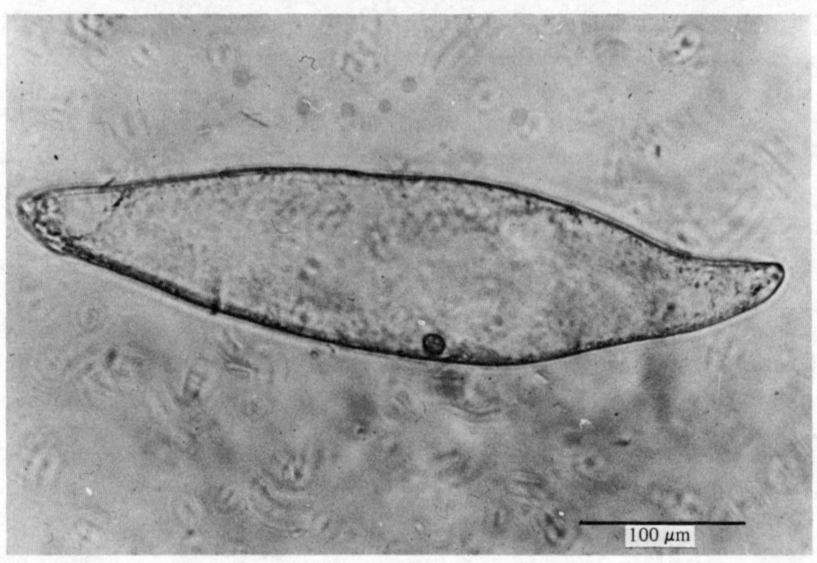

Fig. 2. Cells from a high alkaloid-yielding suspension culture of *Catharanthus roseus*.

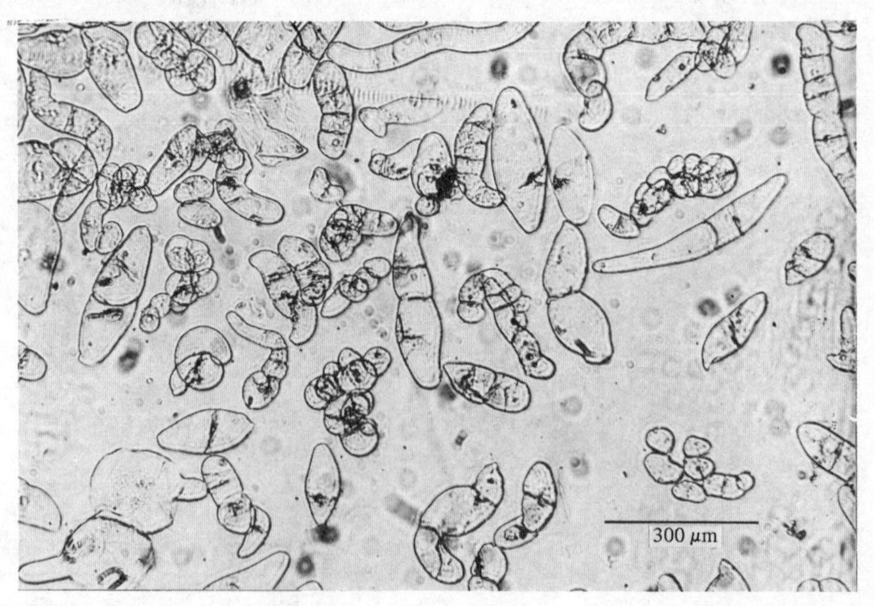

The above comment concerning *C. roseus* cultures raises a further point. In many systems accumulation of secondary metabolites does not become significant until late into exponential growth or the stationary phase of batch culture (Street, 1977). A good example is the accumulation of phenolics in cultures of *Acer pseudoplatanus* (Westcott & Henshaw, 1976). In contrast studies by Zenk *et al.* (1977) and work in our laboratories (A. Stafford & M. W. Fowler; unpublished observations, 1982) with *C. roseus* cultures has shown that instances do occur of product synthesis accompanying growth. In this case synthesis of catharanthus alkaloids (e.g. serpentine) parallels growth as defined by increasing dry weight accumulation (Fig. 3). This is by no means an isolated example (see for instance Zenk, 1978; Wilson, 1980).

From this variety of situations in which secondary product synthesis may or may not occur, it is difficult to see a unifying theme which may be applied as a general strategy for the development of high-yielding cell lines. Certainly several factors have been identified which affect secondary product synthesis. Strategies which take into account some of these factors are outlined below since they have in certain cases provided useful results.

Factors affecting product yield

Progress in selecting high-yielding cell lines has been constrained by a number of factors, not the least of which has been the slow growth of cell cultures, the difficulty of isolating true single-cell clones and the availability

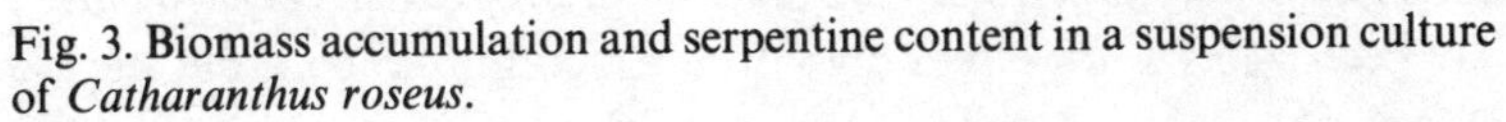

Fig. 3. Biomass accumulation and serpentine content in a suspension culture of *Catharanthus roseus.*

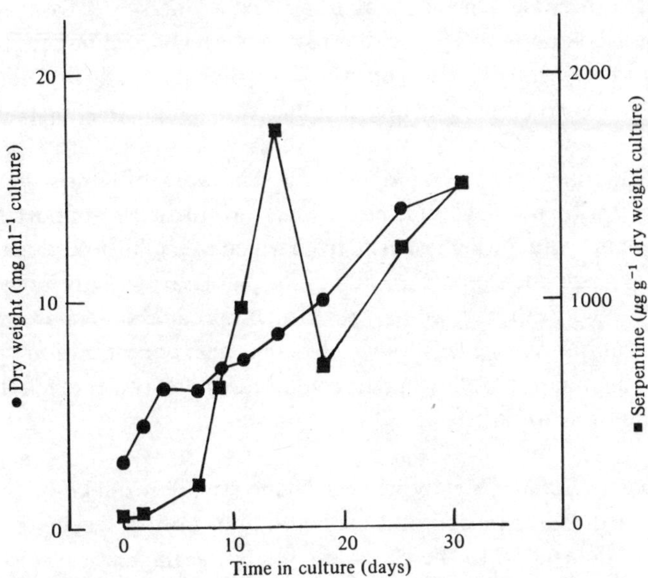

of rapid screening methods. In addition, knowledge of the physiology and biochemistry of cultures, particularly in relation to natural product synthesis is seriously lacking. Some of these points are considered in the following sections and in subsequent chapters of this volume.

Tissue origin – genetic characteristics. Plant cells are generally considered to be totipotent. It should be possible therefore, given the appropriate environmental conditions, to induce any cell to produce any substance that is characteristic of the parent plant. Traditionally the approach has been to initiate cultures from high-yielding plants in the hope that this characteristic will be transferred to the cultures. That high-yielding plants will give high-yielding cultures and that low-yielding cells from low-yielding plants will not give a higher yield has been a major point of controversy for some years. From an extensive study of alkaloid production by plants and cell cultures of *C. roseus*, Zenk *et al.* (1977) found that in general cultures established from high-yielding parent plants tended themselves to have a high alkaloid content in comparison with cultures established from low-yielding plants. In contrast Röller (1978), working with the same plant, *C. roseus*, found no correlation between serpentine production by callus cultures and the plants from which the cultures were derived. To try and gain more definitive information on this type of problem Kinnersley & Dougall (1981) have carried out experiments in which they studied nicotine synthesis in callus cultures that were established from two different pairs of *Nicotiana tabacum* plants. The two pairs of plants had been selected for either high or low nicotine content, but the individuals of each pair were otherwise isogenic for all other loci. Kinnersley & Dougall's data unambiguously show that high-yielding plants give high-yielding cultures. At no time did they observe a situation where a callus from a low-yielding plant gave a higher yield of nicotine than a callus from a high-yielding plant. In other words the nicotine yield in the cultures was in this case clearly genotypically determined from the parent plant. The work of Kinnersley & Dougall together with that of Zenk *et al.* therefore provides support for selecting those plants with the highest yield from which to establish cultures. It does not, however, provide support for taking the plant part with the highest concentration of the desired product. Such an approach suffers from a number of reservations not the least of which is that high concentration may reflect tissue accumulation following directed transport rather than the location of a site of synthesis.

Culture conditions – chemical. Steady progress has been made in our knowledge of culture and media preparation and media composition in recent years. There are two basic aspects to these studies: firstly the influence of media composition on biomass, i.e. the amount of potential synthetic machinery formed, and secondly its influence on secondary product synthesis. In those situations where biomass production is uncoupled from secondary product synthesis (e.g. phenol synthesis (Westcott & Henshaw, 1976)), a careful balance has to be achieved between producing sufficient biomass and not reducing the secondary product yield too much. Where secondary product synthesis accompanies biomass increase, then a key aspect is to achieve as high a contribution of substrate to the desired product as possible relative to the biomass.

Of fundamental importance to biomass production is the nature and quantity of the supplied carbohydrate. To date there have been few successful reports of photoautotrophic growth, and even where it has been observed, growth rates are very low. Consequently the vast majority of cultures are grown heterotrophically, typically with sucrose as the carbon source. However, during the last few years a wide range of carbon sources have been investigated for their ability to support growth and product synthesis (Maretzki, Thom & Nickell, 1974; Fowler, 1978; 1982*b*). These sources range from glucose, fructose, galactose and related mono-and disaccharides to complex non-refined sources such as molasses, milk whey and potato and cereal starch (Fowler 1982*b*). With our own cell lines, covering a variety of species including *C. roseus*, *N. tabacum*, *P. somniferum* and *Cinchona ledgeriana*, there is some indication that glucose may be the most effective carbon source from three standpoints: high growth rate, carbon conversions in excess of 60% and biomass yield. Specific growth rates in excess of 0·2 day^{-1} are regularly achieved from *C. roseus* cells using glucose, with final biomass yields of the order of 30 g dry weight l^{-1}. (We have observed higher biomass yields with other cell lines, occasionally in excess of 50 g dry weight l^{-1}. At these levels mixing of biomass and nutrient becomes a severe problem). Growth using some of the less familiar carbon sources such as galactose and lactose, as well as the non-refined sources, often requires a 'conditioning' period. This may be as short as the time taken for two or three passages after transfer from glucose or sucrose, or typically may take much longer, perhaps 12–15 passages over six months to a year, growth and cell division occurring at a very low rate during the interim period (I. Lyons & M. W. Fowler; unpublished observations, 1981).

Nitrogen supply has also been shown to have an effect on biomass yield, but perhaps more importantly there appears to be a close interrelation between nitrogen loading and growth rate. Noguchi *et al.* (1977) were able to effect a substantial increase in growth rate in tobacco cell cultures with an increased nitrogen supply. Similar effects were also noted with phosphate.

To turn next to product yield, it is interesting to note that the concentration of the carbon source may also influence this directly. Zenk *et al.* (1977) have observed substantial increases in alkaloid yield with *C. roseus* cell cultures when the carbon source (sucrose in this case) was increased above 3% (w/c). Phosphate levels have also been found to have important effects on product yields. Dougall and Weyrauch (1980) have been able to control anthrocyanin accumulation in carrot suspension cultures by using phosphate as the limiting nutrient. Similar observations have been made by Wilson (1980) on anthraquinone biosynthesis by *Galium molugo* cells.

Perhaps the most investigated components of the nutrient medium for their effect on product synthesis are the various plant growth regulators (see Zenk *et al.*, 1977). Variation in the concentrations or composition of some of these may often have spectacular effects, e.g. when naphthaleneacetic acid (NAA) was substituted for 2,4-dichlorophenoxyacetic acid (2,4-D) (2 mg l^{-1}) in suspension cultures of *Morinda citrifolia*, there was a thirty fold increase in

Table 5. *Effect of hormone variation on alkaloid production in cell suspension cultures of* Catharanthus roseus

Medium code		Maximum alkaloid yield (% dry wt)		Numbers of other alkaloids detected by TLC+HPLC
Growth	Production[a]	Serpentine	Ajmalicine	
B5	B5	0.00	0.00	0
B5	B5-H	0.08	0.06	1
B5	IB_5	0.33	0.08	10
B5	NI_{20}	0.04	0.02	2
B5	Z	0.30	0.45	2
M_3-CM	M_3-CM	0.58	0.14	14
NB5	NB5	0.24	0.06	12
M_3-CM	Z	1.00	0.04	9
NB5	Z	0.42	0.32	5

[a] Note that many of the production media to which the cultures are transferred after the growth media do not support growth and the cultures die within 1 or 2 subcultures.

Medium supplements

B5; 1.0 mg l^{-1} 2,4-D, 0.1 mg l^{-1} kinetin
M_3-CM; M&S with 1.0 mg l^{-1} NAA, 0.1 mg l^{-1} kinetin
IB5; B5 with 1.0 mg l^{-1} IAA, 0.1 mg l^{-1} kinetin
NB5; B5 with 1.0 mg l^{-1} NAA, 0.1 mg l^{-1} kinetin
B5-H; B5 no hormones
NI_{20}; modified M&S, 20 mg l^{-1} IAA, 0.2 mg l^{-1} kinetin
} Including 2% sucrose
Z; modified M&S, 0.125 mg l^{-1} IAA, 1.125 mg l^{-1} with 6BAP and 5% sucrose
M&S, Murashige & Skoog (1962) basal medium; 6BAP, 6-benzylaminopurine.
(Previously unpublished data of Dr P. Morris, 1981.)

anthraquinone production (Zenk, El-Shagi & Schulte, 1975). In slight contrast Furuya, Kojima & Syono (1971) observed the synthesis of nicotine by tobacco cell cultures in the presence of indole-3-acetic acid (IAA) at 1 mg l^{-1}, but not with 2,4-D at the same concentration. Different regulators and levels of regulators may also cause major qualitative changes. Table 5 illustrates the variation in yield of serpentine and ajmalicine, as well as the number of alkaloids observed by high pressure liquid chromatography (HPLC) and thin layer chromatography (TLC) in a cell line (C11C) of *C. roseus* supplied with different nutrients and growth regulators.

Optimisation of the nutrient regime is obviously a key factor in enhancing product yield. Observations of the effect of nutrients on product yield have led to the development of so-called 'production media', perhaps the most well known being the production medium of Zenk *et al.* (1977). These media tend to be characterised by low growth and little or no cell division, but enhanced product yield.

In addition to the factors mentioned above, there are, scattered about the literature, references to many other factors which may have some effect on product yield. Readers are referred to Dougall (1979*a*) for a more complete reference list. These other factors include oxygen, carbon dioxide, ethylene, pH, amino acids and other organic supplements and are further discussed by Mantell & Smith (this volume).

Culture conditions – physical. Physical manipulations of the culture environment in order to effect an increase in product yield have yet to receive a systematic investigation. That there are undoubtedly effects of light, temperature, pH, etc. is fully supported by various observations in the literature. This general area has recently been reviewed by Seibert & Kadkade (1980) and Martin (1980).

Selection and screening. The problems associated with isolating single-cell clones from plant cell cultures make strain selection more difficult than with microbial systems. A further problem has been the development of screening systems that are able to detect the very small amount of the desired product that is present in single cells or small populations of cells.

That certain cells in a population synthesise a greater amount of the desired substance than others is well established (see for example Tabata & Hiraoka, 1976; Zenk *et al.*, 1977). Where a pigment is the desired product, microscopic, and in some cases visual, inspection reveals intense colouring of single or small groups of cells. A number of workers have been able to isolate 'clones' from these high pigment areas and develop cell lines in which the cell population is almost totally pigmented and apparently stable. (See for example Kinnersley

& Dougall, 1980; Colijn, Johnsson, Schramm & Kool, 1981.) Similar results have been obtained in our own laboratory in callus, and lately suspension, cultures of *C. roseus* with a deep red pigment (I. Lyons & M. W. Fowler; unpublished results, 1981).

The apparent instability of many cell lines in terms of product yield has been a cause of concern in maintaining high-yielding cultures. There is, however, a positive side to this situation in that, as well as producing low-yielding variants it may also bring forward yet higher-yielding cells, which may then themselves be selected out.

The application of mutagenic techniques as a selection pressure to develop high-yielding lines is one area which seems to raise particular emotion in cell culture circles; the response ranges from that of those who believe that tens of mutants have been induced using classical chemical and physical techniques to that of others who claim that no more than one or two cases have actually been substantiated. This is a vast area of controversy, too big for the present review. Readers are referred to Widholm (1980) for a recent discussion.

Screening has been greatly enhanced since the application of radioimmunoassay (RIA) techniques to the catharanthus alkaloids by Zenk and co-workers in the mid-1970s (Zenk *et al.* 1977). This system has now been applied to a variety of cultures and products with much success. A further development has come through the use of enzyme immunoassay systems (ELISA), which are currently under study in a number of laboratories. These have the advantage of speed and the possibility of autoanalysis when compared with RIA. The need to resort to such sophisticated screening systems emphasises the problems of dealing with very low concentrations of metabolites in slow-growing cells.

Biochemical manipulation. Precursor feeding experiments designed to enhance product synthesis have not been as widely attempted as might have been anticipated. Where this approach has been used, the results have been variable. Perhaps the most successful examples are those of Yeoman, Miedzybrodzka, Lindsey & Milauchlan (1980), who observed the conversion of supplied phenylalanine to capsicum by cultures of *Capsicum frutescens*, and Boulanger, Bailey & Steck (1973), who observed large increases in quinoline alkaloids in *Ruta graveolens* cultures on the addition of 4-hydroxy-2-quinoline. A major stumbling block in attempts at 'diagnostic biochemical manipulation' is again our comparative lack of knowledge of the biosynthetic pathways for many of the secondary products. The major exceptions are the flavonoid pathways (Hahlbrock, Schröder & Vieregge, 1980) and to a lesser extent the synthesis of the catharanthus alkaloids (see for example Stöckigt, 1979).

Novel products

Perhaps one of the most exciting aspects of cell culture technology is the potential for producing novel structures not observed in the parent plant. As noted above (p. 10), cell culture is a form of selection pressure in itself and we might anticipate that cells in culture may be channelled into biochemical activities which date from their evolutionary history. Plant cells possess large amounts of DNA, far more than is probably required for their 'normal' functioning. Does this DNA contain 'silent genes', genes which are structural in nature, and which may code for enzyme proteins able to catalyse a whole range of chemical reactions not yet elucidated (Fowler, 1981)? Synthesis of novel products by cell cultures is already well established, e.g. rutacultin by cultures of *R. graveolens* and sesquiterpene lactones by cultures of *Andrographis paniculata* (for review see Nickell, 1980). The major problem with this approach from a research and development standpoint is the cost of chemical, clinical and toxicological screening for and of the novel product. Undoubtedly though there is great potential here (Misawa, 1977).

Biotransformation

Biotransformation is another means through which the synthetic ability of the plant cell may be channelled to good effect in industrial processes. The range of potential transformations using either plant cells or isolated enzymes is large, and includes hydrolyses, hydrogenations, hydroxylations, oxidations, reductions and so on. A number of biotransformations have been demonstrated, the most advanced towards industrial application being the hydroxylation of β-methyldigitoxin to β-methyldigoxin. This biotransformation has been developed by Reinhard and Alfermann in Tübingen and is reviewed later in this volume (see also Reinhard & Alfermann, 1980). The main restriction on the development of biotransformation systems is again, as noted above (Products from cell cultures), our lack of fundamental knowledge of biosynthetic pathways and their associated enzyme systems.

Mass growth and production systems

Although it is extremely doubtful that plant cell cultures will ever compete with microbial systems as a biomass source, except in certain highly specialised cases like tobacco, any process for chemical synthesis has to go through an initial phase of biomass production purely to obtain a sufficiency of material with synthetic capability. That this phase should be rapid and occur at as low a cost as possible is axiomatic. The financial return made on the end product has to cover not only the period of time when it is being synthesised but also the initial grow-up period. If this initial period has to cover three or four seed stages, for instance

1 l → 10 l → 100 l → 1000 l → 10000 l at, say, 10-day intervals, then that 40-day period can account for a significant part of the total cost. In this section, then, we are concerned not just with approaches to product synthesis but also with the need for optimisation of biomass production as part of the total process.

Factors affecting mass growth

Before we consider such aspects as vessel size and configuration, we need to take account of those fundamental properties of plant cells which affect our approach to mass growth systems. Plant cells are of the order of ten to one hundred times larger than bacteria or fungal cells, being generally between 20 μm and 150 μm in diameter. They are more dense and exhibit a wide variety of shapes and sizes in, and during, culture. Cell volume during culture may vary by as much as 10^5, cells being typically small and densely cytoplasmic at the beginning of log growth and large with peripheral cytoplasm during the stationary phase. A large vacuole in older cells poses particular problems in its own right in relation to both turgor pressure and physical handling. In addition it often serves as a repository for substances normally toxic to the cell. Severe handling may rupture the tonoplast membrane thus releasing the vacuole contents, which may then result in cell death.

Few plant cell suspensions are composed of free cells. They usually contain cell clumps which vary in size from two cells up to as many as 200. These clumps arise in two major ways. Firstly, they may result from incomplete separation following cell division; this is particularly the case in early log growth. Secondly, they may arise as a result of cell surfaces becoming 'sticky' so that the cells aggregate together, a condition frequently observed in late log growth and in cultures excreting polysaccharide into the external medium. Several methods of reducing cell clumping have been tested with varying degrees of success (see Fowler, 1982*b*).

One of the most crucial features of the plant cell from the point of view of mass growth is the cellulose cell wall. The cell wall has a high tensile strength but low shear resistance. This much reduces the use of conventional microbial vessels which are usually designed to develop high levels of shear to aid gas transfer.

Plant cells, although much larger than microbial cells, generally have a much lower physiological and metabolic activity. Doubling times are of the order of 25–100 h (Noguchi *et al.*, 1977) and respiratory rates are in the region of 1 $\mu mol\ O_2\ h^{-1}\ 10^6\ cells^{-1}$. The low respiratory rate in turn results in low oxygen demand which is an important factor in vessel design. Unfortunately, recent studies (Smart & Fowler, 1981; Fowler, 1982*a*) have shown this to be a complicated area much in need of study.

A whole variety of vessels in terms of shape, size and mixing systems has now been used to grow plant cells (see Martin, 1980; Fowler, 1982*a* for reviews). Vessels range from 2 l to 20000 l in size and comprise a wide range of construction materials and mixing systems. Many of the early vessels were based on microbial systems with impellar (turbine) mixing, either in the form of paddles or with plastic or glass-coated spinning bars. A major departure from traditional vessels was marked by Wagner and Vogelmann (1977) who used various combinations of airlift and stirred tank vessels in experiments with *M. citrifolia* and *C. roseus*. Most of these studies were carried out with airlift vessels constructed on the draught-tube principle (Fig. 4). More recently in our own laboratory we have carried out studies with airlift loop vessels (Figs. 4 and 5) of between 10 and 100 l in size. Airlift vessels have the advantage of low shear, generally good mixing (although see below) and the avoidance of complex seals and glands which accompany turbine-driven systems and which may provide routes of infection for bacterial and fungal spores.

Mixing and aeration are two fundamental aspects of mass growth. In microbial systems, vessel designs are biased towards high shear and high oxygen transfer rates. Depending upon the vessel configuration and biomass load, mixing is usually effective at these levels, although doubts have recently been raised and expressed by Finn and Feichter (1979). In general high-shear

Fig. 4. Outline diagram of (*a*) a draught-tube airlift vessel and (*b*) an airlift loop vessel.

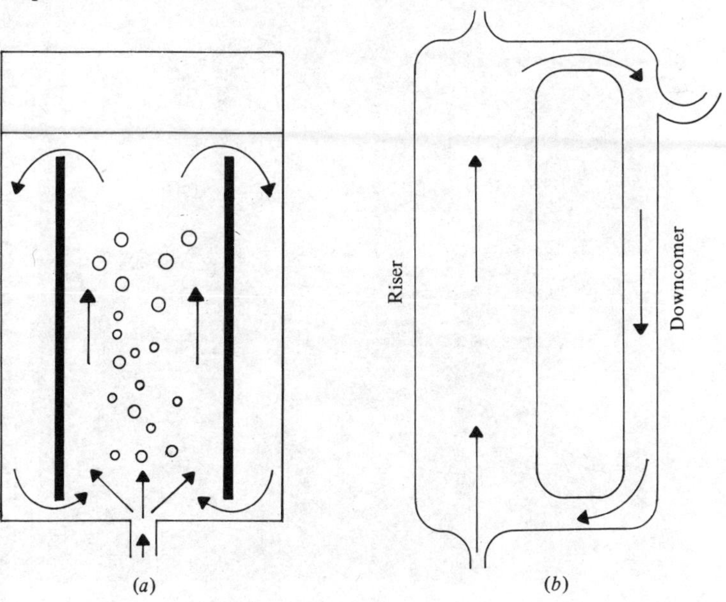

systems are deleterious to plant cells. Wagner & Vogelmann (1977) observed that with *M. citrifolia* cultures, major cell lysis occurred in turbine-stirred vessels and that natural product synthesis dropped by as much as 60% (as compared with air-driven systems). An effect was noticed with turbines

Fig. 5. A 100-l (total volume) airlift loop bioreactor (configuration shown in Fig. 4*b*) used to grow plant cell suspensions.

rotating as slowly as 20 r min^{-1}. They also noticed that the older the cells, the greater the damage. There is a marked variation between different cell lines in their response to turbine stirring and the effects of shear. Tobacco cells generally will remain viable at speeds of up to about 150 r min^{-1} (Kato *et al.*, 1977; Fowler 1982*a*) while it is possible to grow *C. roseus* cells in turbine-stirred vessels at speeds of up to 300 r min^{-1} before there is a major loss of viability (Fowler, 1982*a*).

The need to reduce shear in plant cell systems brings with it the problem of how to provide effective mixing. Plant cells are large and heavy in comparison with microbial cells and sediment rapidly. Consequently, if the level of mixing in the vessel is inadequate, 'dead' zones may result in which rapid accumulation of cells occurs. Senescence soon follows and the culture becomes multiphasic. Furthermore the senescing cells may release substances into the culture broth which may have deleterious effects on the remaining suspended cells. A further complication with plant cell cultures is that nutrient broths tend to be viscous, and that this viscosity rises rapidly as the biomass increases. Part of the viscosity increase, particularly in the later stages of batch growth, is due to the excretion of polysaccharide, which in turn appears to be related to the nature and level of the carbon source (N. J. Smart & M. W. Fowler; unpublished observations, 1981).

An alternative approach to mixing through mechanical stirring has been mentioned earlier (p. 21), that is to use the incoming air-stream in a bubble column or airlift reactor. Such systems have been applied successfully to microbial cultures (Schugerl, Lucke, Lehmann & Wagner, 1978). While having certain desirable properties, such as low shear, in order to achieve adequate mixing an airlift vessel requires careful design. Unfortunately merely increasing the venting rate to give more bubble agitation and nutrient flow is not a ready answer because of problems associated with overgassing (see below).

The rheology, i.e. the flow and deformation of matter, of plant cell cultures also causes particular problems (Wagner & Vogelmann, 1977). The nutrient broth and cell biomass possess non-Newtonian characteristics about which relatively little is known. In a non-Newtonian system apparent viscosity increases exponentially, so with a real increase in biomass load mixing may again become a problem. Wagner & Vogelmann (1977) reported good mixing in a draught-tube vessel of 10-l capacity at biomass levels of 15 g dry weight l^{-1} with cultures of *M. citrifolia*. Proportional scale-up to a 200-l vessel resulted in only a small decline in biomass productivity. Aeration rates were at a level of 0·5 vvm. Kato *et al.* (1977) have operated draught-tube reactors of 65-l capacity containing tobacco cells, at rates of up to 1.0 vvm and biomass levels approaching 15 g l^{-1}; again mixing was good. With draught-tube vessels there are indications of an upper working limit of about 20 g l^{-1} biomass. Above this the rheology of the system comes to exert an influence and stagnant zones develop. Operation at higher biomass levels is, however possible using a different design of airlift reactor, an airlift loop system. Studies in our own laboratory with 10- and 100-l vessels (Fig. 5, N. J. Smart & M. W. Fowler; unpublished observations, 1981) have shown that it is possible to operate at biomass levels of up to 30 g l^{-1} without the development of major stagnant zones. With this system problems of mixing and maintaining culture flow tend to lessen, up to a point, with scale-up.

Another approach, used by Wagner & Vogelmann (1977) has been to combine draught-tube with impellar systems. In a comparison between this novel approach and conventional turbine and draught-tube airlift systems they found that the airlift system on its own tended to give the best productivity. Similar observations have been made in our own laboratory. Although the data tend to point to airlift vessels as being a system particularly suited to mass plant cell growth, it should be emphasised that much information is required on other systems and on using a wider range of cultures to allow a valid comparison to be made.

Where the incoming air-stream is used both to aerate and to mix the cultures, as in airlift reactors, the relation between aeration and mixing takes on a particular importance. Aiba, Humphrey & Millis (1965) have reviewed the relation between biomass, specific product yield and oxygen supply in microbial systems. Unfortunately, few data are available for plant cells. Observations by Kato, Shimizu & Nagai (1975) on tobacco cells indicate a close relation between the initial gas transfer coefficient (K_La) and biomass productivity. At initial K_La values of 5 h^{-1} and less Kato *et al.* (1975) observed a marked limitation on biomass yield. Between K_La values of 5 h^{-1} and 10 h^{-1} there appeared to be a linear relation between initial K_La and final biomass yield. Similar data to the above were recorded in work of our own (N. J. Smart & M. W. Fowler; unpublished observations, 1981) from studies with *C. roseus* cells. We also found that with our *C. roseus* cell lines at least,

Table 6. *Effect of high* K_La *values on the growth of* Catharanthus roseus *cells*

Initial K_La (h^{-1})	Aeration rate (l min^{-1})	Growth rate (day^{-1})	Biomass yield (g l^{-1}/days)	Sucrose conversion (%)
20.5	1.0	0.34	12.4/8	56
25.2	1.5	0.41	11.4/9	51
27.3	2.0	0.33	9.5/11	42

the relation between initial $K_L a$ and final biomass yield held up to a $K_L a$ value of about 20 h^{-1}. At initial $K_L a$ values much above this a decline in final biomass yield was observed together with a fall in growth rate (Table 6). In addition, when the aeration rate was raised in order to increase the $K_L a$ value, the log phase became extended (Fowler, 1982*a*). $K_L a$ values of 20 h^{-1} are comparatively low by microbial standards, and given that limitation, probably due to oxygen starvation, the effect of low $K_L a$ begins to be serious at values of around 5 h^{-1}; we may therefore be faced with a constraint of balancing adequate mixing against a need to keep within a relatively narrow $K_L a$ range for a reasonable biomass and product return. This range of $K_L a$ possibly represents a spread of only some 10–15% of the total dissolved oxygen range available. We need to know much more about oxygen starvation, and, equally importantly, how to control it while maintaining good mixing throughout the vessel. Our data also revealed (N. J. Smart & M. W. Fowler; unpublished observations, 1981) that there may be effects additional to oxygen starvation or low aeration, caused by the gas stream. Preliminary experiments at high $K_L a$ values with tobacco cell cultures which had been provided with a carbon dioxide bleed of 50 ml min^{-1} gave significantly enhanced biomass yields, suggesting that one effect of overgassing may be to strip off key volatiles. Since, in the case of carbon dioxide, none of the above cultures are photosynthetic it suggests a not insubstantial level of non-photosynthetic carbon dioxide fixation. This is obviously an area urgently in need of further study.

Surface adhesion and foaming are problems which individually do not affect plant cell cultures as much as animal cells or microbial systems respectively (Fowler, 1982*a*). Together, however, they can cause major problems, again in relation to mixing. Fig. 6 shows a tobacco culture running in excess of 20 g dry weight biomass l^{-1}. At the top of the vessel is a 'meringue' or 'crust' composed of foam, protein and excreted polysaccharide. As the culture grows, so the level of foaming increases, the cells become sticky and the whole reactor contents become more viscous. Cells begin to stick together and collect in the foam at the top of the vessel. As the biomass increases further so the meringue or crust at the top of the vessel grows down into the culture. This further reduces mixing which gradually becomes inadequate to maintain a homogeneous culture. Eventually 'dry' zones result in the cell accretions and the culture slowly dies. It is to some extent possible to control 'meringue' formation by manipulation of the carbohydrate substrate.

In addition to 'meringue' formation some adhesion to the walls and probes of the vessel also occurs. It is seldom as serious as occurs with some micro-organisms but may still affect the proper functioning of the vessel. Growth on the walls is another aspect of adhesion; little is known of its nature in plant cell cultures.

Other parameters which affect mass growth are inoculum size, pH and temperature. These have recently been reviewed (Fowler, 1982*a*) and will not be further discussed here.

Fig. 6. A draught-tube bioreactor system (configuration shown in Fig. 4*a*) with a tobacco cell culture at high biomass levels.

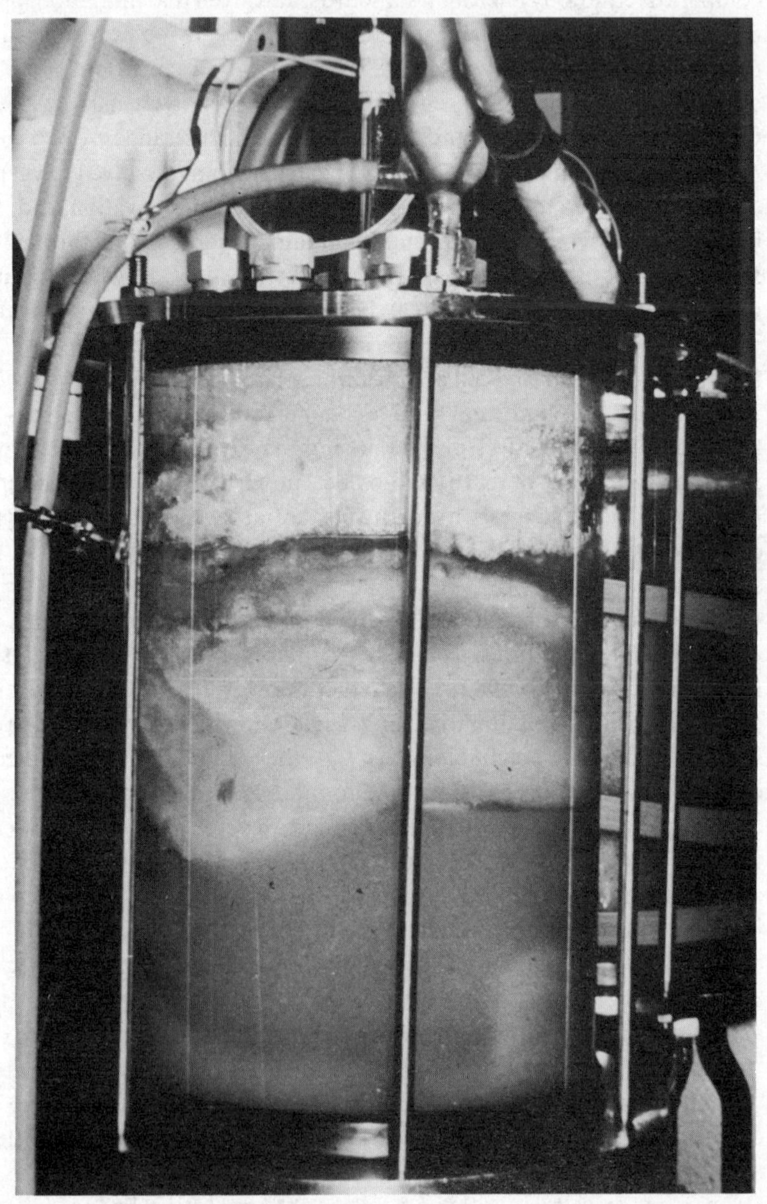

Biomass and productivity

Biomass yields in the range 5–20 g dry weight l^{-1} have been recorded by a number of workers. In particular Kato *et al.* (1977) have obtained yields of 20 g l^{-1} and more from tobacco cell cultures grown in vessels of 15-, 360- and 1500-l capacity with sucrose as the carbon source. In recent work we have recorded yields in excess of 50 g dry weight l^{-1} for tobacco cells grown in shake flasks. This level of biomass, with its very high viscosity, poses major mixing problems and is probably not viable as an operating level for large-scale production. In contrast, biomass levels of 25 g l^{-1} are regularly achieved with *C. roseus* in our 10-l airlift loop vessels without major problems, although we are probably near the maximum biomass load feasible before effective mixing breaks down. An important aspect in terms of process economics is the time taken to reach these high biomass yields. In the cases quoted above they have typically been achieved over a period of about 8–15 days.

Relatively few data are available for biomass productivity from plant cell cultures, but those that are available are encouraging, e.g. 3.82 g l^{-1} d^{-1} in a 1500-l vessel (Kato, Kawozoe, Iizima & Shimizu, 1976) and 6.9 g l^{-1} d^{-1} in a 35-l vessel (Noguchi *et al.*, 1977), both with tobacco cells. The carbon conversion efficiency of plant cell cultures is generally high, in the region of 30–60% (Smart & Fowler, 1981), and is comparable with that of micro-organisms.

Biomass in batch, semi-continuous and continuous culture systems

Batch culture has been the traditional way of growing plant cells in liquid culture, and has been amply reviewed over the years (see for instance Street, 1977; and articles in Thorpe, 1978).

The history of semi-continuous and continuous (chemostat) culture of plant cells is comparatively short compared with batch culture; it is only since the mid-1970s that a determined effort has been made in this area and then only within a few laboratories. Almost all the semi-continuous culture work has been essentially batch culture where large volumes of culture are removed from the bioreactor and replaced with fresh medium (see Kato *et al.*, 1977). Continuous culture has been reviewed recently (Fowler, 1977; Wilson, 1980). Numerous systems have been devised for continuous culture, some more successful than others (Martin, 1980). Early work indicated that chemostat plant cell cultures may conform to classical Michaelis – Menten kinetics (Fowler & Clifton, 1974; Fowler, 1977; Wilson, 1980). However, more recent studies (Dougall & Weyrauch, 1980) suggest that the position may not be so straightforward and that kinetic models developed by Nyholm (1976) may be more applicable in certain circumstances. Because of the possible applications of plant cell cultures as production systems much more information is required here.

Production systems

That plant cell culture as an industrial route to natural product synthesis is in an embryonic state is all too obvious. It is equally obvious, given the variety of situations in which natural products may be synthesised, that no one production system will be applicable to all situations. The ever-widening horizon of approaches to product synthesis using plant cells is therefore to be welcomed and encouraged.

Mass culture. Mass synthesis of natural products by cells grown in large vessels is perhaps the most obvious approach to product synthesis, but is one which in its simplest form of batch growth is possibly the most limited, apart from, as suggested earlier, for the production of biomass. In most circumstances natural product accumulation occurs at or towards the end of log growth when cell division and growth are slowing. (A major exception has been referred to earlier, see Fig. 3). The word 'accumulation' is used advisedly; measurements of a natural product in a cell mass provide little information concerning the synthetic potential of a cell at a particular time and even less about levels of degradation or turnover rates. In other words, the actual synthetic ability of the cell mass may be much greater at points in the batch culture cycle other than that indicated by the amount of accumulated product at any one point. This is an important distinction, particularly if we wish to manipulate synthetic versus degradative biochemistry in order to enhance product yield. Again we come back to a much-repeated earlier comment about lack of detailed information.

In those circumstances where synthesis apparently accompanies growth, then, continuous culture may be an appropriate system for an industrial process. Continuous culture is also applicable to two-stage processes where the first stage is concerned with continued biomass production and the second stage is a batch process for product synthesis under different environmental conditions (see for instance Alfermann *et al.*, 1977).

Immobilised cells, fluidised beds and callus systems. In recent years we have seen great progress in the immobilisation of plant cells on a variety of supports including agarose, starch and polyacrylamide. Brodelius and colleagues (Brodelius, Dwus, Mosbach & Zenk, 1979; Brodelius & Nilsson, 1980) have immobilised *C. roseus* cells on a variety of supports and showed not only continued, but initially enhanced, product synthesis. Cell viability on immobilisation is good and will often exceed 100–150 days. Lambe *et al.*

(1981) have also reported successful immobilisation of *C. roseus* cells, this time on polyacrylamide. Fluidised beds form an alternative approach to immobilised columns. Here the cells are attached in a layer to the outside of beads which are maintained in a fluidised state by a recycling nutrient stream (Morris & Fowler, 1981). Another approach is to use a callus system over which a constant stream of nutrient flows, a system developed by Yeoman and colleagues at Edinburgh (see Lindsey & Yeoman, this volume). All these systems are particularly applicable to biotransformations, and indeed immobilised cell columns are already under investigation for the digitoxin–digoxin biotransformation (Alfermann, this volume).

All the above systems have two key features in common. Firstly the need to generate biomass initially and as rapidly as possible, and secondly the need to be permeable to offered transformation substrates as well as 'leaky' to products. Very few natural products from plant cells of commercial interest are released into the surrounding medium. This is a key area of interest, not just from the point of view of product recovery but also that of enhanced product synthesis. It is possible that end-product inhibition may well restrict the attainable yields of a desired product in quite a number of cases. If some way could be found of releasing the end product from the cell into the surrounding medium without affecting product synthesis and cell viability, this would mark a major contribution to this aspect of cell technology.

No major development in the immobilisation of enzymes from plant cell cultures has so far been reported, although undoubtedly there is great potential here for some of the complex transformations of which plant enzymes are capable.

Downstream processing

This is an area to which relatively little attention has so far been paid, yet given the nature of cell culture technology, there are various ramifications to be taken into account as compared with traditional processes. Cell culture technology suffers from the classical problem of all fermentation technology, that is typically a concentration of desired product which is often very low indeed and either present in a large volume of liquid or surrounded by biomass of great chemical diversity. To a certain extent cell culture technology may have an advantage over whole plant material insofar as extraction from cell material is concerned. Where the product is present in the nutrient broth, or distributed between biomass and broth, different criteria apply. So far classical approaches of organic solvent extraction have been applied to cell cultures. On the horizon is supercritical extraction which could well transform downstream processing. Downstream processing is an important, if not very glamorous, part of the total process; it may account for up to 50% of the total process costs. As an area for assessment and development it must not be neglected.

Economic considerations

No matter how good the technology or elegant the science any new plant cell biotechnological process will be judged by industry solely on one criterion, commercial viability. To attempt to gain some indication as to whether or not a process may be commercially viable is at present difficult, mainly because we lack the technical and financial data base of an operating process with which to carry out a cost analysis. A possible approach is to use information derived from microbial fermentations, the general features of which are similar to large-scale plant cell growth, and take into account key aspects of plant cells such as low growth rate and product yield. Such an approach was used by Goldstein, Ingle & Lasure (1980), who have produced the most comprehensive costing analysis to date. Goldstein and colleagues presented possible costs ranging from about US \$20 kg^{-1} refined product, at a market of 10^6 kg p.a. and technology much advanced from its present level, to US \$1600 kg^{-1} with present technology and a market of 10^4 kg p.a. Some of the assumptions made in the calculations, given our present knowledge and the state of the art, seem a little extreme, for instance hydrated biomass yields of 500 g l^{-1}, water content of 80% (more usually 90–95%) and product yields of 0·5% of hydrated biomass. However, the analyses do give a very good account of the factors which must be taken into consideration when costing a mass culture process. Rather than repeat the detail, I have chosen to carry out a 'sensitivity' analysis based on two simple calculations, both using present costings and technology. Certain general points arise from such an analysis and will be outlined at the end of this section.

Vessel running costs

The costings shown in Table 7 are based on vessels of 20 m^3 and 50 m^3, by no means large by microbial standards but probably in the size range for a plant cell process. (Note that the figures are derived from experience with operational microbial systems.)

If we assume that it takes 10 days for a batch run and there is 15% 'down time' for servicing, lost runs, etc. this gives a figure of approximately 30 runs p.a. For a 20-m^3 vessel we have a daily cost of £14156 and for a 50-m^3 vessel, £25000. Note that these costings do *not* take into account the possibilities of nutrient recycling, and are only for batch runs, not continuous or semi-continuous operation. In other words, they represent a fairly pessimistic view.

The production of serpentine

Serpentine is a well established plant-derived antihypertensive. Its end-use selling price is £50000 kg^{-1} (Sigma Ltd) and it has a market in the region of 3500 kg p.a. A number of laboratories have been able to establish suspension cultures of *C. roseus* which produce the alkaloid at levels as high as the intact plant. The following data have been taken from my own group's work with *Catharanthus* cultures; serpentine, yield 1.0% dry weight; biomass yield, 25.0 g dry weight l^{-1}; run time, 10 days. This gives a final serpentine yield of 0.25 g l^{-1}; thus for a 20-m^3 vessel we have a yield of 0.25×20000 g = 5 kg per run. For 1 year (30) runs) we have a total yield of 150 kg. If we assume a purified product price of £20000 kg^{-1} (n.b. *not* £50000 which is the final packaged price), then we have a product value of £3m. This compares with a running cost of £4.25m (see above). For a 50-m^3 vessel the figures would be 12.5 kg per run, an annual yield of 375 kg and a product value of £7.5m, i.e. a break-even against a running cost of £7.5m (see above). If we achieve a product yield of 2% as opposed to 1%, this transforms the situation; e.g. for a 20-m^3 vessel we can obtain 300 kg p.a. serpentine at a value of £6m, against a running cost of £4.25m. At 50-m^3 vessel size the equivalent figures are £15m product value against £7.5m running costs.

As already stressed above, these simple calculations are very basic and serve as a guideline only. From them we can, however, make certain general points. Firstly, even using a batch culture system, and at the present level of technology, projected costings are very close to a break-even position. Secondly, for an increase in product yield not too far above the levels reported so far (in excess of 1% in a number of cases), serpentine production is potentially commercially viable. (This point also reinforces the total dependence of the process on product yield.) Thirdly, the calculations do not take into account various aspects of technology sophistication, such as nutrient

Table 7. *Vessel running costs*

	Vessel size	
	20 m^3 (£m p.a.)	50 m^3 (£m p.a.)
Running costs	2.00	3.00
Raw material costs	1.00	2.50
(Capital cost)	(5.00)	(8.00)
Capital cost amortised over 10 years	1.25	2.00
Total	4.25	7.50

recycling, semi-continuous culture or immobilisation techniques, all of which could yield substantial cost savings on a batch process. Fourthly, the analysis does emphasise the need to have relatively short grow-up periods (i.e. fast biomass accumulation) prior to 'production' periods in order to keep the total operating cost low. Total process costs are very susceptible to time spent in the pre-production phase. This again emphasises the importance of looking at semi-continuous culture or immobilised cell technologies. Fifthly, a major point of concern must be raised about low-market-volume products. These will only be viable if produced in general-purpose plant not dedicated to one product. Equally, if two or three products can be produced from the same cell line at the same time in the same process vessel then the economic basis changes dramatically.

The development of plant cell biotechnology has now reached a point when such economic considerations as those outlined above, albeit in simple form, have to be taken into account. Funding for enlarged programmes of research and development will not be forthcoming from industry and related sources without some form of economic assessment, as well as consideration of the 'state of the art'.

Future horizons

It is undoubtedly true that the plant kingdom has great potential as a chemicals resource. To what extent plant cell biotechnology will enable the further development of that resource is more difficult to assess. During the last few years major developments have occurred in a number of areas of plant cell biotechnology which point towards industrial application. These areas include enhanced product yield, improved methods of strain selection and screening, scale-up of mass growth systems with new vessel designs and successful application of immobilised cell technology and related technologies, albeit as yet on a small scale. The key aspect is of course product yield, and there are indications that a two to three fold increase in the yield of some products would be commercially viable. Given the level of product enhancement achieved in recent years, such increases in yield should be possible.

We now have a framework of information available to use which enables us to lay out a plant cell biotechnological process and to pinpoint those areas urgently in need of attention if we are to achieve industrial application. Key areas include biosynthetic pathways and their mode of regulation, enhanced and rapid methods of strain selection and screening, biochemical mechanisms for the enhancement of product synthesis and improved ways of extracting and recovering product, including methods of producing 'leaky' cells. The progress made with large-scale growth systems in a number of countries suggests that this will not pose insuperable problems as was perhaps once thought, although undoubtedly difficulties will occur.

The initial products to come from plant cell biotechnology will most likely be in the high-cost, medium- to low-market-volume area. A phosphodiesterase from tobacco cells, marketed by Bethesda Maryland, is probably the first product as such from a plant cell system. We can pin-point others which are not far off, the biotransformation of digitoxin being a good example, and Boerhinger of West Germany are now at pilot plant scale with this process. Developments in recent years with the catharanthus alkaloids suggest that a commercial process here is not too far into the future. (*Note added in proof*. Mitsui & Co. of Japan announced in July, 1982 a process for the industrial production of shikonin by plant cell culture.)

Given present interest we should perhaps include a comment about gene manipulation. I see this technology affecting plant cell biotechnology in only two ways. Firstly, where a plant product is a protein (e.g. thaumatin), there is every chance that given the appropriate vectors the information coding for the protein may be transferred to a microbial host which is more appropriate to industrial production. Secondly, it may also be of assistance in enhancing the level of product yield, possibly through gene duplication or initiation of fresh synthetic sequences. This latter area will require much more information regarding the plant cell genome than we have at present, but is an area which could be of immense importance. I do not think it likely that in the next 20 years we shall see the wholesale transfer from plant cells to micro-organisms of gene sequences coding for typically medium to low molecular weight natural products. Many of the pathways for these products are polycistronic in gene location. In addition, many comprise 30–40 steps, quite a number of which may be spontaneous rather than enzymic.

In summary the next 10 years should witness the coming on stream of a few plant cell biotechnological processes, typically in the high-added-value speciality chemical sector, although tobacco biomass, for instance, could be an exception, being a low-added-value bulk product. Whether or not plant cell biotechnology will then become a more general tool in the exploitation of the plant kingdom for chemicals production will depend upon the success of these processes, developments in those areas of constraint mentioned above and confidence in the chemical industry that plant cell biotechnology is a *realistic* alternative technology.

I wish to express my deep gratitude to members of the Wolfson Institute staff, particularly Dr P. Morris, Dr A. Stafford and Dr N. Smart, for their help in the preparation of this manuscript and for providing data and other material for it. In addition my thanks go to my secretary, Mrs Anne Gibbs for coping with the many last-minute alterations.

I wish to acknowledge the very generous support of the Wolfson Foundation and many industrial organisations which have directly or indirectly supported work in the Institute that is mentioned in this review.

References

Aiba, S., Humphrey, A. E. & Millis, N. F. (1965). *Biochemical Engineering*. New York and London: Academic Press.

Alfermann, A. W., Boy, H. M., Doller, P. C., Hagedorn, W., Heins, M., Wahl, H. & Reinhard, E. (1977). Biotransformation of cardiac glycosides by plant cell cultures. In *Plant Tissue Culture and its Biotechnological Application*, ed. W. Barz, E. Reinhard & M. H. Zenk, pp. 125–41. Berlin, Heidelberg and New York: Springer-Verlag.

Boulanger, D., Bailey, B. K. & Steck, W. (1973). Formation of edulinine and furoquinoline alkaloids from quinoline derivatives by cell suspension cultures of *Ruta graveolens*. *Phytochemistry*, **12**, 2399–405.

Brodelius, P., Deus, B., Mosbach, K. & Zenk, M. H. (1979). Immobilised plant cells for the production and transformation of natural products. *FEBS Letters*, **103**, 93–7.

Brodelius, P. & Nilsson, K. (1980). Entrapment of plant cells in different matrices. A comparative study. *FEBS Letters*, **122**, 312–26.

Butcher, D. N. (1977). Secondary products in tissue cultures. In *Plant Cell, Tissue and Organ Cultures*, ed. J. Reinert & Y. S. Bajaj, pp. 668–93. Berlin: Springer-Verlag.

Colijn, C. M., Johnsson, L. M. W., Schramm, A. W. & Kool, A. J. (1981). Synthesis of malvidin and petunidin in pigmented tissue cultures of *Petunia hybrida*. *Protoplasma*, **107**, 63–8.

Constabel, F., Gamborg, O. L., Kurz, W. G. W. & Steck, W. (1974). Production of secondary metabolites in plant cell cultures. *Planta Medica*, **25**, 158–65.

Dougall, D. K. (1972). Cultivation of plant cells. In *Growth, Nutrition and Metabolism of Cells in Culture*, vol. 2, ed. G. H. Rothblat & V. J. Cristofalo, pp. 371–406. New York: Academic Press.

Dougall, D. K. (1979*a*). Factors affecting the yields of secondary products in plant tissue cultures. In *Plant Cell and Tissue Cultures – Principles and Applications*, ed. W. R. Sharp, P. D. Larsen, E. F. Paddock & V. Raghaven, pp. 727–43. Columbus, Ohio: Ohio State University Press.

Dougall, D. K. (1979*b*). Production of biologicals by plant cell cultures. In *Cell Substrates*, ed. J. C. Petricciani, H. E. Hopps & P. J. Chapple, pp. 135–52. New York: Plenum Publishing Corporation.

Dougall, D. K. & Weyrauch, K. W. (1980). Growth and anthocyanin production by carrot suspension cultures grown under chemostat conditions with phosphate as the limiting nutrient. *Biotechnology and Bioengineering*, **22**, 337–52.

Finn, R. K. & Feichter, A. (1979). The influence of microbial physiology on reactor design. In *Microbial Technology, Current State, Future Prospects*, ed. A. T. Bull, D. E. Ellwood & C. Ratledge, pp. 83–106. Cambridge: Cambridge University Press.

Fowler, M. W. (1977). Growth of cell cultures under chemostat conditions. In *Plant Tissue Culture and its Biotechnological Application*, ed. W. Barz, E. Reinhard & M. H. Zenk, pp. 253–65. Berlin, Heidelberg and New York: Springer-Verlag.

Fowler, M. W. (1978). Regulation of carbohydrate metabolism in cell suspension cultures. In *Frontiers of Plant Tissue Culture 1978*, ed. T. A. Thorpe, pp. 443–52. Calgary: University of Calgary.

Fowler, M. W. (1981). Plant cell biotechnology to produce desirable substances. *Chemical and Industry, 229–33.*

Fowler, M. W. (1982*a*). The large scale cultivation of plant cells. *Progress in Industrial Microbiology*, **17**, 207–29.

Fowler, M. W. (1982*b*). Substrate utilisation by plant cell cultures. *Journal of Chemical Technology and Biotechnology*, **32**, 338–46.

Fowler, M. W. & Clifton, A. (1974). Activities of enzymes of carbohydrate metabolism in cells of *Acer pseudoplatanus* L. maintained in continuous (chemostat) culture. *European Journal of Biochemistry*, **45**, 445–50.

Furuya, T. & Ishii, T. (1972). The manufacturer of Panax plant tissue culture containing crude saponins and crude sapogenins which are identical with those of natural Panax roots. *Japanese Patent No.* 48–31917.

Furuya, T., Kojima, H. & Syono, K. (1971). Regulation of nicotine biosynthesis by auxins in tobacco callus tissues. *Phytochemistry*, **10**, 1529–32.

Goldstein, W. E., Ingle, M. B. & Lasure, L. (1980). Product cost analysis. In *Plant Tissue Culture as a Source of Biochemicals*, ed. E. J. Staba, pp. 191–234. Boca Raton: CRC Press.

Haberlandt, G. (1902). Kultinversuche mit isollerten Pflanzellen. *Sber. Akad. Wiss. Wien.* **111**, 69–92.

Hahlbrock, K., Schröder, J. & Vieregge, J. (1980). Enzyme regulation in parsley and soybean cell cultures. In *Advances in Biochemical Engineering*, vol. 18, Plant Cell Cultures II, ed. A. Feichter, pp. 39–60. Berlin: Springer-Verlag.

Kato, A., Kawozoe, S., Iizima, M. & Shimizu, Y. (1976). Continuous culture of tobacco cells. *Journal of Fermentation Technology*, **54**, 52–7.

Kato, A., Shimizu, Y. & Nagai, S. (1975). Effect of initial $K_L a$ on the growth of tobacco cells in batch culture. *Journal of Fermentation Technology*, **53**, 744–8.

Kato, A., Shiozawa, Y., Yamada, A., Nishida, K. & Noguchi, M. (1977). A jar fermentor culture of *Nicotiana tabacum* L. cell suspensions. *Agricultural and Biological Chemistry*, **36**, 899–904.

Kaul, B., Stohs, S. J. K. & Staba E. J. (1969). *Dioscorea* tissue culture. III. Influence of various factors on diosgenin production by *Dioscorea deltoidea* callus and suspension culture. *Lloydia*, **32**, 347–59.

Khanna, P. K. & Mohan, S. (1973). Isolation and identification of diosgenin and sterols from fruits and *in-vitro* cultures of *Momordica charantia L. Indian Journal of Experimental Biology*, **11**, 58–60.

Kinnersley, A. M. & Dougall, D. K. (1980). Increase in anthocyanin yield from wild-carrot cell cultures by a selection system based on cell-aggregate size. *Planta*, **149**, 200–4.

Kinnersley, A. M. & Dougall, D. K. (1981). Correlation between nicotine content of tobacco plants and callus cultures. In *W. Alton Jones Cell Science Centre Annual Report*, pp. 7–8. Lake Placid: W. Alton Jones Cell Science Centre.

Lambe, C. A., Reading, A., Roe, S., Rosevear, A. & Thomson, A. R. (1981). Alkaloid production by *Catharanthus roseus* cells in suspension culture. In *Abstracts of Communications*, Second European Congress of Biotechnology, Eastbourne, England, p. 196. London: Society of Chemical Industry.

Maretzki, A., Thom, M. & Nickell, L. G. (1974). Utilisation and metabolism of carbohydrates in cell and callus cultures. In *Tissue Culture and Plant Science 1974*, ed. H. E. Street, pp. 239–61. London: Academic Press.

Martin, S. M. (1980). Environmental factors: B. Temperature, aeration, and pH. In *Plant Tissue Cultures as a Source of Biochemicals*, ed. E. J. Staba, pp. 143–60. Boca Raton, Florida: CRC Press.

Misawa, M. (1977). Production of natural substances by plant cell cultures described in Japanese Patents. In *Plant Tissue Culture and its Biotechnological Application*, ed. W. Barz, E. Reinhard & M. H. Zenk, pp. 17–26. Berlin, Heidelberg and New York: Springer-Verlag.

Morris, P. & Fowler, M. W. (1980). Growth and alkaloid content of cell suspension cultures of *Papaver somniferum*. *Planta Medica*, **39**, 284–5.

Morris, P. & Fowler, M. W. (1981). A new method for the production of fine plant cell suspension cultures. *Plant Cell Tissues and Organ Culture*, **1**, 15–24.

Murashige, T. (1978). The impact of plant tissue culture on agriculture. In *Frontiers of Plant Tissue Culture 1978*, ed. T. A. Thorpe, pp. 15–26. Calgary: The Bookstore, University of Calgary.

Murashige, T. & Skoog, F. (1962). A revised medium for rapid growth and bioassays with tobacco tissue cultures. *Physiologia Plantarum*, **15**, 473–97.

Nickell, L. G. (1980). Products. In *Plant Tissue Culture as a Source of Biochemicals*, ed. E. J. Staba, pp. 256–69. Boca Raton, Florida: CRC Press.

Noguchi, M., Matsumoto, T., Hirata, Y., Yamamoto, K., Katsuyama, A., Kato, A., Azechi, S. & Kato, K. (1977). Improvement of growth rates of plant cell cultures. In *Plant Tissue Culture and its Biotechnological Application*, ed. W. Barz, E. Reinhard & M. H. Zenk, pp. 85–94. Berlin, Heidelberg and New York: Springer-Verlag.

Nyholm, N. (1976). A mathematical model for microbial growth under limitation by conservative substrates. *Biotechnology and Bioengineering*, **18**, 1043–56.

Reinhard, E. & Alfermann, A. W. (1980). Biotransformation by plant cell cultures. In *Advances in Biochemical Engineering*, vol. 16, Plant Cell Cultures I, ed. A. Fiechter, pp. 49–84. Berlin: Springer-Verlag.

Reinhard, E., Corduan, G. & Volks, O. H. (1968). Tissue culture of *Ruta graveolens*. *Planta Medica*, **16**, 9–16.

Röller, U. (1978). Selection of plants and plant tissue cultures with high content of serpentine and ajmalicine. In *Production of Natural Compounds by Cell Culture Methods*, ed. A. W. Alfermann & E. Reinhard, pp. 95–108. Munich: Gesellschaft für Strahler & Umweltforschung mbH.

Routier, J. B. & Nickell, L. G. (1956). Cultivation of Plant Tissue. *US Patent* 2, 747., 334.

Schugerl, K., Lucke, J., Lehmann, J. & Wagner, F. (1978). Application of tower bioreactors in cell mass production. In *Advances in Biochemical Engineering*, vol. 8, Mass Transfer in Biotechnology, ed. T. K. Ghose, A. Fiechter & N. Blakebrough, pp. 63–132. Berlin: Springer-Verlag.

Seibert, M. & Kadkade, P. G. (1980). Environmental factors: A. Light. In *Plant Tissue Cultures as a Source of Biochemicals*, ed. E. J. Staba, pp. 123–42. Boca Raton, Florida: CRC Press.

Smart, N. J. & Fowler, M. W. (1981). Effect of aeration on large scale cultures of plant cells. *Biotechnology Letters*, **3**, 171–6.

Staba, E. J. (1977). Tissue culture and pharmacy. In *Plant Cell, Tissue and Organ Culture*, ed. J. Reinert & Y. S. Bajaj, pp. 694–702. Berlin: Springer-Verlag.

Stöckigt, J. (1979). Cell-free biosynthesis of ajmalicine and related indole alkaloids. *Lloydia*, **41**, 655–8.

Street, H. E. (Ed.) (1977). *Plant Tissue and Cell Culture*, 614 pp. Oxford: Blackwood Scientific Publications.

Tabata, M. & Hiroaka, N. (1976). Variation of alkaloid production in *Nicotiana rustica* callus cultures. *Physiologia Plantarum*, **38**, 19–23.

Teuscher, E. (1973). Problems der Produktion sekundarer Pflanzenstroffe mit Hilfe von Zellkulture. *Pharmazie*, **28**, 6–18.

Thorpe, T. A. (Ed.) (1978). *Frontiers of Plant Tissue Culture*, 556 pp. Calgary: University of Calgary.

Wagner, F. & Vogelmann, H. (1977). Cultivation of plant tissue cultures in bioreactors and formation of secondary metabolites. *In Plant Tissue Culture and its Biotechnological Applications*, ed. W. Barz, E. Reinhard & M. H. Zenk, pp. 245–52. Berlin, Heidelberg and New York: Springer-Verlag.

Westcott, R. J. & Henshaw, G. G. (1976). Phenolic synthesis and phenylalanine ammonia lyase activity in suspension cultures of *Acer pseudoplatanus* L. *Planta*, **131**, 67–73.

Widholm, J. M. (1980). Selection of plant cell lines which accumulate compounds. In *Plant Tissue Cultures as a Source of Biochemicals*, ed. E. J. Staba, pp. 99–114. Boca Raton, Florida: CRC Press.

Wilson, G. (1980). Continuous culture of plant cells using the chemostat principle. In *Advances in Biochemical Engineering*, vol. 16. Plant Cell Cultures I, ed. A. Fiechter, pp. 1–25. Berlin: Springer-Verlag.

Yeoman, M. M., Miedzybrodzka, M. B., Lindsey, K. & Milauchlan, W. R. (1980). The synthetic potential of cultured plant cells. In *Plant Cell Cultures: Results and Perspectives*, ed. F. Sala, B. Parisi, R. Cella & O. Ciferi, pp. 327–44. Amsterdam: Elsevier.

Zenk, M. H. (1978). The impact of cell culture on industry. In *Frontiers of Plant Tissue Culture 1978*, ed. T. A. Thorpe, pp. 1–13. Calgary: University of Calgary.

Zenk, M. H., El-Shagi, H., Arens, H., Stöckigt, J., Weiler, E. W. & Deus, B. (1977). Formation of the indole alkaloids serpentine and ajmalicine in cell suspension cultures of *Catharanthus roseus*. In *Plant Tissue Culture and its Biotechnological Application*, ed. W. Barz, E. Reinhard, & M. H. Zenk, pp. 27–43. Berlin, Heidelberg and New York: Springer-Verlag.

Zenk, M. H., El-Shagi, H. & Schulte, U. (1975). Anthraquinone production by cell suspension cultures of *Morinda citrifolia*. *Planta Medica* Supplement, pp. 79–101.

K. LINDSEY and M. M. YEOMAN

Novel experimental systems for studying the production of secondary metabolites by plant tissue cultures

Introduction

A large number of the chemicals that are used in the industries of pharmaceuticals, food, flavourings and perfumes are of vegetable origin. These include compounds such as alkaloids, steroids, oils and pigments, many of which are still obtained from plant extracts. As has been described in this volume by Fowler, plant cells and tissues cultured in vitro have been considered, for many years, to be a potential source of specific secondary metabolites. Unfortunately, however, the success of the fermentation industry in harnessing microorganisms to produce particular chemicals in abundance has not been repeated by plant cell culturists. With few exceptions, such as diosgenin production by *Dioscorea* cell cultures (Kaul & Staba, 1968), callus and cell suspension cultures tend to produce particular secondary metabolites in much lower quantities than are found in the whole plant, and the chemicals produced are often structurally different from, or present in different proportions to, those of the whole plant (Forrest, 1969; Boulanger, Bailey & Steck, 1973; Hiraoka & Tabata, 1974; Ikuta, Syono & Furuya, 1974; Hirotani & Furuya, 1977; Jalal, Overton & Rycroft, 1979; Sejourne *et al.*, 1981). We believe it to be significant, however, that as a rule callus cultures accumulate higher levels of secondary metabolites than do cells grown in liquid suspension. Fowler and colleagues (Fowler, this volume) have demonstrated, in this respect, that cells cultured in fermentor vessels can accumulate considerable quantities of biomass, but the yield of designated secondary compounds usually remains relatively low.

In this chapter, the use of immobilised plant cells as a means of producing enhanced yields of secondary metabolites will be considered. We shall describe, in particular, two novel cell culture techniques used in our laboratory in which physically stationary cells are grown in or on an inert support.

Firstly, however, the concept and techniques of cell immobilisation will be introduced.

An historical perspective

The first reported instance of the immobilisation of whole cells was in 1966, when cells of the lichen *Umbilicaria pustulata* were entrapped in a polyacrylamide gel (Mosbach & Mosbach, 1966). In the following year, van Wezel (1967) grew embryonic animal cells immobilised on DEAE-sephadex microbeads, and since then a variety of organisms, but mostly microbial cells, have been immobilised in a range of substrata. As can be seen from the information presented in Table 1, the methods of immobilisation may be placed into one of four categories. The first may be described as the immobilisation of cells (and, indeed, subcellular organelles) *in an inert substratum*. This technique involves the entrapment of cells either in one of

Table 1. *Examples of the immobilisation of whole cells*

Source of cells	Immobilisation substratum	Reference
Immobilisation in an inert substratum		
Morinda citrifolia, *Catharanthus roseus*, *Digitalis lanata*	Alginate beads	Brodelius, Deus, Mosbach & Zenk (1979)
Catharanthus roseus	Alginate, agarose, agar, carrageenan, alginate + gelatin, agarose + gelatin, gelatin, polyacrylamide	Brodelius & Nilsson (1980)
Saccharomyces uvarum	Alginate	Cheetham, Blunt & Bucke (1979)
Bacillus stearothermophilus spores	Alginate	Dallyn, Falloon & Bean (1977)
Candida tropicalis	Alginate, polyacrylamide, polystyrene	Hackel, Klein, Megnet & Wagner (1975)
Saccharomyces cerevisiae, *Kluyveromyces marxianus*, chloroplasts, mitochondria	Alginate	Kierstan & Bucke (1977)
HeLa, K562 erythroleukaemic human skin fibroblasts human kidney carcinoma rat colon carcinoma	Alginate, agarose,	Nilsson & Mosbach (1980)
Arthrobacter simplex	Alginate	Ohlson, Larsson & Mosbach (1979)
Escherichia coli	Polyacrylamide	Chibata, Tosa & Sato (1974)
Corynebacterium simplex	Polyacrylamide	Larsson, Ohlson & Mosbach (1976)

Table 1. (*cont.*)

Source of cells	Immobilisation substratum	Reference
Immobilisation in an inert substratum (cont.)		
Curvularia lunata	Polyacrylamide	Mosbach & Larsson (1970)
Umbilicaria pustulata	Polyacrylamide	Mosbach & Mosbach (1966)
Lactobacillus bulgaricus, Escherichia coli, Saccharomyces lactis	Polyacrylamide	Ohmiya, Ohashi, Kobayashi & Shimizu (1977)
Corynebacterium simplex	Collagen	Constantinides (1980)
Streptomyces venezuelae	Collagen	Saini & Vieth (1975)
Aspergillus and *Penicillium* spores	ECTEOLA-cellulose	Johnson & Ciegler (1969)
Human diploid cells	DEAE-gels	van Wezel (1976)
Escherichia coli, Saccharomyces cerevisiae, Serratia marcescens, Acetobacter sp.	Metal hydroxide precipitates	Kennedy, Barker & Humphrey (1976)
Saccaromyces pastorianus	Agar	Toda & Shoda (1975)
Micrococcus denitrificans	Liquid-surfactant membranes	Mohan & Li (1975)
Aspergillus foetidus, Bacillus subtilis, Saccharomyces cerevisiae	Steel wool nets	Atkinson, Black, Lewis & Pinches (1979)
Adsorption to an inert substratum		
Fetal calf buccal cells, chick embryo fibroblasts, He_p-2 and L cells	Dextran microbeads	Beaudry *et al.* (1979)
Chick embryo fibroblasts, human fibroblasts	Dextran microbeads	Levine, Wong, Wang & Thilly (1977)
BHK cells	Glass beads	Spier & Whiteside (1976*a*)
BHK cells	DEAE-sephadex microbeads	Spiers & Whiteside (1976*b*)
Embryonic rabbit skin cells, human embryonic lung cells	DEAE-sephadex microbeads	van Wezel (1967)
Filamentous fungi	Rotating metal disc	Blain, Anderson, Todd & Divers (1979)
Adsorption to an inert substratum via biological macromolecules		
Various human cell lines, sheep erythrocytes	Protein-coated agarose	Carlsson *et al.* (1979)
Covalent bonding to an otherwise inert substratum		
Micrococcus luteus	Carboxymethylcellulose	Jack & Zajic (1977)

a number of matrices, such as alginate, agar, polyacrylamide or collagen, or in a combination of gels. The second method is the *adsorption* of cells *to an inert substratum*. Usually, animal cells are allowed to adhere to charged microspheres or glass beads. The third immobilisation method, which is used only rarely, involves the adsorption of cells to an inert substratum (such as a gel) *via* biological macromolecules (such as lectins). The fourth technique, again not commonly employed, requires the *covalent bonding* of cells to an otherwise inert substratum such as carboxymethylcellulose.

In the majority of the cases described in the literature, the immobilisation of microbial cells was merely a natural progression from the immobilisation of enzymes for one- or two-step biotransformations (reviewed by Mosbach, 1976) and indeed the technology devised for enzyme immobilisation was adopted for whole cell entrapment. The use of whole cells rather than isolated enzymes is advantageous in many ways as it allows relief from the technical difficulties of enzyme isolation and purification, enhanced enzyme stability, intracellular cofactor availability and the potential for multistep (multienzyme) reactions.

Until recently, there has been little interest in the immobilisation of whole plant cells. Certainly, plant cells can be used, in an analogous way to immobilised enzymes or microbial cells, for biotransformation reactions, and such work is described in this volume by Alfermann. However, there are few reports of the immobilisation of plant cells as a means of producing, *de novo*, high yields of particular secondary metabolites. We believe that immobilised cells have specific advantages over callus and liquid-suspended cell cultures for this purpose, and the rationale behind this concept is now discussed.

Why use immobilised cells?

The synthesis and accumulation of secondary metabolites can be considered to be a facet of differentiation (Yeoman, Lindsey, Miedzybrodzka & McLauchlan, 1982). The laying down of lignin by tracheids and xylem vessel elements, for example, is characteristic of the termination of a complex differentiation process (O'Brien, 1974), and attempts to quantify differentiation have, for instance, involved determinations of the activities of enzymes involved in lignin biosynthesis (Haddon & Northcote, 1975). Furthermore, there are numerous reports in the literature that describe a positive correlation between secondary metabolite accumulation and the incidence of differentiation in cell cultures (ranging from the formation of specialised cells to cell organisation and organ formation); this evidence has been reviewed by Yeoman, Miedzybrodzka, Lindsey & McLauchlan (1980) and Yeoman *et al.* (1982), and the details need not be repeated here. An inverse relation between differentiation and accumulation, and the growth of cell cultures has also been reported. In general, differentiation and secondary metabolite accumulation occur to the greatest extent towards the end of the growth cycle, and the use of conditions in which growth is slowed down, such as low temperature (Meyer-Teuter & Reinert, 1973) and inhibitors of protein and RNA synthesis (reviewed by Yeoman *et al.*, 1982), has been shown to cause increases in both processes.

A survey of the levels of alkaloids accumulated by a large number of species in vitro (Lindsey, 1982) similarly suggests that compact or organised and slow-growing cell cultures accumulate higher quantities than do friable and rapidly-growing cultures. It is suggested that the organisation of cells is essential for normal cell metabolism to proceed. Such organisation, and its consequent effects on physical and chemical gradients, is the most obvious difference between high- and low-yielding cultures, and it seems axiomatic that, the closer a cell, or rather, a group of cells are to the whole plant (in their level of organisation), the more likely they are to carry out the metabolic pathways of the whole plant. It is the cytoplasm of the cell which is responsible for the 'cueing' of developmental and other metabolic sequences, and suspended cells suffer from the disadvantage of being subjected to a particularly unnatural environment. Such a point has recently been made nicely by Zeleneva & Khavkin (1980), who discovered that, whereas the patterns of activity of a variety of enzymes investigated were similar in developing maize (*Zea mays*) calluses to those found in maize roots, the patterns of activity during the stationary phase of the growth of suspended cells differed widely from those in callus and intact plant tissue.

With these considerations in mind, it will be apparent that the immobilisation of cells will provide conditions that are conducive to cell differentation, thereby encouraging production of high yields of secondary metabolites. The potential advantages, therefore, of cell immobilisation are now discussed, and will then be evaluated in the light of available experimental evidence.

Slow growth of cells

Cells immobilised in or on an inert substratum would be expected to produce biomass at a slower rate than when grown as liquid suspension cultures. As indicated above, there is much evidence for an inverse relation between growth and secondary metabolite accumulation; let us consider the relation between the two. It is known that changes in the activity of a number of enzymes are associated with changes in the growth rate of cell cultures (Forrest, 1969; Hahlbrock, Kuhlen & Lindl, 1971; Davies, 1972; Thorpe & Meier, 1973; Mäder, Münch & Bopp, 1975; Westcott & Henshaw, 1976; Nato, Bazetoux & Mathieu, 1977; Zeleneva & Khavkin, 1980). Furthermore, Mäder *et al.* (1975) demonstrated that by artificially inhibiting the growth of *Nicotiana tabacum* callus cultures it was possible to induce a peroxidase isoenzyme pattern characteristic of that found at the onset of differentiation. What, then, are the interrelations between growth and metabolism, and how can these be explained in relation to the apparently important role of cell organisation and differentiation? There is evidence in the literature for the involvement of at least two types of mechanism in these interrelations. The first mechanism is based on the concept that growth regulates secondary metabolite production only indirectly, by determining the extent of cell aggregation. Organisation which results from aggregation of cells also seems to be essential for the production of high-yielding cultures (reviewed by Yeoman *et al.*, 1982, and also suggested by Carceller, Davey, Fowler & Street, 1971), and it is suggested that an adequate degree of aggregation may be possible only in slow-growing cultures. The possible role of aggregation in the production of increased yields of secondary metabolites is discussed below. The second hypothetical mechanism for the role of growth rate is a kinetic one in which it is suggested that, broadly, 'primary' and 'secondary' pathways compete differentially for precursors in fast- and slow-growing cells. If environmental conditions that are favourable to rapid cell growth prevail, then, obviously, the metabolic pathways required for rapid growth will operate. If rapid growth is blocked for any reason, then some 'primary' pathways will become inoperative and 'secondary' pathways will come into play. As discussed by Davies (1972), Phillips & Henshaw (1977), Yeoman *et al.* (1980, 1982) and Lindsey (1982), such a change in metabolism might be a relatively simple diversion of precursors from one set of enzymes, leading to one type of product, to another set of enzymes, leading to a second type of product. Furthermore, both enzyme systems may be continuously present in the cell and not synthesised *de novo*, whether or not they are in use (Neumann & Meuller, 1971, 1974; Mizukami, Konoshima & Tabata, 1977; Ramawat & Arya, 1979; P. A. Aitchison; unpublished data, 1977 and D. V. Banthorpe; unpublished data, 1979, 1980, quoted in Yeoman *et al.*, 1982). Thus, although the precise relation between growth and accumulation is not well understood, the low growth rate of immobilised cells certainly would be conducive to high yields of metabolites.

High cell–cell contact

The method of immobilisation *per se* allows the cells to grow in close physical association with each other, and, together with the slow growth, ideally results in chemical and physical communication between cells.

The establishment of chemical and physical gradients

In order that uninhibited secondary metabolism can proceed, it is crucial that the cultured cells be permitted to develop internal gradients. It is suggested that such a phenomenon may be the key to the question of the relation between cell organisation and the accumulation of high yields of secondary products.

Since the regulation of alkaloid and other secondary metabolite production may be under epigenetic control, i.e under extranuclear control, any *cytoplasmic* changes could result in qualitative or quantitative changes in the production of secondary metabolites. A consideration of the environment of the cell is therefore appropriate.

All cells in the whole plant are surrounded by other cells, but the relative position of a cell with respect to others may vary throughout its life history. Moreover, the position of the cell in the plant is determined ultimately by the rate and pattern of division of both it and the surrounding cells, and the degree and form of differentiation of that cell is characteristic of its position in the plant body. Thus, the question of the role of cell organisation in the control of metabolism can be rephrased to ask: how does the physical environment of a cell affect its metabolism? This is, of course, an enormous topic, but some examples are immediately obvious.

The cytoplasmic environment of the genome is a dynamic system and is itself under the influence of a wider environment, namely the body of the plant, or, in the case of cell cultures, simply the surrounding cells, and the ambient atmosphere or the liquid nutrient medium. Of the 'external' environment, two important factors which affect metabolism are the levels of oxygen and carbon dioxide, and light. The roles of light in the control of cell growth and metabolism are diverse; they include the control of photosynthesis in some cultured cells (Nishida, Sato & Yamada 1980; Yamada, Imaizumi, Sato & Yasuda, 1981). Even in heterotrophic, non-green cultures, light will have an important role in a number of physiological processes such as cell division and microfibril orientation (see Roberts, 1976) and enzyme activation (reviewed by Zucker, 1972). The intensity and wavelength of the light reaching a cell will be determined in part by the position of the latter within the cell mass, and the response of the cell to the stimulus will therefore be affected by the organisation of the tissue.

Gradients of oxygen and carbon dioxide concentrations will exist from the inside to the outside of an organised structure, and the importance of these gradients in differentiation has been studied by Dalton & Street (1976) and Bradley & Dahmen (1971) respectively. Thus the secondary metabolism of a large cell clump, which has a small surface area: volume ratio, may differ from that of isolated cells or smaller groups, which have larger surface area:volume ratios, as a direct result of the effect of gas concentration gradients. Similarly, gradients of growth-regulating substances and nutrients and mechanical pressure (Yeoman & Brown 1971) will also be set up, again providing potential for differential metabolism between organised and unorganised cells.

This short consideration of the possible roles of organisation in the control of metabolism attempts to bring attention to the fact that cells grown dispersed (as in liquid suspension cultures) and cells grown as aggregates (as callus or immobilised cells) will have quite different environments, and as a consequence different patterns of metabolism.

Facilitated manipulation of the chemical environment

In the past, the manipulation of the environment of cultured cells as a means of increasing the yields of secondary metabolites has been shown to be a promising approach, but has often been an empirical one (Butcher, 1977; Staba, 1980). Usually, cultures which had previously been shown to synthesise and accumulate to some extent a specific secondary compound were supplied with a range of mineral salts, different concentrations of growth-regulating substances (usually auxins and cytokinins) and primary carbon sources (usually sucrose) in the hope that a particular combination would induce the cells to accumulate high levels of the secondary compound(s). There are, however, a number of problems associated with the traditional culture techniques with regard to chemical manipulation. For example, if the nutrient medium supplying callus or liquid-suspended cells is to be altered, then physical manipulation of the cells is inevitable, and may result in the damage and contamination of the cultures. Furthermore, the addition of possibly toxic precursors to the nutrient medium is limited to low concentrations if the viability of the cultures is to be maintained. The chemostat may be used with relative ease for the manipulation of the chemical environment, but is not ideal for the purpose of increasing yields simply because the cells are suspended and dispersed and are not therefore subjected to the gradients which seem to be essential for the production of high yields of metabolites. All these drawbacks are overcome by the use of immobilised cell cultures by virtue of the circulation of perhaps large volumes of nutrient medium across physically stationary cells. This permits the feeding of large quantities of precursors at low concentrations, and allows sequential chemical treatments to be performed.

Facilitated harvest of exported secondary products

In order to produce a continuous production system, it is necessary that the desired metabolite be exported by the cells into the surrounding nutrient medium, from which it can be easily isolated. Cultured cells of some species, such as *Capsicum frutescens* (Yeoman *et al.*, 1980, 1982) and *Trigonella foenum-graecum* (Radwan & Kokate, 1980), do release secondary compounds, and the immobilised cell culture system allows the harvest of the products without damage to the cultures. Moreover, chemical treatments to induce the release of desired products may be relatively easily performed on immobilised cells which fail to naturally export them. It is felt that this is essential for maximising production as it relieves feedback inhibition mechanisms which would limit the synthesis of the compounds due to intracellular accumulation.

The foregoing has sought to justify the use of immobilised cells for the production of increased yields of secondary metabolites. We have suggested specific advantages of this technique over the use of suspended cells, and we shall now describe and discuss two immobilised cell culture systems that have

Fig. 1. The flatbed culture apparatus. R, Nutrient medium reservoir; PP, peristaltic pump; CV, culture vessel containing cells seated on fabric substratum (F); B, foam bung.

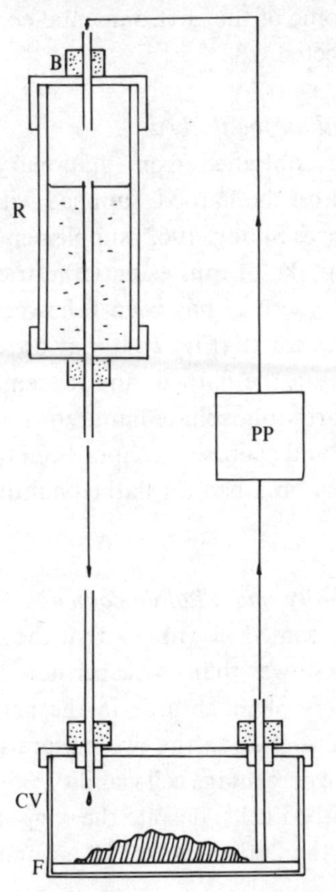

been developed and characterised in our laboratory. The first is a flatbed culture system in which cells are grown while seated in a horizontal culture vessel, and the second is a column culture system in which cells are grown in the vertical mode. In both, liquid nutrient medium is circulated across physically stationary cells.

The flatbed culture system

The construction of the flatbed apparatus is illustrated in Fig. 1. Nutrient medium drips, under the influence of gravity, from a cylindrical reservoir, of a volume of 70 ml, into a glass culture vessel, of a volume of 350 ml. In the culture vessel are contained cells (usually 45–50 g fresh weight) seated on a substratum of non-toxic polypropylene fabric matting. The nutrient medium moves by capillary action across the fabric, thereby supplying the cells, is thence pumped from the culture vessel back into the reservoir by means of a peristaltic pump and recirculated. Subaseals (Gallenkamp) are used to seal the ends of the reservoir and the nutrient medium entrance and exit ports in the culture vessel.

Attempts have been made to define some of the environmental conditions to which flatbed-cultured cells are subjected.

Nutrient uptake by cells cultured on the flatbed

Cells of *Datura innoxia* Mill., obtained from stationary phase suspension cultures, have been cultured on the flatbed, supplied with 50 ml Murashige & Skoog medium (Murashige & Skoog, 1962) supplemented with 10^{-5} M 2,4-D and 10^{-5} M kinetin. The uptake of four essential nutrients, i.e. orthophosphate, ammonia, nitrate and sucrose, has been followed over a 10-day culture period as an indication of growth (Fig. 2). It may be seen that the cells are capable of nutrient uptake on the flatbed, and it seems likely, indeed, that they become subjected to orthophosphate limitations in a very short period of time. Ammonia, nitrate and sucrose have not been found to be limiting. Similar results have also been obtained for flatbed-cultured cells of *Solanum nigrum* L.

Fresh weight increases, cell viability and alkaloid content

It can be seen from the data presented in Table 2 that the increase in fresh weight of cells on the flatbed is slower than in suspension cultures. There may be a difference of two orders of magnitude in the percentage increase in the fresh weight of cells of *S. nigrum* in the two culture systems. Nevertheless there is little difference in the percentage cell viability, of between 70 and 80% for both suspended and flatbed cells, despite the suggestion by the nutrient uptake data of cell lysis on the flatbed. Alkaloid accumulation,

Table 2. *Growth and viability of cells of* Solanum nigrum *grown in suspension culture and on the flatbed*

	Suspended cells of *S. nigrum*		Flatbed cells of *S. nigrum*	
Replicate	Increase in fresh weight (%)	Cell viability (%)	Increase in fresh weight (%)	Cell viability (%)
1	844	72.4	11.3	68.2
2	929	78.3	5.8	72.2
3	826	71.1	3.9	74.6
Mean	866	73.9	7.0	71.7

The analyses were performed after 18 days for suspended cells and after 7 days for flatbed cells.

Fig. 2. Nutrient uptake by cells of *Datura innoxia* cultured in the flatbed culture apparatus. P (triangles), orthophosphate; N (circles), nitrate; S (diamonds), sucrose; A (squares), ammonia.

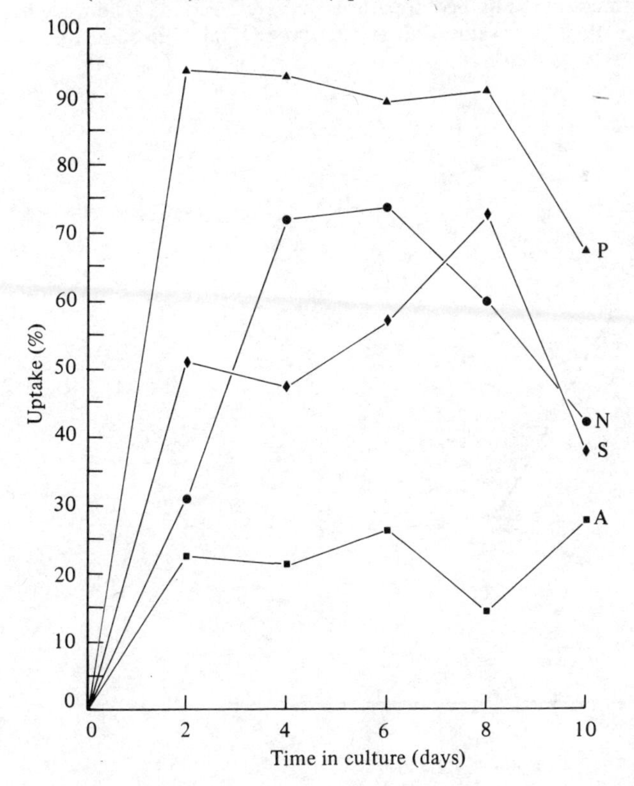

measured using a semi-quantitative technique based on the cell-squash method of Ogino, Hiraoka & Tabata (1978), is also possible on the flatbed and has been found to reach a slightly higher level after a 7-day culture period than after an 18-day culture period in suspension culture (approximately 12 mg alkaloids g^{-1} dry weight cells on the flatbed compared with approximately 10 mg alkaloids g^{-1} dry weight cells in suspension culture).

Oxygen uptake

The availability of oxygen to the cells may be limited since the only oxygen available is (a) that present in the air of the culture vessel at the time of sealing, (b) that dissolved in the nutrient medium and/or obtained from the air in the medium reservoir and (c) that from any diffusion of air into the apparatus through seals, joints and rubber tubing.

The data illustrated in Fig. 3 show the results of the analysis for aeration of the nutrient medium which was supplied to flatbed cultures of *S. nigrum* over seven days. It may be seen that by day 7 *c.* 30% of the oxygen originally present in the medium has been removed by the cells; the mean wet weight

Fig. 3. Oxygen levels in the nutrient medium supplying cells of *Solanum nigrum* cultured on the flatbed apparatus (squares) and in calcium alginate (triangles). 100% air saturation ≡ 0.25 μmol O_2 ml^{-1} medium (n.b. air contains 20.9% oxygen).

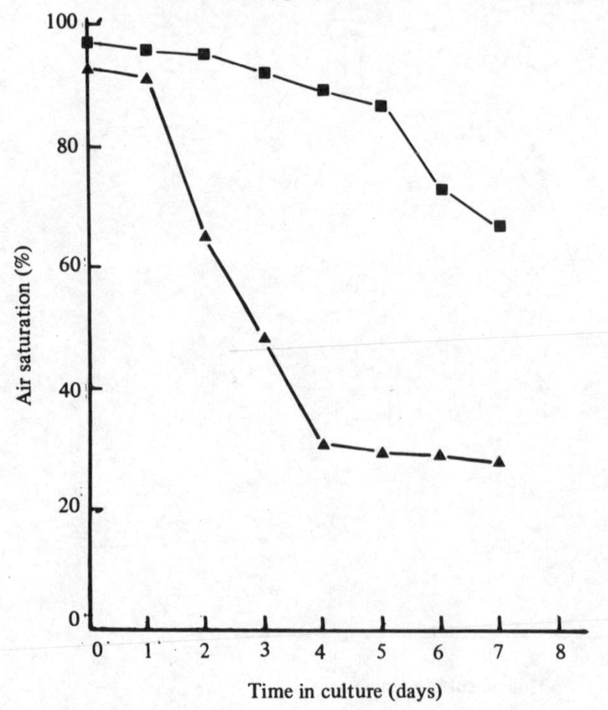

has been found to increase by 30% over this period. Thus the growing cells make use of the oxygen in the nutrient medium, but would also obtain the gas from the air in the culture vessel. It seems likely that enough oxygen is available to the cells to allow further growth, but this is limited by the slow rate of diffusion of oxygen in water.

Formation of 'drip areas'

In flatbed systems in which nutrient medium drips directly onto the cells (see Fig. 1), it has been noticed that after three or four days a change occurs in that particular region of the culture. In these so-called 'drip areas', the tissue invariably turns dark or rusty brown, and is thereby distinct from the rest of the culture, which, for *S. nigrum* cells, for example, is usually of a pale creamy colour. Moreover, the cells of the drip areas are often more compacted or clumped than the cells of the remainder of the culture. This apparent induction of differentiation in the cells of the drip areas has promoted an investigation into the levels of alkaloids accumulated by them.

The results of a series of experiments designed to investigate this phenomenon are summarised in Table 3. When cells isolated from the drip areas are cultured in liquid suspension and on agar plates, their alkaloid contents are significantly lower (a fall of *c.* 5 mg alkaloids g^{-1} dry weight cells when transferred to liquid culture); when the suspended cells are returned to the flatbed, the 'second generation' drip area cells accumulate relatively high levels of alkaloids once again.

These data demonstrate not only that a simple manipulation of the cells' environment (such as the direct dripping of nutrient medium) can cause a

Table 3. *The alkaloid content of cells of* S. nigrum *on the flatbed (drip areas), in liquid suspension and on agar plates*

Alkaloid content (mg glycoalkaloids g^{-1} dry wt cells)				
			Second generation flatbed	
Drip area cells (Day 0)	Suspended cells (Day 15)	Agar cells (Day 21)	Drip area	Non-drip area
15	10	12	15	12

Cells were cultured on three flatbeds for 10 days. Drip areas were formed and isolated (day 0 drip area).
The isolated 'drip area' cells were grown either in suspension culture (for 15 days) or on agar (for 21 days).
The suspended cells were returned to the flatbed for 10 days, and 'second generation' drip areas were formed.

dramatic change in the expression of the metabolism, but also that the flatbed culture system is particularly amenable to this type of treatment.

Precursor feeding

Further manipulations using the flatbed system have involved the feeding of precursors of specific secondary metabolites to cells. We have previously reported that capsaicin production by cells of *Capsicum frutescens* cultured on the flatbed can be increased from nanogram quantities to hundreds or exceptionally thousands of micrograms, by supplying 5 mM isocapric acid (Yeoman *et al.*, 1980, 1982). The capsaicin, furthermore, was exported by the cells into the nutrient medium, as indicated above, and was not accumulated intracellularly.

The main points to come from this work on the flatbed culture system using cells from several species may be summarised as follows. It would appear that this principle of cell culture, namely the movement of a solution of nutrients across a callus-like mass of stationary cells, does allow cell growth and alkaloid production to proceed. It has been established that the cells are able to take up nutrients (including oxygen) from the medium, and in fact probably suffer from phosphate limitation. The relatively slow growth of the cells (compared to their growth rate in suspension culture, from which they were derived) is probably due, at least in part, to the nutrient stress, and, together with the callus-like habit of the cells, leading to more cell–cell contact than in suspension culture, may contribute to the generally higher levels of secondary metabolites produced in the flatbed cells than in the suspended cells. Despite these advantages, however, the flatbed system has one considerable disadvantage when scaling-up for industrial use is contemplated. This is that a cell culture apparatus in a horizontal mode is awkward to design and operate and would require a considerable floor area. In order to resolve this problem it was therefore decided to design and develop a column system of culture in which cells would be attached to a biologically inert substratum inside a vertical column and which would preserve the basic advantages of the flatbed system. In addition to offering the important advantage of 'scaling-up' to an industrial level of operation, column culture also offers certain advantages to the laboratory worker, such as better control of nutrient flow rate over the cells, more even illumination of the cells and saved space. It also ensures that a greater proportion of cells (than in a flatbed culture vessel) would be dripped onto by the medium, thereby creating an enlarged 'drip area' of cells in which secondary metabolite accumulation might be enhanced.

The development and characterisation of such a culture system will now be described.

The column culture system

The development of the column culture system was, in effect, a simple step from the flatbed; the culture vessel was transformed from an essentially horizontal format to a vertical one. As in the flatbed system, 50 ml of liquid nutrient medium was contained in a reservoir, and was allowed to drip, under the influence of gravity, into a vertical glass column containing immobilised cells. Medium was removed from the bottom of the column, pumped back to the reservoir and recirculated (Fig. 4).

Cell immobilisation procedure

It was considered important to support the cells in some way in the column, because the effect of dripping liquid nutrient medium would be to pack unsupported cells into a dense and largely anaerobic agglomeration of

Fig. 4. The column culture apparatus. R, nutrient medium reservoir; PP, peristaltic pump; CV, culture vessel containing cells immobilised in 'baskets' of nylon netting (hatched areas).

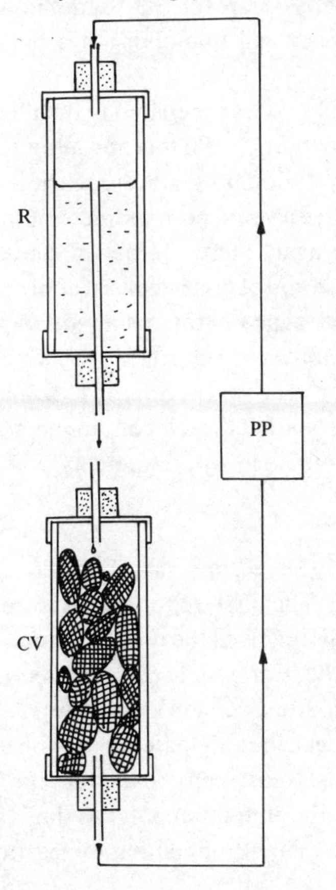

tissue at the bottom of the column. Thus the immobilisation procedure adopted aims to entrap cells in 'baskets' of nylon netting with the aid of a non-toxic, permeable and stable gel. These 'baskets' act as the basic unit within the column, which holds them. Air spaces are formed between the baskets, the netting gives structure to the gel and the cells are in contact with each other and are able to grow through the baskets.

Nylon panscrubbers, or pieces of them, have been found to be a suitable form of netting in which the cells can be embedded. This material is non-toxic, easily cut up and is able to withstand autoclaving. Two gels have been investigated in this laboratory for their suitability as cell entrapment substrata, namely agar and calcium alginate. The methods of immobilisation are now described.

Cell immobilisation using agar. The first substratum used to immobilise cells was agar. It was an obvious first choice, having been used in the culture of tissues for many years as a 'seat' for cells; it is also freely permeable to nutrients. It has been used independently for plant cell immobilisation by Brodelius & Nilsson (1980) and for yeast cell immobilisation by Toda & Shoda (1975).

A 2% solution (w/v) of agar (Oxoid No. 3) was prepared in distilled water, autoclaved at 121 °C (15 lb in^{-2} steam pressure) for 20 min and allowed to cool to 35–40 °C in a water-bath. Cells from stationary phase suspension cultures of different species were sieved (using a sieve of pore diameter 1 mm) and mixed 1:1 (v/v) with the still-molten agar. Sterile pieces of panscrubber netting of approximately cylindrical shape and of dimensions of approximately 2–3 × 1 × 1 cm were dipped into the cell–agar mixture using sterile forceps, and the resultant panscrubber/cell agar units were inserted into glass columns (15 cm × 2.5 cm internal diameter) as the agar was beginning to solidify. The result was a column of about 10 nylon 'baskets' each containing a mass of agar-embedded cells. Each column contained approximately 5–7 g fresh weight of cells.

Cell immobilisation using calcium alginate. The second immobilisation substratum used was a calcium alginate gel. The first reported instance of the immobilisation of whole higher plant cells involved the use of calcium alginate pellets, in which cells of *Morinda citrifolia*, *Catharanthus roseus* and *Digitalis lanata* were embedded (Brodelius, Deus, Mosbach & Zenk, 1979). As can be seen from the information in Table 1, calcium alginate has also been used for the immobilisation of animal cells, yeast cells, bacterial spores and subcellular organelles and enzymes, and the aluminium salt has similarly been used for yeast cell immobilisation. The method of cell immobilisation used in our laboratory is based on that of Kierstan & Bucke (1977). A 2% solution (w/v) of sodium alginate (Sigma London Chemical Co., practical grade type IV) was prepared, for the experiments described here, in distilled water (but for other experiments, in supplemented Murashige & Skoog medium), was autoclaved at 121 °C (15 lb in^{-2} steam pressure) for 20 min and allowed to cool to room temperature. This solution was then mixed 1:1 (v/v) with cells sieved, as for the 'agar' method, from stationary phase cells, and sterile panscrubber netting was dipped into the mixture using sterile forceps. The pieces of netting containing the fluid mixture of sodium alginate and cells were then quickly transferred into a sterile solution of 0.05 M calcium chloride in distilled water, and left there for at least 10 min to allow the calcium alginate gel to solidify within and around the nylon netting, thus entrapping the cells. The calcium alginate is made stable by the cross-linking of alginate molecules by the divalent calcium anions (see Cheetham, Blunt & Bucke, 1979). When solid, the panscrubber/cell-alginate units were washed three times in sterile distilled water and inserted into glass columns as in the agar method of immobilisation.

Once the cell entrapment procedures were established, basic experiments could be performed to characterise the column culture system. This was done by manipulating the environment of the cells and determining the effect of such treatments on culture viability and alkaloid yield.

Nutrient uptake, viability and alkaloid content of immobilised cells

Cells immobilised in agar. The uptake of nutrients by cells of *S. nigrum* immobilised in 2% agar has been investigated as an index of their growth. The results obtained have shown that both the rate and extent of uptake of orthophosphate, ammonia, nitrate and sucrose are less than into cells immobilised on the flatbed (Lindsey, 1982). The percentage cell viability is not severely reduced by agar immobilisation and culturing (typically a mean of 60–65% per column after 10–12 days compared with a viability of 70–80% in a rapidly-growing suspension culture). The cells are also capable of alkaloid accumulation, with a final content of *c.* 11–13 mg alkaloids g^{-1} dry weight cells (again, after 10–12 days). No alkaloids have been detected in the medium, the pH of which falls over an experimental period by *c.* 0.7 units to pH 5.1.

Cells immobilised in calcium alginate. The nutrient uptake data for cells of *S. nigrum* immobilised in 2% calcium alginate are presented in Fig. 5. Orthophosphate, ammonia, nitrate and sucrose are all taken up at more or less steady rates, and these results indicate that cell growth is better in alginate than in agar. The viability of the cells is comparable with that of cells

Fig. 5. Nutrient uptake by cells of *Datura innoxia* immobilised in calcium alginate, in the presence (solid points) and absence (open points) of 5 mM ornithine. P. (squares), orthophosphate; S (triangles), sucrose; A (squares), ammonia; N (triangles), nitrate.

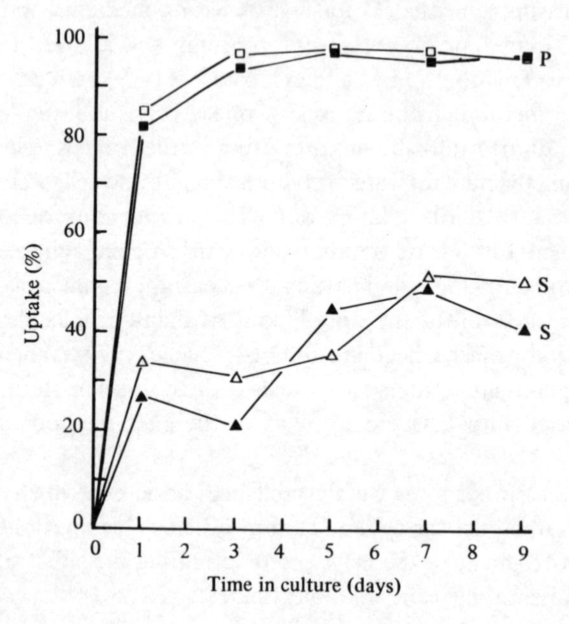

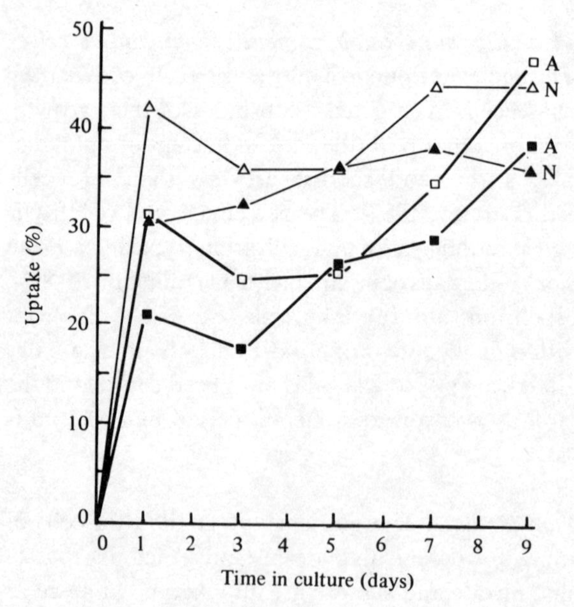

immobilised in agar (typically a mean of 60–65% per column after a culture period of 8–10 days), and the alkaloid content is also similar (*c.* 12 or 13 mg alkaloids g^{-1} dry weight cells). Alkaloids are not released into the nutrient medium, and by day 8 the pH of the medium has fallen by about 0.4 units to pH 5.4.

Fig. 3 illustrates oxygen uptake by cells immobilised in alginate. The increase in the rate of oxygen uptake corresponds approximately to the initiation of steady nutrient uptake, suggesting that the cells are metabolically most active after the first couple of days of culture, followed after days 5 or 6 by a period of reduced activity. Since the oxygen level representing 5% available oxygen (the extinction point, i.e. the level of oxygen below which no aerobic respiration is detectable) is approximately 0.060 μmol O_2 ml^{-1} medium, it seems likely that the immobilised cells suffer from oxygen limitation by between days 4 and 8.

The effects of illumination

Following the discovery that alginate-immobilised cells of *S. nigrum* grow and accumulate alkaloids under the experimental conditions used, the importance of light was investigated.

The results illustrated in Fig. 6 show that cultures illuminated at 15 μmol m^{-2} s^{-1} (fluorescent light) remove phosphate, ammonia, nitrate and possibly sucrose from the medium more quickly and to a greater extent than do dark-grown cultures.

After 8 days of culture the mean percentage cell viability is greater for illuminated cultures (at 65–70%) than for the dark-grown cultures (about 55%), and there is a significant reduction in the alkaloid content of the dark-grown cultures (less than 10 mg alkaloids g^{-1} dry weight cells) compared with illuminated cultures (12–13 mg alkaloids g^{-1} dry weight cells).

These results suggest that cell growth is better in the illuminated cultures, as indicated by the nutrient uptake data. It also appears that light is important in maintaining cell viability and alkaloid accumulation. Although darkened cultures appear to grow more slowly, they contain lower levels of alkaloids than do illuminated cells, and it therefore seems that slow growth *per se* does not automatically determine that alkaloids should be accumulated: light is apparently a contributory factor. Since greening was not observed in the illuminated cultures examined, other effects of light must be of importance in maintaining 'normal' cell metabolism.

Precursor feeding

The effects of precursors of secondary compounds on the metabolism of immobilised cells have been investigated. Cells of *Datura innoxia* have been treated with ornithine, a precursor of tropane alkaloids, to determine the effects on the qualitative and quantitative alkaloid composition, and on cell growth and viability. Furthermore, cells of *C. frutescens* have been treated with a precursor of capsaicin, isocapric acid, to determine (1) whether the

Fig. 6. Nutrient uptake by cells of *Solanum nigrum* immobilised in calcium alginate, in the presence (open points) and absence (solid points) of light (15 μmol m^{-2} s^{-1}). P (triangles), orthophosphate; S (squares), sucrose; A (triangles), ammonia; N (squares), nitrate.

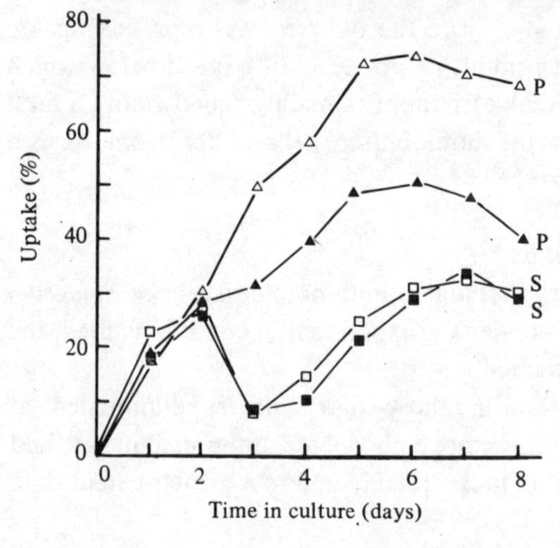

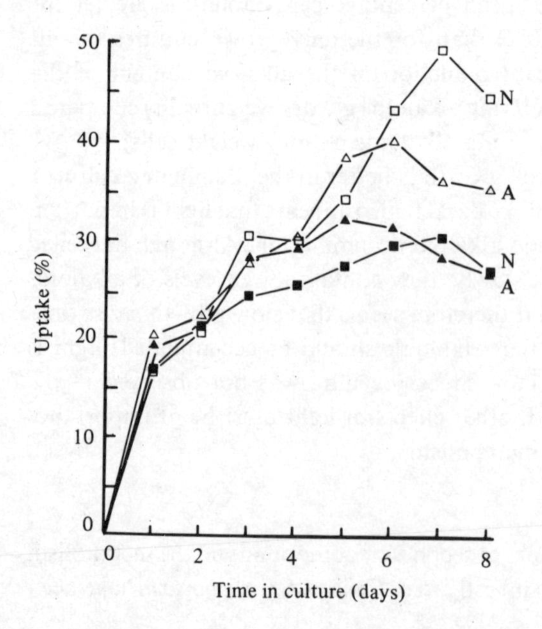

alginate-immobilised cells could produce capsaicin (this was compared with production by cells grown in the flatbed system) and (2) whether the immobilisation procedure inhibits the release of capsaicin from the cells that is observed in the flatbed culture system.

Effect of ornithine on alginate-immobilised cells of Datura innoxia. The effect of 5 mM ornithine on nutrient uptake by the cells is illustrated in Fig. 5. The apparent adverse effect of ornithine on cell growth is reflected in the lower percentage cell viability of treated cultures (a mean of 63.3% for samples from three columns compared with a mean of 71.9% for the untreated control cultures, after 9 days).

Analysis of the cell alkaloid content after 9 days by means of the cell squash technique and thin layer chromatography reveals a small increase, of about 0.5 mg alkaloids g^{-1} dry weight cells, in the ornithine-treated cells over the untreated control cultures. This is accounted for principally by an increase in scopolamine. No alkaloids are detectable in the nutrient medium at the end of such an experimental period.

Thus 5 mM ornithine causes a reduction in the growth and viability of immobilised cells of *D. innoxia*, and has a small effect in increasing the yield of tropane alkaloids. Although cells of *D. innoxia* and *S. nigrum* accumulate and retain alkaloids while immobilised, there is evidence that the immobilisation procedure itself does not prevent the release of the compounds into the nutrient medium.

Effect of isocapric acid on alginate-immobilised cells of Capsicum frutescens. Cells of *C. frutescens* cultured on the flatbed and in alginate columns in the presence of 5 mM isocapric acid for 10 days do not accumulate capsaicin to detectable levels. It has been found, however, that the nutrient medium supplying either the flatbed-cultured cells or the column-cultured cells contains capsaicin at a concentration of approximately 2.5 $\mu g\ ml^{-1}$. Thus, despite the fact that the flatbeds contain five times as much tissue as the columns on a fresh weight basis, similar quantities of capsaicin are released by the cells, presumably uninhibited by the alginate gel. This suggests that, either the alginate-immobilised cells are more efficient at capsaicin synthesis than cells immobilised on the flatbed, or, perhaps, the synthesis of capsaicin is under negative feedback regulation, and the activity of one or more enzymes in the capsaicin biosynthetic pathway is inhibited by concentrations of capsaicin exceeding 2.5 $\mu g\ ml^{-1}$ in flatbed-cultured cells.

Concluding remarks

The evidence presented here and elsewhere demonstrates that the cells of a number of species are able to survive immobilisation. Our results have described two culture systems and have indicated their potential for the production of high-yielding cultures. The 'drip area' data demonstrate well the 'reversible' nature of cells with respect to alkaloid accumulation, for the drip area cells effectively dedifferentiate when grown as cell suspensions but revert (differentiate) to high-yielding, coloured cells when returned to the flatbed. Similarly, despite the fact that the flatbed- and column-grown cells are obtained from rapidly-growing and friable suspensions, they acquire certain physical and chemical similarities to the cells in a callus culture. Firstly, the cells grow in closer association than when in liquid culture. Although the extent of communication between the cells is not yet known, there is nevertheless much more clumping of (in particular) the flatbed-grown cells than of those in fast-growing suspension cultures. This allows the establishment of physical and chemical gradients. Secondly and most importantly, the immobilised cells accumulate higher levels of alkaloids than do fast-growing, friable suspension cultures.

The viability of the cultures is not severely affected by the immobilisation procedures. Growth of the column-grown cells (as indicated by nutrient uptake) is better in alginate than in agar, and this difference may be associated with detrimental effects on the cells of molten agar (which is at a temperature of about 40 °C) during the immobilisation procedure. Results of investigations in other laboratories using an alginate entrapment system indicate that cell metabolism is not adversely affected in a qualitative way, for such reactions as phenol oxidation (Hackel, Klein, Megnet & Wagner, 1975) and glucose–ethanol conversion (Kierstan & Bucke, 1977) by yeast cells, alkaloid biosynthesis (Brodelius *et al.*, 1979; Brodelius & Nilsson, 1980) and respiration (Brodelius & Nilsson, 1980) in whole plant cells and the Hill reaction and ATP synthesis in immobilised chloroplasts and mitochondria respectively (Kierstan & Bucke, 1977) all take place within the environment of a calcium alginate gel. In most of these examples there is, however, an obvious diminution of the rates of the reactions when compared with those in freely-suspended cells, a noticeable exception being the increased efficiency of ajmalicine synthesis from tryptamine and secologanin in entrapped (compared with freely-suspended) cells of *C. roseus* (Brodelius & Nilsson, 1980).

Flatbed-cultured and alginate-immobilised cells appear to suffer from phosphate limitation after a short period of time, and with the possible exception of nitrate in the case of *C. frutescens*, none of the other nutrients are removed to the extent that they could be considered limiting. The uptake of phosphate by the column-cultured cells is not as rapid as in the flatbed system, indicating that the column-cultured cells grow more slowly. This may be a direct result of entrapment, perhaps due to mechanical constraints on cell growth, but it seems more likely to be due to an oxygen limitation in the columns.

This 'nutritional stress' to which the immobilised cells become naturally subjected means that this culture system may be particularly useful, by bringing the cells to an early stationary phase. Interestingly, Nettleship & Slaytor (1974) have found that of a variety of nutrient-limited media, phosphate-free medium was especially conducive to alkaloid and other secondary metabolite production by a callus culture of *Peganum harmala*. Furthermore, Yeoman *et al*., (1980) found that nitrate-free medium enhanced capsaicin production by a callus culture of *C. frutescens*.

Since immobilised cells appear to reach a premature stationary phase, it would be expected that the addition of precursors to the nutrient medium would be particularly effective in enhancing secondary metabolite production, assuming that they are limiting. The results presented here demonstrate that although an increase in tropane alkaloids is effected in *D. innoxia* cells supplied with 5 mM ornithine, the increase is small. It is likely that the increase in yield could be improved in three ways. (1) Increasing the amount of precursor and/or extending the culture period might lead to a direct increase in yield. M. M. Yeoman & E. Forche (unpublished observations) found that ornithine and phenylalanine increased the quantities of tropane alkaloids accumulated by suspended cells of *D. innoxia*, but only after incubation for three to six weeks. (2) The use of precursors closer than ornithine to the desired product would probably have been more effective in increasing yields. Tropine and tropic acid have been found to be particularly useful in promoting tropane alkaloid production in cell cultures (Konoshima, Tabata, Yamamoto & Hiraoka, 1970; Tabata, Yamamoto, Hiraoka & Konoshima, 1972). (3) The relief of feedback inhibition mechanisms would allow improved production. The intracellular accumulation of specific secondary metabolites would drastically limit the potential for a continuous production system, and experiments involving treatments using low temperatures and solvents have been carried out in our laboratory to induce the release of alkaloids from cells.

In conclusion, we can propose a general method of cell culture which would be expected to result in the production of relatively high levels of secondary metabolites. This can be summarised as follows.

(1) Cells should be grown physically stationary, in close association with each other, in order to encourage the development of physical and chemical gradients and allow the partial differentiation of the culture. In the cases of at least some species, the culture should be illuminated and induced to produce chloroplasts and turn green, in order to ensure the development of a metabolism that is close to that of the cells in the intact plant.

(2) The composition of the nutrient medium and/or the oxygen level should be manipulated to slow down the growth of the culture, and sequential chemical treatments involving, in particular, the use of growth-regulating substances should be performed in order to mimic some of the processes which, in vivo, regulate the metabolism associated with differentiation.

(3) Large quantities of precursors should be supplied to the cells at low concentrations. These should be as close as possible, in the biosynthetic pathway, to the desired product.

(4) Cells should be used which either naturally export the desired metabolite(s) into the nutrient medium or can be induced to release them.

We wish to express our thanks to Albright and Wilson Ltd for financial support and to Mrs E. Raeburn for typing and processing the manuscript so efficiently.

References

Atkinson, B., Black, G. M., Lewis, P. J. S. & Pinches, A. (1979). Biological particles of given size, shape and density for use in biological reactors. *Biotechnology and Bioengineering*, **21**, 193–200.

Beaudry, Y., Quillon, J. P., Frappa, J., Deloince, R. & Fontanges, R. (1979). Culture of diploid and heteroploid cells on dextran 'microbeads' in suspension: Studies of growth kinetics. *Biotechnology and Bioengineering*, **21**, 2351–8.

Blain, J. A., Anderson, J. G., Todd, J. R. & Divers, M. (1979). Cultivation of filamentous fungi in the disc fermentor. *Biotechnology Letters*, **1**, 269–74.

Boulanger, D., Bailey, B. K. & Steck, W. (1973). Formation of eduline and furoquinoline alkaloids from quinoline derivatives by cell suspension cultures of *Ruta graveolens*. *Phytochemistry*, **12**, 2399–405.

Bradley, M. V. & Dahmen, W. J. (1971). Cytohistological effects of ethylene, 2,4-D, kinetin and carbon dioxide on peach mesocarp callus cultured *in vitro*. *Phytomorphology*, **21**, 154–64.

Brodelius, P., Deus, B., Mosbach, K. & Zenk, M. H. (1979). Immobilized plant cells for the production and transformation of natural products. *FEBS Letters*, **103**, 93–7.

Brodelius, P. & Nilsson, K. (1980). Entrapment of plant cells in different matrices. *FEBS Letters*, **122**, 312–16.

Butcher, D. N. (1977). Secondary products in tissue cultures. In *Applied and Fundamental Aspects of Plant Cell, Tissue and Organ Culture*, ed. J. Reinert and Y. P. S. Bajaj, pp. 668–93. Berlin, Heidelberg and New York: Springer-Verlag.

Carceller, M., Davey, M. R., Fowler, M. W. & Street, H. E. (1971). The influence of sucrose, 2.4-D and kinetin on the growth, fine structure and lignin content of cultured sycamore cells. *Protoplasma*, **73**, 367–85.

Carlsson, J., Gabel, D., Larsson, E., Ponten, J. & Westermark, B. (1979). Protein-coated agarose surfaces for attachment of cells. *In Vitro*, **15**, 844–50.

Cheetham, P. S. J., Blunt, K. W. & Bucke, C. (1979). Physical studies on cell immobilisation using calcium alginate gels. *Biotechnology and Bioengineering*, **21**, 2155–68.

Chibata, I., Tosa, T. & Sato, T. (1974). Immobilised aspartase-containing microbial cells: Preparation and enzymatic properties. *Applied Microbiology*, **27**, 878–5.
Constantinides, A. (1980). Steroid transformations at high substrate concentrations using immobilised *Corynebacterium simplex* cells. *Biotechnology and Bioengineering*, **22**, 119–36.
Dallyn, H., Falloon, W. C. & Bean, P. G. (1977). The immobilisation of bacterial spores in alginate gel. *Laboratory Practice*, **26**, 773–5.
Dalton, C. C. & Street, H. E. (1976). The role of the gas phase in the greening and growth of illuminated cell suspension cultures of spinach (*Spinacia oleracea* L.). *In Vitro*, **12**, 485–94.
Davies, M. E. (1972). Polyphenol synthesis in cell suspension cultures of Paul's scarlet rose. *Planta*, **104**, 50–65.
Forrest, G. I. (1969). Studies on the polyphenol metabolism of tissue cultures derived from the tea plant (*Camellia sinensis* L.). *Biochemical Journal*, **113**, 765–72.
Hackel, U., Klein, J., Megnet, R. & Wagner, F. (1975). Immobilisation of microbial cells in polymeric matrices. *European Journal of Applied Microbiology*, **1**, 291–3.
Haddon, L. E. & Northcote, D. H. (1975). Quantitative measurement of the course of bean callus differentiation. *Journal of Cell Science*, **17**, 11–26.
Hahlbrock, K., Kuhlen, E. & Lindl, T. (1971). Anderungen von Enzymaktivitäten wahrend des Wachstums von Zellsuspensionkulturen von *Glycine max*: Phenylalanin-ammonia-lyase und *p*-cumurat: CoA-ligase. *Planta*, **99**, 311–18.
Hiraoka, N. & Tabata, M. (1974). Alkaloid production by plants regenerated from cultured cells of *Datura innoxia*. *Phytochemistry*, **13**, 1671–5.
Hirotani, M. & Furuya, T. (1977). Restoration of cardenolide synthesis in redifferentiated shoots from callus cultures of *Digitalis purpurea*. *Phytochemistry*, **16**, 610–11.
Ikuta, A., Syono, K. & Furuya, T. (1974). Alkaloids of callus tissues and redifferentiated plantlets in the Papaveraceae. *Phytochemistry*, **13**, 2175–9.
Jack, T. R. & Zajic, J. E. (1977). The immobilisation of whole cells. *Advances in Biochemical Engineering*, **5**, 125–45.
Jalal, M. A. F., Overton, K. H. & Rycroft, D. S. (1979). Formation of three new flavones by differentiating callus cultures of *Andrographis paniculata*. *Phytochemistry*, **18**, 149–51.
Johnson, D. E. & Ciegler, A. (1969). Substrate conversion by fungal spores entrapped in solid matrices. *Archives of Biochemistry and Biophysics*, **130**, 384–8.
Kaul, B. & Staba, E. J. (1968). *Dioscorea* tissue cultures: 1. Biosynthesis and isolation of diosgenin from *Dioscorea deltoidea* callus and suspended cells. *Lloydia*, **31**, 171–9.
Kennedy, J. F., Barker, S. A. & Humphrey, J. D. (1976). Microbial cells living immobilised on metal hydroxides. *Nature*, **261**, 242–4.
Kierstan, M. & Bucke, C. (1977). The immobilisation of microbial cells, subcellular organelles and enzymes in calcium alginate gels. *Biotechnology and Bioengineering*, **14**, 387–97.
Konoshima, M., Tabata, M., Yamamoto, H. & Hiraoka, N. (1970). Growth and alkaloid production of *Datura* tissue cultures. *Yakugaku Zasshi*, **90**, 370–7.
Larsson, P. O., Ohlson, S. & Mosbach, K. (1976). New approach to steroid conversion using activated immobilised microorganisms. *Nature*, **263**, 796–7.

Levine, D. W., Wong, J. S., Wang, D. I. C. & Thilly, W. C. (1977). Microcarrier cell culture: New methods for research-scale application. *Somatic Cell Genetics*, **3**, 149–55.

Lindsey, K. (1982). Studies on the growth and metabolism of plant cells cultured on fixed-bed reactors. PhD Thesis, University of Edinburgh.

Mäder, M., Münch, P. & Bopp, M. (1975). Regulation of peroxidase patterns during shoot differentiation in callus cultures of *Nicotiana tabacum*. *Planta*, **123**, 257–65.

Meyer-Teuter, H. & Reinert, J. (1973). Correlation between rate of cell division and loss of embryogenesis in long term tissue cultures. *Protoplasma*, **78**, 273–83.

Mizukami, H., Konoshima, M. & Tabata, M. (1977). Effect of nutritional factors on shikonin derivative formation in *Lithospermum* callus cultures. *Phytochemistry*, **16**, 1183–6.

Mohan, R. R. & Li, N. N. (1975). Nitrate and nitrite reduction by liquid-membrane-encapsulated whole cells. *Biotechnology and Bioengineering*, **17**, 1137–56.

Mosbach, K. (Ed.) (1976). *Methods in Enzymology*, vol. 44 (Immobilised Enzymes). New York: Academic Press.

Mosbach, K. & Larsson, P. O. (1970). Preparation and application of polymer-entrapped enzymes and microorganisms in microbial transformation processes with special reference to steroid 11β-hydroxylation and Δ^1-dehydrogenation. *Biotechnology and Bioengineering*, **12**, 19–27.

Mosbach, K. & Mosbach, R. (1966). Entrapment of enzymes and microorganisms in synthetic cross-linked polymers and their application in column techniques. *Acta Chemica Scandinavica*, **20**, 2807–10.

Murashige, T. & Skoog, F. (1962). A revised medium for rapid growth and bioassays with tobacco tissue cultures. *Physiologia Plantarum*, **15**, 473–97.

Nato, A., Bazetoux, S. & Mathieu, Y. (1977). Growth rate of cells affects enzyme activity. *Physiologia Plantarum*, **41**, 116–23.

Nettleship,L. & Slaytor, M. (1974). Adaptation of *Peganum harmala* callus to alkaloid production. *Journal of Experimental Botany*, **25**, 114–23.

Neumann, D. & Mueller, E. (1971). Beiträge zur Physiologie der Alkaloide: V. Alkaloidbildung in Kallus-und Suspensionkulturen von *Nicotiana tabacum*. *Biochemie und Physiologie der Pflanzen*, **162**, 503–13.

Neumann, D. & Mueller, E. (1974). Formation of alkaloids in callus cultures of *Macleaya*. *Biochemie und Physiologie der Pflanzen*, **165**, 271–82.

Nilsson, K. & Mosbach, K. (1980). Preparation of immobilised animal cells. *FEBS Letters*, **118**, 145–50.

Nishida, K., Sato, F. & Yamada, Y. (1980). Photosynthetic carbon metabolism in photoautotrophically and photomixotrophically cultured tobacco cells. *Plant and Cell Physiology*, **21**, 47–55.

O'Brien, T. P. (1974). Primary vascular tissues. In *Dynamics of Plant Ultrastructure*, ed. A. W. Robards, pp. 414–40. London: McGraw-Hill.

Ogino, T., Hiraoka, N. & Tabata, M. (1978). Selection of high nicotine-producing cell lines of tobacco callus by single cell cloning. *Phytochemistry*, **17**, 1907–10.

Ohlson, S., Larsson, P. O. & Mosbach, K. (1979). Steroid transformation by living cells immobilised in calcium alginate. *European Journal of Applied Microbiology and Biotechnology*, **7**, 103–10.

Ohmiya, K., Ohashi, H., Kobayashi, T. & Shimizu, S. (1977). Hydrolysis of lactose by immobilised microorganisms. *Applied and Environmental Microbiology*, **33**, 137–46.

Phillips, R. & Henshaw, G. G. (1977). The regulation of synthesis of phenolics in stationary phase cell cultures of *Acer pseudoplatanus* L. *Journal of Experimental Botany*, **28**, 785–94.

Radwan, S. S. & Kokate, C. K. (1980). Production of higher levels of trigonelline by cell cultures of *Trigonella foenum-graecum* than by the differentiated plant. *Planta*, **147**, 340–4.

Ramawat, K. G. & Arya, H. C. (1979). Effects of amino acids on ephedrine production in *Ephedra gerardiana* callus cultures. *Phytochemistry*, **18**, 484–5.

Roberts, L. W. (1976). *Cytodifferentiation in Plants*. Cambridge: Cambridge University Press.

Saini, R. & Vieth, W. R. (1975). Reaction kinetics and mass transfer in glucose isomerisation with collagen-immobilised whole microbial cells. *Journal of Applied Chemistry and Biotechnology*, **25**, 115.

Sejourne, M., Viel, C., Bruneton, J., Rideau, M. & Chenieux, J. C. (1981). Growth and furoquinoline alkaloid production in cultured cells of *Choisya ternata*. *Phytochemistry*, **20**, 353–5.

Spier, R. E. & Whiteside, J. P. (1976*a*). The production of foot-and-mouth disease virus from BHK 21C13 cells grown on the surface of glass spheres. *Biotechnology and Bioengineering*, **18**, 649–57.

Spier, R. E. & Whiteside, J. P. (1976*b*). The production of foot-and-mouth disease virus from BHK 21C13 cells grown on the surface of DEAE-sephadex A50 beads. *Biotechnology and Bioengineering*, **18**, 659–67.

Staba, E. J. (Ed.) (1980). *Plant Tissue Culture as a Source of Biochemicals*. Boca Raton, Florida: CRC Press.

Tabata, M., Yamamoto, H., Hiraoka, N. & Konoshima, M. (1972). Organisation and alkaloid production in tissue cultures of *Scopolia parviflora*. *Phytochemistry*, **11**, 949–55.

Thorpe, T. A. & Meier, D. A. (1973). Sucrose metabolism during tobacco callus growth. *Phytochemistry*, **12**, 493–7.

Toda, K. & Shoda, M. (1975). Sucrose inversion by immobilised yeast cells in a complete mixing reactor. *Biotechnology and Bioengineering*, **17**, 481–97.

van Wezel, A. L. (1967). Growth of cell strains and primary cells of microcarriers in homogeneous culture. *Nature*, **216**, 64–5.

van Wezel, A. L. (1976). The large-scale cultivation of diploid cell strains in microcarrier culture. Improvement of microcarriers. *Developmental Biology Standard*, **37**, 143–7.

Westcott, R. J. & Henshaw, G. G. (1976). Phenolic synthesis and phenylalanine ammonia-lyase activity in suspension cultures of *Acer pseudoplatanus* L. *Planta*, **131**, 67–73.

Yamada, Y., Imaizumi, K., Sato, F. & Yasuda, T. (1981). Photoautotrophic and photomixotrophic culture of green tobacco cells in a jar-fermentor. *Plant and Cell Physiology*, **22**, 917–22.

Yeoman, M. M. & Brown, R. (1971). Effects of mechanical stress on the plane of division in developing callus cultures. *Annals of Botany*, **35**, 1001–12.

Yeoman, M. M., Lindsey, K., Miedzybrodzka, M. B. & McLauchlan, W. R. (1982). Accumulation of secondary products as a facet of differentiation in plant cell and tissue cultures. In *Differentiation In Vitro*. British Society for Cell Biology Symposium 4, ed. M. M. Yeoman & D. E. S. Truman, pp. 65–82. Cambridge: Cambridge University Press.

Yeoman, M. M., Miedzybrodzka, M. B., Lindsey, K. & McLauchlan, W. R. (1980). The synthetic potential of cultured plant cells. In *Plant Cell*

Cultures: Results and Perspectives, ed. F. Sala, B. Parisi, R. Cella & O. Ciferri, pp. 327–43. Amsterdam: Elsevier North-Holland Biomedical Press.
Zeleneva, I. V. & Khavkin, E. E. (1980). Rearrangement of enzyme patterns in maize callus and suspension cultures: Is it relevant to the changes in the growing cells of the intact plant? *Planta*, **148**, 108–15.
Zucker, M. (1972). Light and enzymes. *Annual Review of Plant Physiology*, **23**, 133–56.

A. W. ALFERMANN, W. BERGMANN, C. FIGUR, U. HELMBOLD, D. SCHWANTAG, I. SCHULLER and E. REINHARD

Biotransformation of β-methyldigitoxin to β-methyldigoxin by cell cultures of *Digitalis lanata*

Introduction

The production of valuable plant products by biotransformation from cheap precursors which cannot be transformed effectively by chemical or microbial means is an interesting field for the practical application of plant cell cultures and has been reviewed recently (Furuya, 1978; Reinhard & Alfermann, 1980). Among the reactions observed, biotransformation of cardenolides is of special pharmaceutical importance because their glycosides (i.e. cardiac glycosides) are widely used in medicine for treatment of heart diseases. Digitoxin and digoxin, both of which are extracted from *Digitalis lanata* plants, are the most interesting ones. Nowadays digitoxin is used in therapy to a lesser extent than digoxin. *D. lanata* plants, however, always contain substantial amounts of digitoxin. This situation has led to increasing stockpiles of digitoxin. Digoxin differs from digitoxin only by an additional hydroxyl function at C-12. Undifferentiated cell cultures of *D. lanata* do not produce cardiac glycosides, but they are able to perform special biotransformations on substrates added into the medium.

Culture maintenance and analytical procedures

The medium, cell strain, culture conditions in shake culture and 20-l airlift reactors as well as the analytical procedures by thin layer chromatography (TLC) are as published elsewhere (Alfermann *et al.*, 1977; Wahl, 1977; Helmbold, 1979). In addition to the analysis by TLC, the cardenolide content of the medium of the 200-l reactor was analysed by high pressure liquid chromatography (HPLC) (Mechler, 1982) using a Hewlett-Packard 1084B Liquid Chromatograph equipped with a stainless steel column (25 cm long, 4.6 mm internal diameter), LiChrosorb RP 8, 7 μm (Merck, Darmstadt) as stationary phase and acetonitrile/water (a gradient moving from 40 to 50% acetonitrile until 5.5 min and thereafter held at 50%) as liquid phase. Oven temperature was 40 °C; temperature of acetonitrile 30 °C, and of water

50 °C. Measuring wavelength was 220 nm, and reference wavelength 360 nm. Preculture (see Fig. 3) for the 200-l reactor was performed in a Giovanola B 20 reactor (Giovanola Frères, Monthey, Switzerland) converted into an airlift reactor with an inner draught tube (area of draught tube: area of outer ring = 1:1.8). A ring sparger at the bottom of the draught tube produced an aeration rate of 0.25–0.42 vvm. For cultivation in 200-l volumes, a 200-l pilot plant of Chemap Company (Männedorf, Switzerland) was converted into an airlift reactor with an inner draught tube (area of draught tube: area of outer ring = 1:1). In this case, aeration was performed outside the draught tube (aeration rate = 0·25–0·5 vvm). Total width: height of the reactor was 1:3, and in the case of the B 20, 1:2. The procedure for immobilization and cultivation of the *Digitalis* cells was as described previously (Alfermann, Schuller & Reinhard, 1980).

Biotransformation of digitoxin

Fig. 1 shows the biotransformation of digitoxin, which would be a most appropriate substrate to use because it is produced during the technical

Fig. 1. Digitoxin is transformed by cell cultures of *Digitalis* to a series of products. Purpureaglycoside A is the main compound, whereas those hydroxylated at C-12 like digoxin, deacetyllanatoside C and lanatoside C are only produced in minor amounts. Dtx, digitoxose molecule.

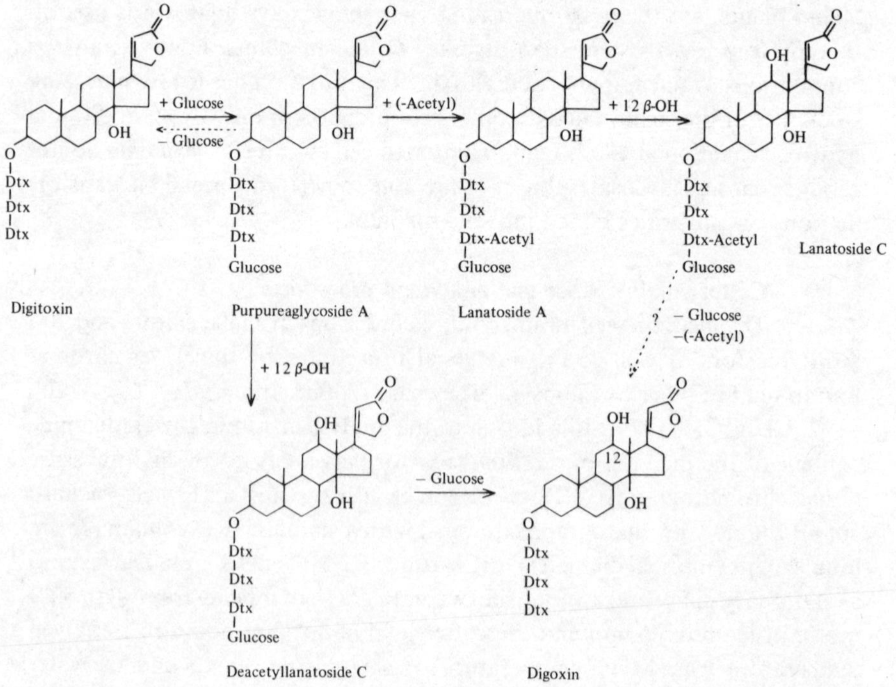

isolation of digoxin. Digitoxin is transformed by cell cultures of *Digitalis* to a series of products, some of which are hydroxylated at C-12. The main product, however, purpureaglycoside A, does not contain a hydroxyl function at C-12. Therefore, we have tested a series of digitoxin derivatives among which β-methyldigitoxin proved to be the most promising.

Biotransformation of β-methyldigitoxin

β-methyldigitoxin is prepared from digitoxin by chemical methylation. *Digitalis* cell cultures hydroxylate this compound to β-methyldigoxin (Fig. 2), a compound which is used widely in medicine and which up to now has been produced by chemical methylation of digoxin. During intensive screening work, we tested a large number of cell strains for their hydroxylating capacity. All the cell strains tested had been started from *D. lanata* plants rich in digoxin content. In this screening programme, however, it was found that different cell strains performed various reactions to a very different extent. A first group of strains was almost incompetent at biotransformation. A second group was only able to demethylate β-methyldigitoxin and to glucosylate it to purpureaglycoside A, eventually accompanied by acetylation to lanatoside A. A third group of strains could hydroxylate β-methyldigitoxin in addition to the reactions mentioned above. A fourth group could achieve the desired reaction and hydroxylated β-methyldigitoxin to β-methyldigoxin at high yields with almost no side reaction.

Development of a bioreactor process

Those strains of *Digitalis* with high hydroxylating capacity have been tested in 20-l airlift reactors. In order to optimise the product yields, various parameters have been manipulated. The manipulations included adding

Fig. 2. Digitoxin can be methylated by chemical means to β-methyldigitoxin which is transformed by selected cell lines of *Digitalis lanata* very rapidly and efficiently to β-methyldigoxin. Dtx, digitoxose molecule.

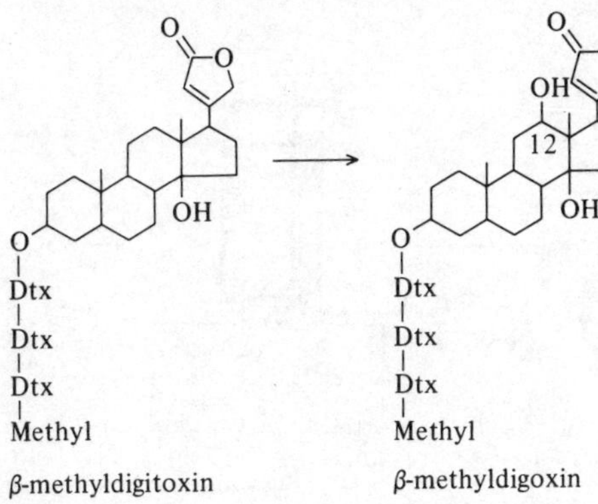

O O OH O | Dtx | Dtx | Dtx | Methyl

β-methyldigitoxin

O O OH 12 OH O | Dtx | Dtx | Dtx | Methyl

β-methyldigoxin

substrate during different growth phases in one or several batches, maintaining a constant level of substrate during cultivation, as well as the absolute level of substrate and varying the pH, cultivation temperature, aeration rate, level of glucose, level of phosphate and the solvent for β-methyldigitoxin, which is quite insoluble in the medium. In addition to highly purified β-methyldigitoxin, more crude preparations have been used to lower the costs of the process. Until now, product yields of up to about 700 mg l^{-1} have been achieved in a 20-l airlift reactor during a cultivation time of 17 days. More than 80% of the substrate was converted into the desired product and *c.* 10% remained untransformed and could be recycled. An interesting feature of this process is that more than 90% of the product can be extracted from the medium, only 10% being stored in the cells. In an eventual industrial process, methyldigoxin would probably be extracted only from the medium, that in the cells being discarded. This should result in an efficiency of the process of between 70 and 80%.

Fig. 3. Scaling-up of *Digitalis* cell cultures to a working volume of 200 l. For stages *2* and *3*, conventional bioreactors converted into airlifts with inner draught tubes are used. *Stage 1*: multiplication of cells in 1-l Erlenmeyer flasks. 300 ml of medium are incubated with 20 g of cells (fresh weight). Time, 7 days. *Stage 2*: multiplication of cells. A total volume of 30 l is incubated with the contents of 20–25 Erlenmeyer flasks of *Stage 1*. Time, 10 days. *Stage 3*: subcultivation and biotransformation. Reactor volume, 200 l; time, 15–20 days. 1, sampling device; 2, air inlet; 3, air outlet; 4, transfer valve; 5, control for temperature, pH, dissoved oxygen; 6, draught tube; 7, stirrer (during sterilization); 8, outlet; 9, sampling device and valve for transfer, respectively; 10, air inlet; 11, air outlet; 12, draught tube.

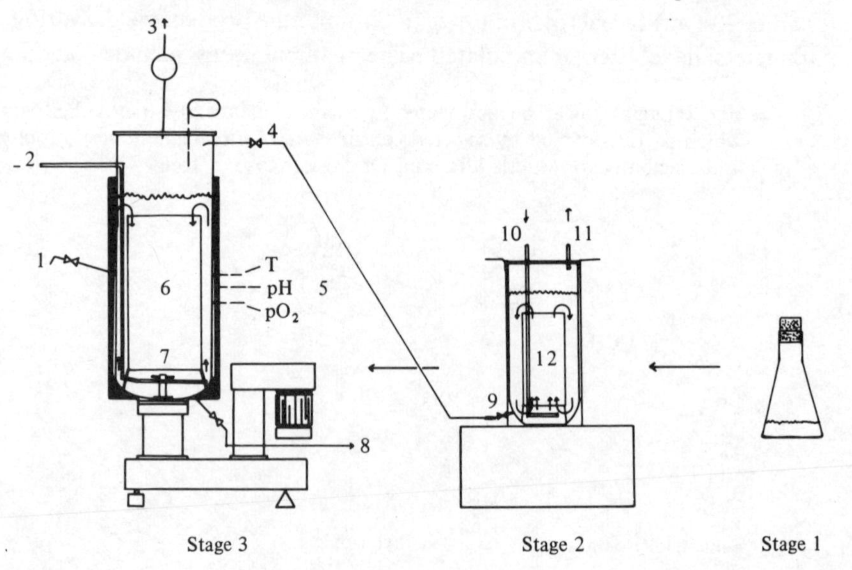

If one wants to use such a process for technical application it is necessary to test whether a scale-up to larger volumes is possible. Recently, we have installed a 200-l pilot plant. In these experiments, we use a normal commercial bioreactor which is converted into an airlift reactor with an inner draught tube. Fig. 3 demonstrates the scaling-up of the *Digitalis* cell cultures from 1-l Erlenmeyer flasks to the 200-l volume. Fig. 4 shows a typical biotransformation of β-methyldigitoxin in this reactor. On day 5 of cultivation, the cells are incubated with the substrate. Product formation parallels growth, in this case reaching a maximum after a further 13 days of biotransformation. After that time, 430 mg of β-methyldigoxin, which is about 70% of the substrate added,

Fig. 4. Biotransformation of β-methyldigitoxin (Mdt) to β-methyldigoxin (Mdg) in as 200-1 airlift reactor. Medium of Murashige & Skoog (1962), 2 mg l^{-1} kinetin, 1 mg l^{-1} IAA (indole-3-acetic acid), 1 g l^{-1} casamino acids (Difco Laboratories), 340 mg l^{-1} KH_2PO_4. Glucose is fed from day 5 to maintain a glucose level of between 4 and 10 g l^{-1}. The β-methyldigitoxin used is a crude preparation containing 90% β-methyldigitoxin and 3% α-methyldigitoxin with the rest unknown.

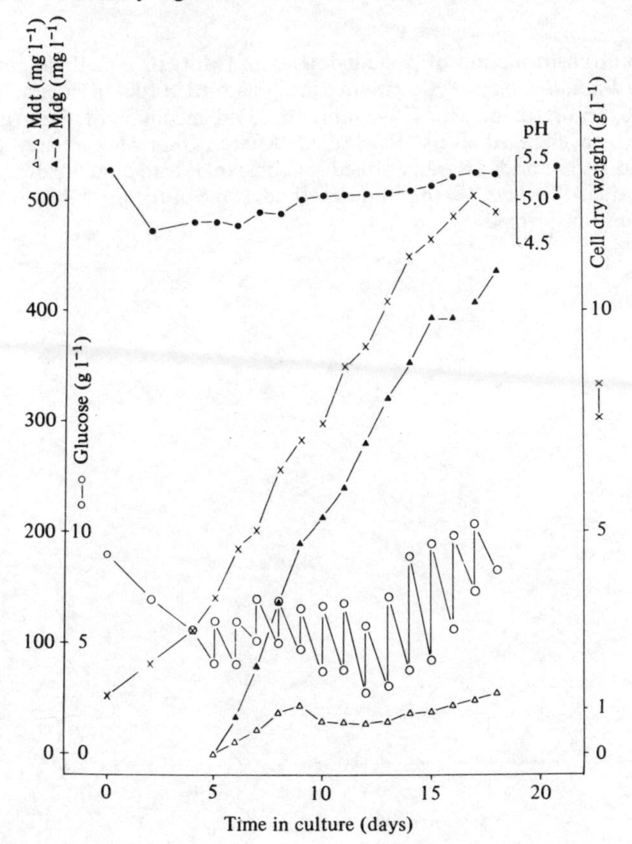

can be extracted from 1 l of medium. About 10% of the substrate remains untransformed. The amount of β-methyldigoxin produced in a 200-l reactor would be sufficient to prepare more than 800000 cardiovascular tablets, e.g. 'Lanitop', to treat 1000 patients with heart disease for more than one year.

Biotransformation by immobilised *Digitalis* cells

An interesting and fruitful development in recent years has been the immobilisation of bacterial, fungal, animal and plant cells in various matrices (Brodelius, 1978; Brodelius *et al.*, 1979) for the production or biotransformation of various compounds (see also Lindsey & Yeoman, this volume). Not only can immobilised cells be used for the study of fundamental aspects, but when appropriately applied they can also function over a long time, thus lowering the overall costs of the biocatalyst used. As with other plant cell culture systems, Figs. 3 and 4 show that, as compared with microbial processes, the slow growth rate of *Digitalis* cells results in long growth times for scaling-up to industrial fermentation volumes. It can be calculated from Fig. 3 that it takes at least 40 days to scale up to a working volume of 50 m³,

Fig. 5. Biotransformation of β-methyldigitoxin (Mdt) to β-methyldigoxin (Mdg) by *Digitalis lanata* cells entrapped in alginate gel in 100-ml Erlenmeyer flasks, 50 beads (diameter of 4–5 mm) in 25 ml medium. Medium and substrate are changed every third day. After a short lag phase, the β-methyldigitoxin added is transformed at a constant rate to β-methyldigoxin for more than 170 days. Owing to the method, some of the methyldigitoxin remains untransformed.

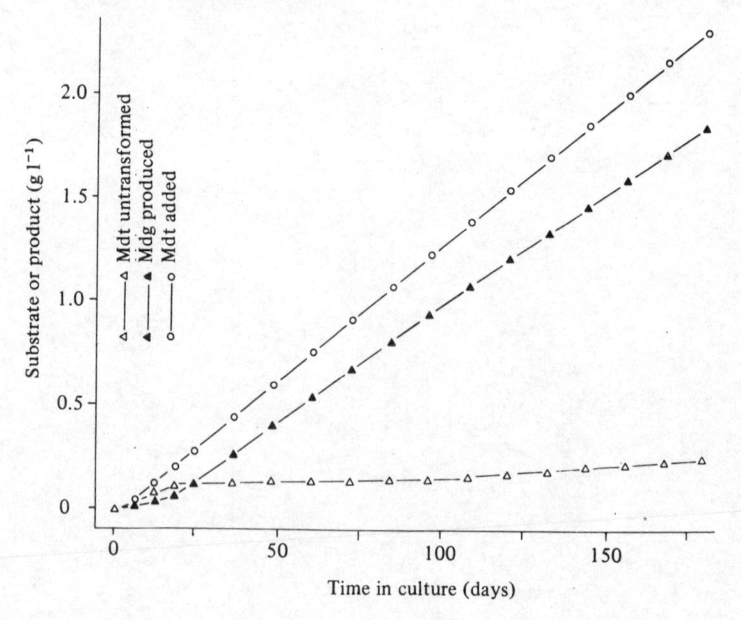

an economically viable working volume. This long growth time has until now been an important obstacle which prevents the practical use of our process. Therefore, we have tested biotransformation of cardiac glycosides by immobilised *Digitalis* cells (Alfermann *et al.*, 1980). Fig. 5 shows the biotransformation of β-methyldigitoxin by these cells when immobilised in alginate gel. It can be seen that in the small batch culture system, after a short lag phase, the immobilised cells proceed to hydroxylate β-methyldigitoxin at a constant rate for more than 170 days.

Perspectives

Our data show that the principal problems surrounding the biotransformation of cardiac glycosides by *Digitalis* cells are now solved. High product yields can be achieved after selection of specialised cell strains and optimisation of the culture conditions. Further improvements with regard to shortening the time for the scaling-up to industrial fermentation volumes will be most important for achieving a practical application of this plant cell culture process in industry. An eventual use of immobilised cells over a long period of operation is of great interest, especially because the product is found in the culture medium. Transferring these first encouraging results with immobilised cells into much larger reactor volumes, is, however, still necessary.

We thank the Federal Ministry for Research and Technology, Bonn, for sustaining this work by a research grant (PTB 03-8425-2). We also thank Dr S. H. Mantell for help in preparing the English version of this chapter.

References

Alfermann, A. W., Boy, H. M., Döller, P. C., Hagedorn, W., Heins, M., Wahl, J. & Reinhard, E. (1977). Biotransformation of cardiac glycosides by plant cell cultures. In *Plant Tissue Culture and Its Bio-technological Application*, ed. W. Barz, E. Reinhard & M. H. Zenk, pp. 125–41. Berlin, Heidelberg and New York: Springer-Verlag.

Alfermann, A. W., Schuller, I. & Reinhard, E. (1980). Biotransformation of cardiac glycosides by immobilized cells of *Digitalis lanata. Planta Medica*, **40**, 218–23.

Brodelius, P. (1978). Industrial application of immobilized biocatalysts. *Advances in Biochemical Engineering*, **10**, 75–129.

Brodelius, P., Deus, B., Mosbach, K. & Zenk, M. H. (1979). Immobilized plant cells for the production and transformation of natural products. *FEBS Letters*, **103**, 93–7

Furuya, T. (1978). Biotransformation by plant cell cultures. In *Frontiers of Plant Tissue Culture 1978*, ed. T. A. Thorpe, pp. 191–200. Calgary: Calgary University.

Helmbold, U. (1979). Versuche zur Steigerung der Hydroxylierung von β-Methyldigitoxin durch Fermenterkulturen von *Digitalis lanata* Zellstämmen. Doctoral Thesis, Fakultät für Chemie und Pharmazie, Universität Tübingen.

Mechler, E. (1982). Quantitative determination of β-Methyldigoxin by HPLC in fermenter broth. *Planta Medica*, **45**, 164.

Murashige, T. & Skoog, F. (1962). A revised medium for rapid growth and bioassays with tobacco tissue cultures. *Physiologia Plantarum*, **15**, 473–97.

Reinhard, E. & Alfermann, A. W. (1980). Biotransformation by plant cell cultures. *Advances in Biochemical Engineering*, **16**, 49–83.

Wahl, J. (1977). Fermentation von pflanzlichen Zellkulturen und 12β-Hydroxylierung von β-Methyldigitoxin durch Zellkulturen von *Digitalis lanata*. Doctoral Thesis, Fachbereich Pharmazie, Universität Tübingen.

S. H. MANTELL and H. SMITH

Cultural factors that influence secondary metabolite accumulations in plant cell and tissue cultures

Introduction

An extensive range of secondary metabolites, i.e. compounds without any clear function or role in the vital primary process of plant cells, which includes alkaloids, isoprenoids, plant phenolics, volatile oils and specialised proteins, has been isolated from cell and tissue cultures (Butcher, 1977; Staba, 1980) and *de novo* biosynthesis of these compounds in vitro is now well established (Kurz & Constabel, 1979). In the case of alkaloid biosynthesis in vitro, cultures derived from explants of various parts of an alkaloid-producing plant are capable of synthesising alkaloids and contain a spectrum of these substances which is identical to that found in the whole plant (Böhm, 1980). Some alkaloid-producing cultures may even contain raised levels of alkaloids in vitro, e.g. aposcopolamine predominates in culture extracts of several *Datura* spp. whereas this alkaloid is present in only trace amounts in vivo (Corduan, 1975). Furthermore, certain cultures produce novel alkaloids to those found in vivo, e.g. edulinine and aromorine production in *Ruta graveolens* and *Stephania cepharantha* cell cultures, respectively (Böhm, 1980). This phenomenon of altered secondary metabolism in cell cultures is by no means restricted to alkaloids but also appears to apply to the other groups of secondary plant metabolites produced by cell cultures, and has been attributed to various factors.

The first of these is the lack of tissue differentiation in some callus and liquid cell suspension cultures. For example, root-differentiating calluses of *Atropa belladonna* are capable of producing tropane alkaloids whilst non-differentiated cultures of the same plant material are not (Thomas & Street, 1970). Similarly, differentiated tissue cultures of *Papaver bracteatum* and *Nicotiana tabacum* tend to produce more thebaine and nicotine respectively than do comparable batches of undifferentiated calluses (Kamimura & Nishikawa, 1976; Pearson, 1978). In the latter case, however, although loss of nicotine-producing capability in seedling callus lines of cv. Maryland Mammoth is correlated with the decline in shoot number, similar cultures of other cultivars, such as NC 2512, held under identical conditions do not show this correlation between declining alkaloid yield and reduced shoot formation. The lack of specialised cell structures in some cultures may be a further reason for the absence (or at least for much reduced levels) of accumulated secondary metabolites (particularly volatile oils, resins and latex (Krikorian & Steward, 1969)). Despite this, evidence that organogenesis in tissue cultures is not an essential prerequisite for secondary metabolite biosynthesis in vitro is now accumulating (Staba, 1980; Fowler, this volume). In fact, decreased secondary metabolite yields have sometimes been reported for cultures as a consequence of organogenesis. For example, undifferentiated cultures of *Dioscorea deltoidea* and *Agave wightii* yield 1–2% (*dry weight basis*) steroidal sapogenins, but when cultures differentiate to produce roots or bulbils respectively, only trace amounts of sapogenins are produced (Kaul & Staba, 1968; Sharma & Khanna, 1980). Also, Böhm (1978) has observed that the latex cells scattered throughout cell aggregates of *Macleaya microcarpa* suspension cultures have a distinct storage function and that other morphologically uniform cells within the same cultures actively synthesise alkaloids. Organ and tissue specialisation in the morphological sense would therefore not appear to be a prerequisite for biosynthesis of alkaloid in these cultures. The number of reports on the establishment of high metabolite-yielding dedifferentiated cell cultures increases yearly, some cultures producing up to 50–60 times the amount of specific secondary metabolites present in the differentiated parent plant, e.g. the high levels of indole alkaloids produced by *Rauwolfia serpentina* cell cultures (Stöckigt, Pfitzner & Firl, 1981).

A second factor may be that the biosynthetic potential of cell cultures derived from one part of a plant differs from that of cultures derived from another part of the plant. There is little support for this idea and in general the source of an explant does not influence the secondary metabolite production capacity of cells and tissues cultured for prolonged periods. This reflects the phenomenon of the biochemical totipotency of plant cells referred to by Zenk (1978), i.e. that all the necessary genetic and physiological potential for secondary metabolite formation is present in an isolated cell and that cultured cells, irrespective of the part of the plant from which they are excised, can be expected to yield similar secondary metabolites when held under stable cultural conditions.

A third possible factor is that structural rearrangements of the genomes of cultured cells caused by endoreduplication and/or nuclear fragmentation processes (features inherent to in vitro cell growth; see D'Amato, 1977; Yeoman & Forche, 1980) might lead to significant alterations in the genotypes of a portion of a cell population, thereby causing altered secondary metabolism in these cells. Since the proportion of cells containing altered genotypes might increase during prolonged culture periods (especially in cases where these genotypes produce fast-growing cell types), a gradual decline in secondary metabolite productivity may be one outcome. It is significant, however, that the biosynthetic capabilities of plant cell cultures can be restored upon regeneration of plants from dedifferentiated cultures. For example, Kartnig (1977) observed that the productivity and cardenolide spectra of *Digitalis* cell cultures declined to non-detectable levels after 16–18 serial passages of 5–6 weeks each. However, when plants were regenerated from these apparently non-productive cultures, the original cardenolide spectra were restored in regenerated plants. Therefore, the original genotype for controlling cardenolide biosynthesis had been retained through many subcultures. Situations of this type lead one to suppose that the expression of genes controlling the biosynthesis of secondary metabolites is most likely under some form of derepression/repression control mechanism. This hypothesis is supported by the findings of Hirotani & Furuya (1977) which showed that the genes responsible for the control of the biosynthesis of secondary metabolites (in this case the biosynthesis of cardenolides in *Digitalis purpurea*) are in fact still present in the cells of unproductive cultures. Evidence in the literature now tends to support the view that cultured plant cells derived from plants which produce detectable levels of secondary metabolites do themselves contain the necessary genetic information for the biosynthesis of these same metabolites.

As well as to genotypic perturbations, the phenotype of cultured cells may be subject to extranuclear heritable alterations, i.e. to epigenetic variability of the type described by Meins & Binns (1978) and reviewed in Dougall (1980). Epigenetic phenomena may result in the accumulation of auxin- or cytokinin-independent cells in cultures, and the presence of these cells may drastically reduce the proportion of productive cells in cultures (for more detailed discussion see under Growth regulators).

Improvements in the secondary metabolite productivity of plant cell cultures can be made by taking explants from parent plants which accumulate high levels of a particular secondary metabolite. Indeed, Zenk *et al.* (1977*a*) and Kinnersley & Dougall (1980*b*) have clearly demonstrated this. Cultures derived from parent plants of *Catharanthus roseus* and *N. tabacum* which produce high levels of serpentine and nicotine respectively, produce higher amounts of these secondary metabolites than do comparable cultures derived from parent plants which produce low levels of these metabolites. Variant cells in these cultures can be further selected for by using special screening techniques (see Tabata *et al.*, 1978). These variants are believed to possess enhanced biochemical potential by which they have the capacity to produce and accumulate secondary metabolites. Provided that stable cultures of these cells are exposed to appropriate physiological stimuli (i.e. to the 'triggers' or 'effectors' of secondary metabolic pathways), it is possible to induce the biosynthesis of significant amounts of secondary metabolites in vitro (Zenk, 1978). The extent to which cultural conditions influence secondary metabolite accumulation in vitro is therefore worthy of review because of their likely role as triggers of secondary metabolism (Yeoman *et al.*, 1980).

It is a characteristic of many microbial and plant cell cultures that, after a phase of rapid cell division in a batch culture system, the growth rate slows down and secondary product formation and other features of cell specialisation begin (Fig. 1). This is the so-called trophophase–idiophase development described by Bu'Lock (1975) and Luckner, Nover & Böhm (1977). Generally, a primary metabolite is synthesised as a direct result of the metabolic processes which keep the cells alive and growing and accumulates in tandem

Fig. 1. A comparison of primary (*a*) and secondary (*b*) metabolite accumulation in microbial fermentations with secondary metabolite accumulation in plant cell cultures (*c*). (*a*) Ethanol production by yeast cells; (*b*) penicillin production by fungal cells; (*c*) nicotine production by tobacco cells. The relative durations of trophophase (dotted areas) and idiophase (clear areas) are shown for cultures producing secondary metabolites.

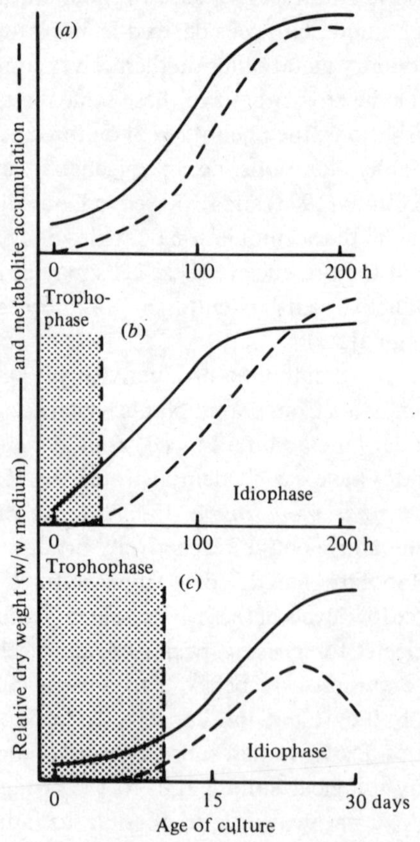

with increases in cell dry weight (Fig. 1 *a*). In contrast, a secondary metabolite is generally not formed as a direct result of metabolism that maintains cells in an actively dividing state and the accumulation of these metabolites tends to lag behind cell growth (Fig. 1 *b*, *c*). Therefore, cultural conditions favouring rapid growth are seldom those that are best for the biosynthesis of secondary metabolites (Luckner, 1980). Consequently, in a batch culture process designed to produce a particular plant product, a compromise between two sets of optimum conditions, i.e. those for the accumulation of biomass and those for the biosynthesis and accumulation of secondary metabolites, is best striven for. A two-stage strategy has been suggested as the most suitable way of increasing the productivity of plant cell cultures raised in batch systems (Alfermann & Reinhard, 1978).

The accumulation of secondary metabolites at any one stage of culture is the result of a dynamic balance between biosynthetic, biotransformational and biodegradative processes. This situation is complex and is undoubtedly the reason why mainly empirical approaches have been adopted so far in studies on cultural factors which influence the accumulation of secondary metabolites in plants. Cultural factors are considered below under three main categories: those pertaining to the external and internal cultural environments and those brought about by the use of different preculture treatments of cells. The degree of influence of these factors is illustrated by reference to the production of nicotine and steroids by cell cultures of *N. tabacum* and *S. aviculare* respectively, which has been studied in our laboratory.

Experimental systems

Methods used for establishing and maintaining plant cell and tissue cultures are those comprehensively described by Street (1977 *a*) and more recently by Seabrook (1980) and Dougall (1980). Three major types of in vitro cultures have been regularly used in studies on secondary plant metabolites. These are organ, callus (usually grown on solid media) and liquid cell suspension cultures. The latter, though containing aggregates of anything up to 200 cells, normally consist of dispersed cells and in this state it has been possible for these to be maintained in large fermentor vessels (King, 1980). For metabolic studies, cell suspension cultures are generally used only after a stabilisation period of culture (5–10 generations) in order to avoid the possibility of carry-over of secondary metabolites from parent tissue. Qualitative and quantitative analysis of secondary metabolites is normally carried out on culture media as well as on the cells since some metabolites (particularly alkaloids) are thought to be excreted from the cells into the media. Metabolite yields are usually compared on a product weight per unit weight of cells or volume of medium basis. Further details on these methods and their application to studies on secondary plant metabolism can be found in Street (1977*b*), Martin (1980) and in previous chapters of this book.

For our work on the production of nicotine, suspension cultures of *N. tabacum* were obtained by transferring callus from a 3–5-week-old stock culture, grown on solidified Murashige & Skoog (MS) (1962) medium supplemented with 10^{-5} M naphthaleneacetic acid (NAA) and 10^{-6} M kinetin, into 70 ml of liquid stock culture medium and culturing on an orbital shaker (110–120 r min^{-1}). Routine subculture of cell suspensions was carried out every two weeks using a dilution of one stock culture (previously filtered through 500 μm wire sieves to remove large cell aggregates) to 500 ml fresh medium (*c*. 1:7). For nicotine production, cultures were grown in a special production medium (MS minus casein hydrolysate and supplemented with 10^{-6} M NAA and 10^{-7} M kinetin). Cultures of *Solanum aviculare* were established and maintained in a similar manner except that stock media contained 10^{-5} M NAA and 5×10^{-6} M kinetin. Sucrose was maintained at 30 g l^{-1} in stock cultures. Continuous diffuse light (5.38 W m^{-2}) was supplied to cultures from 25-W 'Cool White' tubes. The 4-l batch fermentors used in the nicotine and steroid work were similar to those described by Wilson, King & Street (1971) with the following modifications. Internal heating coils were omitted and as a replacement for a magnetic stirrer (which induced deleterious shear) the air inlet tube was extended to within 1 cm of the culture base and the aeration of the culture (at 3 l min^{-1}) served to keep the cells suspended.

Identification of individual groups of secondary metabolites in cultures described can be found in the cited references. Nicotine was determined using either gas liquid chromatographic (GLC) or UV spectrophotometric methods on ether extracts of at least three replicate batches of *N. tabacum* cells. GLC separations were carried out in 180-cm glass analytical columns packed with 10% carbowax 20 M with 5% potassium hydroxide supported on 80–100 mesh chromosorb W. The column temperature was 160 °C, the temperature of the injection port was 200 °C and the temperature of the detectors was 250 °C. Gas flow rates were: nitrogen (carrier), 30 ml min^{-1}; hydrogen, 30 ml min^{-1}; air, 300 ml min^{-1}. The amount of nicotine was measured in a 5-μl diethyl ether sample and quantified using quinaldine as an internal standard. UV spectrophotometry was carried out using a modification of the method of Willits, Swain, Connelly & Brice (1950). Steroidal constituents of *S. aviculare* were separated by GLC on a 3% OV-17 Chrom Q (100–200 mesh) column at a temperature of 275 °C using nitrogen carrier gas with a flow rate of 45 ml min^{-1}. Cell samples from at least four replicate cultures of *S. aviculare* were prepared for GLC by refluxing for 2 h in 3 M sodium hydroxide followed by extraction in ether.

Factors in the external cultural environment

Light

The characteristics of radiation which influence plant development in vivo are also those which affect plant tissues and cells in vitro. As cultures do not normally grow photoautotrophically (except in a few special cases as reviewed by Yamada, Sato & Hagimori, 1978), the majority of plant tissue culture work has been carried out using low radiance levels of broad spectral quality. The behaviour of cultures is influenced by photoperiodicity, light quality and light intensity (Seibert & Kadkade, 1980). The activities of the enzymes involved in the biosynthesis of cinnamic acids, coumarins, lignins, flavones, flavonols, chalcones and anthocyanins are influenced significantly by light (Hahlbrock, Schröder & Vieregge, 1980). For instance, the activities of the Group 1 and Group 2 enzymes of the flavonoid pathway in cultured cells of *Petroselinum hortense* (parsley) show an increase when cells are exposed to light for 2 and 4 h respectively. The activities of the Group 1 enzymes, phenylalanine ammonia lyase (PAL) included, can be increased independently of light by transferring dark-grown parsley cells to distilled water. Further, the extent of this PAL-activity change is dependent on the degree of cell dilution. A second increase in PAL activity is observed 5 h after dilution but this is light induced. In determining the active spectral region of PAL induction by light, several investigators have shown that blue light increases both the activity of this enzyme and subsequent anthocyanin production in callus and cell cultures, e.g. those of *Haplopappus gracilis* (Stickland & Sunderland, 1972). The effect of the light induction of PAL and other flavonoid pathway enzymes in parsley cell cultures is shown by the degree to which secondary metabolites accumulate as a result of exposing cultures to continuous 'Cool White' fluorescent light. More than 20 flavone and flavonol glycosides are produced after such light treatment (Grisebach & Hahlbrock, 1974). Flavone glycoside synthesis is most sensitive to UV light at wavelengths below 320 nm and subsequent post-UV irradiation red/far-red modulations in glycoside synthesis have been demonstrated (Wellmann, 1975), indicating that a low-energy phytochrome system is most likely operating in such cultures. The induction of anthocyanin biosynthesis in cell cultures by light mentioned above has been demonstrated repeatedly in different cell systems. An action spectrum for this induction with peaks at 372 nm and 438 nm has been reported (Seibert & Kadkade, 1980). Moreover, cultures of *H. gracilis* left in the dark for three days prior to light exposure accumulated more anthocyanin than did cultures kept in the dark for one day prior to light exposure (Fritsch, Hahlbrock & Grisebach, 1971). Interestingly, Alfermann & Reinhard (1971) were able to replace this light requirement for anthocyanin biosynthesis with auxin treatment in carrot cell cultures. However, the time dependence of light-induced and auxin-induced synthesis was quite different with anthocyanins being detectable only three days after light induction but not until 6 days after the addition of auxin. Other work (Davies, 1974) showed that cycloheximide and 2-thiouracil inhibited the latter induction but promoted the former. Such results indicate that there may be at least two different mechanisms involved in the regulation of anthocyanin accumulation in carrot cultures.

Other stimulatory effects of light induce increased polyphenol biosynthesis in cultures from the tea plant (Forrest, 1969) and Paul's Scarlet Rose (Davies, 1972). Upon illumination, tea callus cultures showed several-fold increases in catechin, epicatechin and leucoanthocyanin synthesis. In Paul's Scarlet Rose cell suspension cultures, the initial rate of polyphenol synthesis was influenced by several factors, the most important of these being light intensity. High intensities of light could partially reverse the inhibition of polyphenol synthesis caused by 5×10^{-5} M 2,4-dichlorophenoxyacetic acid (2,4-D) and could stimulate accumulation at lower (5×10^{-7} M) auxin levels. Biosynthesis of other metabolites such as volatile oils in stem tissue cultures of *R. graveolens* also depends on both light intensity and light quality. Volatile-oil accumulation in cultures grown under continuous 'Cool White' light (250 lx) resembled that of photosynthesising parent plants (Corduan & Reinhard, 1972). However, the relative composition of the oils could be significantly altered by administering different spectral treatments to cultures. Cells grown under continuous red or far-red light (1.25 W m^{-2}) produced the same major oil components as those found in dark-grown cultures but others grown under blue light produced an oil composition comparable to that grown under 'Cool White' fluorescent light (250 lx) on long days (either 15- or 24-h photoperiods). On short days (6-h photoperiod) in the same light quality conditions, cultures produced oil accumulations which were a mixture of those obtained separately under continuous light and continuous dark conditions (Nagel & Reinhard, 1975). 'Cool White' light also stimulates the biosynthesis of numerous other secondary metabolites including steroidal sapogenins (e.g. diosgenin in tuber-derived callus and cell suspensions of *Dioscorea*), steroidal alkaloids (e.g. solasodine and solamargine in *Solanum* (Seibert & Kadkade, 1980)) and some alkaloids (e.g. serpentine in *C. roseus* cell cultures (Röller, 1978)). In these cases, the promotory effects of light may possibly be due to its known influence on the rate of uptake of sugars and nutrients into plant cells. Enhanced uptake of both [^{14}C]sucrose and nitrate is known to be stimulated by light in etiolated plant tissues, both of these responses being under the control of phytochrome (e.g. Goren & Galston, 1966; Jones & Sheard, 1975; respectively) through its mediation of increases in intracellular ATP levels that are necessary for the active uptake of nutrients by cells (see White & Pike,

1974). In contrast to these triggering effects of light on secondary metabolism in vitro, several reports indicate that light can have inhibitory effects, e.g. that of Tabata, Yamamoto, Hiraoka & Konoshima (1972) on suppressed alkaloid production in *Scopolia parviflora*, that of Brain (1976) on suppressed L-dopamine (L-DOPA) accumulation in static cultures of *Mucuna pruriens* and that of Ohta & Yatazawa (1978) on a 70% suppression of nicotine production in *N. tabacum* callus by high light intensities. Although low light intensities did not completely inhibit nicotine synthesis in *N. tabacum* cv. NC 2512, continuous darkness proved to be a more suitable condition for supporting high levels of nicotine accumulation (Fig. 2). These findings bear a close resemblance to the effect of light and dark treatments on the nicotine content of tobacco seedlings in vivo: seeds germinated in the light contain lower levels of nicotine than do those germinated in the dark (Weeks, 1970). Higher levels of accumulation of secondary metabolites under dark rather than light conditions suggest that photodegradation of certain metabolites and/or enzymes may occur although there is no direct evidence to substantiate this. It should, nevertheless, be borne in mind since in the case of nicotine

Fig. 2. Effect of light on the fresh weight and nicotine content of suspension cultures of *N. tabacum* cv. NC 2512. (From Pearson, 1978.)

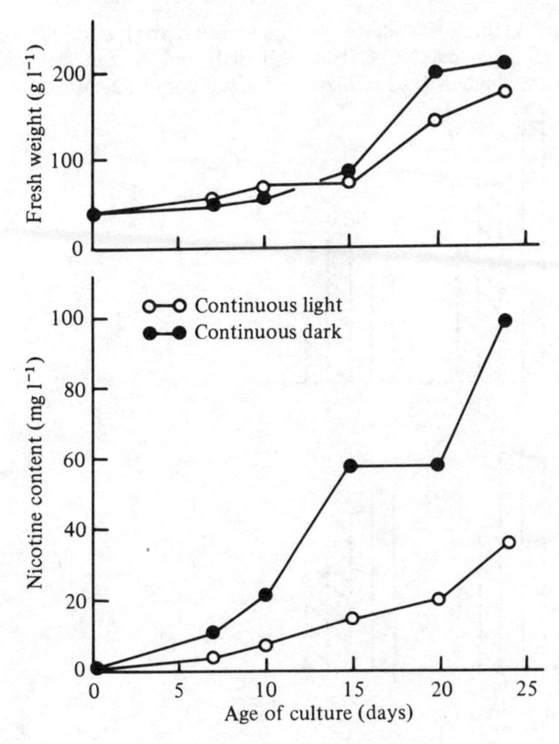

photodegradation in vitro has been demonstrated (Doumery & Chouteau, 1975). Also, increased alkaloid accumulation in darkness could be due to the repression, described by Barz *et al.* (1978) and Böhm (1978), of biodegradative processes which regulate the turnover of plant alkaloids. Finally, there are also reports in the literature which indicate that light/dark treatments have no significant effect on the accumulation of some secondary metabolites, e.g. those of Zenk, El-Shagi & Schulte (1975) and Ikeda, Matsumoto & Noguchi (1977) on anthraquinone production in *Morinda citrifolia* and ubiquinone production in *N. tabacum* suspension cultures, respectively.

Temperature

Little information is available in the literature on temperature optima for the growth of cell cultures, let alone for secondary metabolite production in these cultures. This is probably due to the fact that, traditionally, in vitro studies have been carried out as a matter of routine at temperatures of around 25 °C. Investigations may prove rewarding since the effects of temperature on alkaloid levels in vivo are well known. For instance, 100–200% more nicotine accumulates in tobacco seedlings held at 27 °C than in those held at either 21 or 32 °C (Weeks, 1970). The data shown in Fig. 3 demonstrate

Fig. 3. Effect of temperature on fresh weight and nicotine production in two callus lines of *N. tabacum* cv. NC 2512, unshaded, and *N. tabacum* cv. Maryland Mammoth, shaded. (L. P. Hazell, personal communication.)

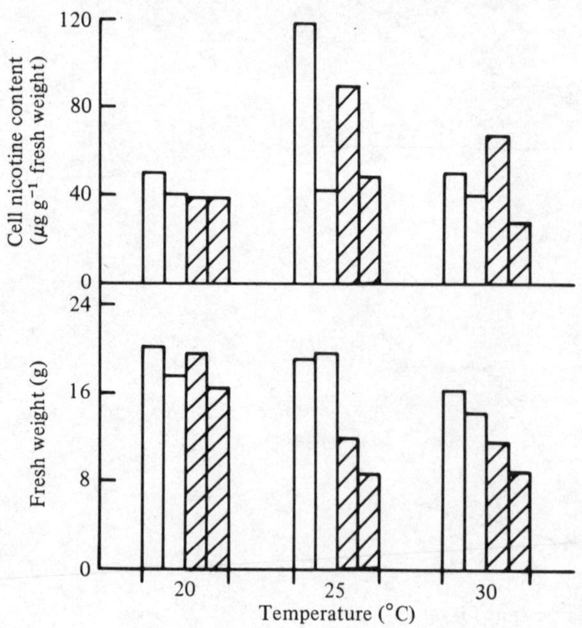

the influence of temperature on growth and nicotine production by tobacco callus cultures. A previous report on growth optima for such cultures by Matsumoto, Ohunishi, Nishida & Noguchi (1972) indicated that optimal growth occurs at 32 °C rather than at 28 °C or 24 °C. Similar studies on alkaloid production in *Peganum* callus cultures by Nettleship & Slaytor (1974) showed that optimal growth of callus occurred at 30 °C while maximum alkaloid production was attained at 25 °C with levels of production decreasing rapidly at higher temperatures. The basis of such observations may be temperature-dependent shifts in the metabolism of cultured cells. For instance, the rate of both sucrose and amino-nitrogen utilisation in *Ipomoea* cell suspension cultures is maximal between 30 and 32 °C and both of these rates decline by about 25% when cells are moved from 30 to 25 °C whereas growth rate declines little (Rose & Martin, 1975). The fatty acid content of cultured plant cells is much increased in cultures grown at suboptimal growth temperatures such as 15 °C (MacCarthy & Stumpf, 1980).

Culture vessel agitation

The special case of rotary shake-flask cultures, a most widely used system for experimentation on secondary metabolite production, is considered here. In this system, the rotation speed of a shaker (normally 90–120 r min^{-1}) can have an important effect on growth and metabolite accumulation. Rajasekhar, Edwards, Wilson & Street (1971), in a study of the effect of the shaking rate on the growth of *Atropa* and *Acer* cultures, concluded that reduced growth at suboptimal shaking speeds was not due to oxygen deficiency nor to accumulation of carbon dioxide but rather to either an unknown volatile toxic factor (not ethylene) or to restricted nutrient uptake resulting from a stationary liquid-phase boundary surrounding cells. Similar studies by Matsumoto *et al.* (1972) on *N. tabacum* cell growth demonstrated that this was not greatly different at 80 r min^{-1} or 110 r min^{-1}. However, more recent results on growth and nicotine production by tobacco cells (Pearson, 1978) indicate that raised culture agitation (150 r min^{-1}) induces high nicotine yields whereas normal agitation (110 r min^{-1}) causes slight inhibition of growth but an almost total suppression of nicotine production. There is clearly need for further clarification of aspects of culture vessel agitation if data on secondary metabolite accumulation in shake-flask cultures is to be meaningful.

Factors in the internal cultural environment

In comparison to those cultural factors which can be relatively easily controlled outside cultures, those factors associated with the culture itself, i.e. the growth medium, the metabolic state of stationary and dividing cells, dead cells with their associated detritus and the gaseous environment above the culture surface, are undoubtedly complex multicomponent entities. Consequently, the approach to the study of these remains an empirical one. For convenience, this multicomponent system is divided here into four main parts on the understanding that consideration of one parameter does not preclude the direct and/or indirect influences of the others on its own role in secondary metabolite accumulation.

Medium components

Media formulations suitable for the culture of callus and cell suspensions of plants have been given by Seabrook (1980). The medium most widely used is that of Murashige & Skoog (1962), which was devised originally for *rapid growth* of tobacco calluses. This medium, although capable of supporting the growth of most plant cells, is by no means optimal for inducing secondary metabolism in these cells. Generally, medium conditions which most frequently support active secondary metabolism are those which limit rapid cell division and lead to a comparatively early cessation of exponential growth. Cultural conditions which support protracted periods of decelerating and stationary phases are generally conducive to secondary metabolite biosynthesis and accumulation in plant cell cultures.

Growth regulators. Plant growth regulators are effective triggers of secondary metabolism in vivo (see Böhm, 1980). Similarly, in vitro both the quality and quantity of auxins initially present in media or administered during the course of culture development have a marked effect on primary (Everett, Wang & Street, 1978; Everett, Wang, Gould & Street, 1981) and secondary metabolism (Gamborg *et al.*, 1971). With the exception of habituated and oncogenic cultures, plant cell cultures require the addition of growth regulators, i.e. auxins and cytokinins, to media for consistent growth by cell division. Growth by differentiation or morphogenesis, on the other hand, can usually be induced by lowering the auxin concentration or by supplying less active growth substances. Since the production of secondary metabolites in plant cell cultures is a function of both cell multiplication and division, growth regulators have a major role in determining the potential productivity of a given culture. There are numerous examples in the literature of these types of growth regulator effects on secondary metabolism in vitro, so for more extensive coverage of these aspects, reference should be made to recent reviews such as those by Kurz & Constabel (1979) and Staba (1980).

The effect of auxin type on secondary product synthesis has been investigated in cultures of numerous species. For production of thebaine by suspension cultures of *Papaver bracteatum*, indole-3-acetic acid (IAA) was found to be better than NAA or 2,4-dichlorophenoxyacetic acid (2,4-D) for stimulating alkaloid synthesis (Kamimura & Nishikawa, 1976). Further, Zenk *et al.* (1975) observed even greater differences, with NAA stimulating the production of anthraquinones in suspension cultures of *M. citrifolia* and 2,4-D totally suppressing anthraquinone production in these cultures. Both auxins were examined over a wide range of concentrations, thus ruling out any possibility that the observed results were due to differences in auxin activity. These cells also showed a remarkably sensitive response to 200 synthetic effectors (of an auxin type) as far as anthraquinone production was concerned (Zenk, 1978). The quality of the substituent in the *para*-position of the phenoxyacetic acid series had a drastic influence on the triggering of secondary metabolite formation under conditions where the various growth regulators all supported growth and multiplication of cells. Similar observations were made in studies on the effects of 33 phenoxyacetic acids (again with different substitution patterns) on serpentine production by *C. roseus* cells (Böhm, 1978). Of these, six compounds triggered higher serpentine concentrations than did the standard auxin NAA. These compounds contained no substituents or methyl groups in position four of their rings. In contrast, phenoxyacetic acids which inhibited serpentine biosynthesis were all halogen-substituted in this position and the lowest alkaloid levels were recorded in 2,4-D-supplemented media. Auxin-quality effects on the production of other alkaloids such as nicotine are also well known. Furuya, Kojima & Syono (1971) found that in callus cultures of *N. tabacum* cv. Bright Yellow cultured for five years in the presence of 2,4-D no alkaloids were detected, while nicotine, anatabine and anabasine were readily found in calluses growing in media supplemented with IAA. Upon transfer of calluses from 2,4-D- to IAA-supplemented media or vice versa it was demonstrated that nicotine biosynthesis was activated by IAA and suppressed by 2,4-D. In conclusion, 2,4-D has generally been found to be less suitable for triggering secondary metabolism in plant cell cultures than either IAA or NAA. In spite of this, 2,4-D continues to be a most widely used auxin and as such may account for many of the low yields and absences of secondary metabolites observed in some cultures. The possible reason for 2,4-D inhibition of alkaloid biosynthesis has been suggested by Kurz & Constabel (1979) to be the much lower pools of glutamate and aspartate (important amino-acid precursors of many alkaloids) that are present in cells exposed to this auxin. A marked effect of initial auxin (NAA) concentration on nicotine biosynthesis in tobacco cell cultures has been observed (Fig. 4). A stimulatory as well as an inhibitory effect of NAA concentration on nicotine biosynthesis is seen from these data.

In exceptional cases, the auxin type (see Tabata *et al.*, 1972) and concentration (Hiraoka & Tabata, 1974) have been reported to have little or no effect on secondary metabolite (nicotine) yield. However, these observations were made on cultures in which alkaloid content was consistently low, so that there was a possibility that other factors were also limiting biosynthesis. Instances have even been reported where maximum yields of secondary metabolites were obtained when cells were raised on media devoid of auxin, e.g. in cardenolide production by *Digitalis* cultures (Büchner & Staba, 1964) and alkaloid production by *Peganum* callus (Nettleship & Slaytor, 1974) and *C. roseus* (Carew & Krueger, 1977). In these three situations, however, omission of auxin resulted in culture differentiation and morphogenesis.

There are only a few reports of metabolite levels being increased by raising auxin levels, e.g. maximum carotenoid and ubiquinone contents of carrot cell cultures were induced by the comparatively high 2,4-D level of 10 mg l^{-1} (Ikeda, Matsumoto & Noguchi, 1976).

In most studies auxins have been applied to cultures in combination with cytokinins. Thus, results of such combinations have to be assessed accordingly though there are several instances where distinct cytokinin effects on secondary

Fig. 4. Effect of NAA concentration on the fresh weight and nicotine content of suspension cultures of *N. tabacum* cv. NC 2512. (From Pearson, 1978.)

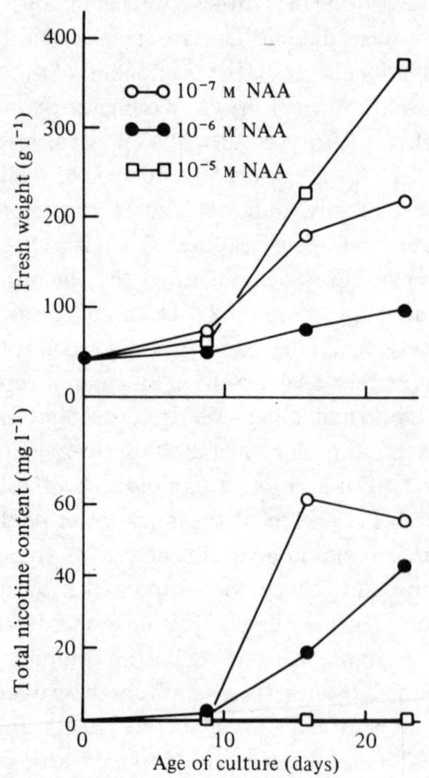

metabolite accumulation have been recorded. In callus cultures of *Datura tatula*, kinetin showed no noticeable effect on growth yet it was inhibitory to alkaloid production at high concentrations (Tabata, Yamamoto & Hiraoka, 1971). In *Scopolia maxima* cultures, a comparatively high concentration of kinetin promotes alkaloid production. Constabel, Shyluk & Gamborg (1971) found an apparent stimulatory effect of a high level of kinetin (5×10^{-5} M) on anthocyanin production, but subsequent study revealed that this promotory effect was only transitional since by extending culture by a few days, cells grown in the presence of lower levels of kinetin could accumulate similar levels of anthocyanin. In *N. tabacum* callus and suspension cultures, kinetin levels in excess of 10^{-5} M totally suppress nicotine production (Shiio & Ohta, 1973; Pearson, 1978).

It should be stressed that the combined effects of auxins and cytokinins on secondary metabolite production are difficult to assess particularly since there is a lack of data on relative endogenous levels of these growth regulators in

Fig. 5. GLC separation of steroids of *Solanum aviculare*. Steroidal constituents, with retention times relative to cholesterol given in parentheses, are: 1, cholesterol (1.00); 2, campesterol (1.34); 3, stigmasterol (1.41); 4, sitosterol (1.62); 5, tigogenin (1.80); 6, diosgenin (1.83); 7–9, other steroidal sapogenins/alkaloids (1.92, 2.08, 2.29 respectively); 10, solasodine (2.66); 11–13, unidentified metabolites (3.20, 3.70, 5.30 respectively).

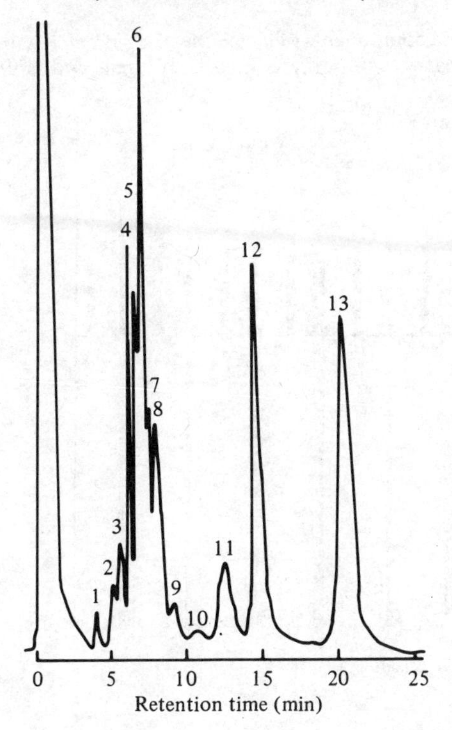

the cultured cells. However, there is little doubt that certain combinations are more beneficial than others. One example of this is illustrated by their effects on steroid metabolite production by *Solanum aviculare* cultures. Reduction of both auxin and cytokinin levels in the medium leads to an increase in the steroid spectrum (Figs. 5 and 6). Primary steroids such as cholesterol, stigmasterol, campesterol and β-sitosterol are the most prominent metabolites in cells grown on rapid growth medium (10^{-5} M NAA and 5×10^{-6} M kinetin). By applying an auxin/cytokinin stepdown subculturing procedure, the same cell lines are induced to accumulate a wider range of steroid constituents which include diosgenin and other steroidal alkaloids. This treatment also causes earlier cessation of exponential phase growth. Similarly, Fig. 7 shows the more predominant effect that auxin has, compared with cytokinin, on steroidal sapogenin (diosgenin) yields and biomass in *S. aviculare* cultures even in the presence of a wide range of cytokinin treatments.

Before passing on to consider other medium components, it should be noted that auxin and cytokinin habituation may be more widespread phenomena in plant cell cultures than has hitherto been realised. Evidence that the difference between cytokinin-dependent and cytokinin-independent lines is a heritable alteration in phenotype (i.e. an epigenetic alteration) but not an alteration in the genotype (i.e. a mutagenic alteration) is particularly

Fig. 6. Growth and steroid components of a cell line of *S. aviculare* cultured on MS medium supplemented with different levels of auxin/cytokinin.

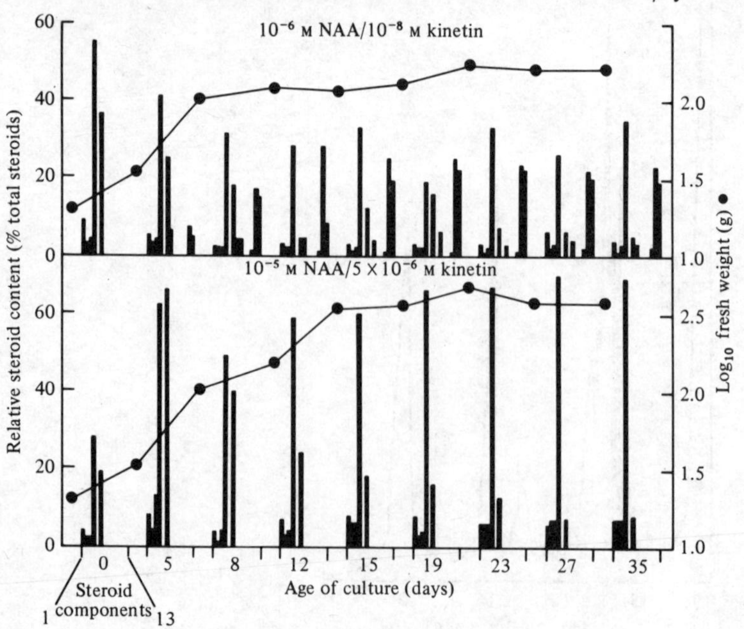

strong (see Meins & Binns, 1978). Growth regulator requirements of non-habituated cultures and of cultures in which a portion of the cell population is habituated may therefore differ. Moreover, the habituation phenomenon may explain why cultured cells gradually lose their ability to produce secondary metabolites when grown in media which contain standard concentrations of growth regulators. Only by single-cell selection or filtration is the recovery of non-habituated (productive) cells from habituated cultures possible. Kinnersley & Dougall (1980*a*) propose that marked improvements in the anthocyanin-producing capacity of carrot cell cultures that were obtainable by repeated filtering of suspensions to remove cell aggregates of more than 63 μm in diameter could possibly be attributed to the gradual removal of cytokinin-habituated cells from cultures. High levels of endogenous cytokinins appear to be associated with large cell aggregates. Indirect evidence for this is the fact that the raising of kinetin concentrations to 0.1 mg l^{-1} in media causes marked inhibition of anthocyanin accumulation and a concomitant increase in aggregate size.

Fig. 7. Fresh weight and diosgenin levels of *S. aviculare* cells cultured on MS medium supplemented with different levels of cytokinins and auxins for 28 days. BAP, 6-benzylaminopurine.

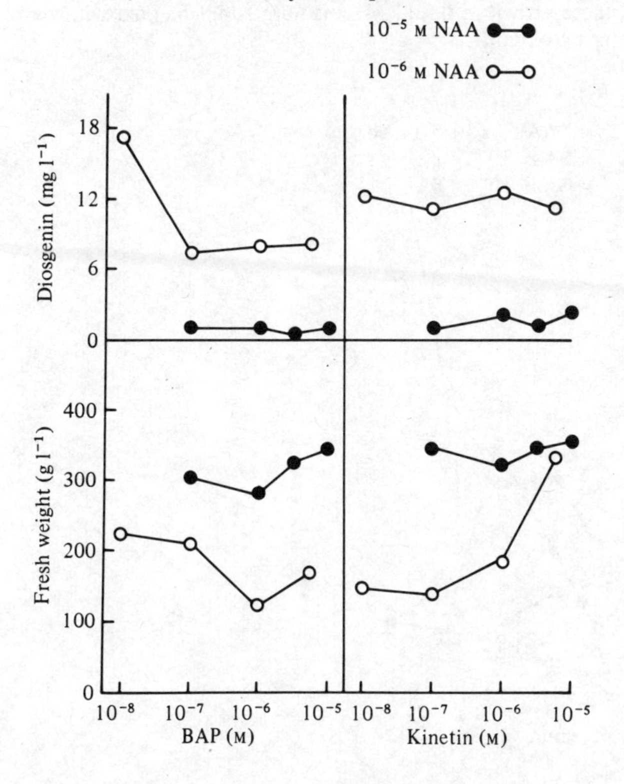

The effect of contrasting levels of the cytokinin 6-benzylaminopurine (BAP) on the accumulation of steroidal sapogenins in *S. aviculare* is shown in Fig. 8. Highest accumulations were achieved in 60-ml rotary shake-flask cultures grown in media containing 10^{-8} M BAP. Culture of the same cell stocks in 4-l systems did not result in the same pattern of sapogenin accumulation. Associated with culture in larger vessels was a marked increase in cell aggregation, possibly caused in part by the absence of culture vessel agitation and in part by poorer macromixing. These observed increases in cell aggregation may also have led to increases in endogenous levels of cytokinin resulting in depressed sapogenin level during normally beneficial cytokinin treatments.

The likelihood that excessive levels of endogenously produced cytokinin might be a frequent occurrence in plant cell cultures raises the question of whether maturation and ageing processes (biochemical characteristics of which are apparently essential components of secondary metabolism) are being inhibited by these high levels of growth regulator. It is well known that senescence of whole organs and tissues is suppressed by applications of

Fig. 8. Production of steroidal sapogenins in 60-ml stock cultures of *S. aviculare* (squares) and in 60-ml (filled symbols) and 4-l (open symbols) subcultures of these grown in liquid MS medium supplemented with two different levels of cytokinin.

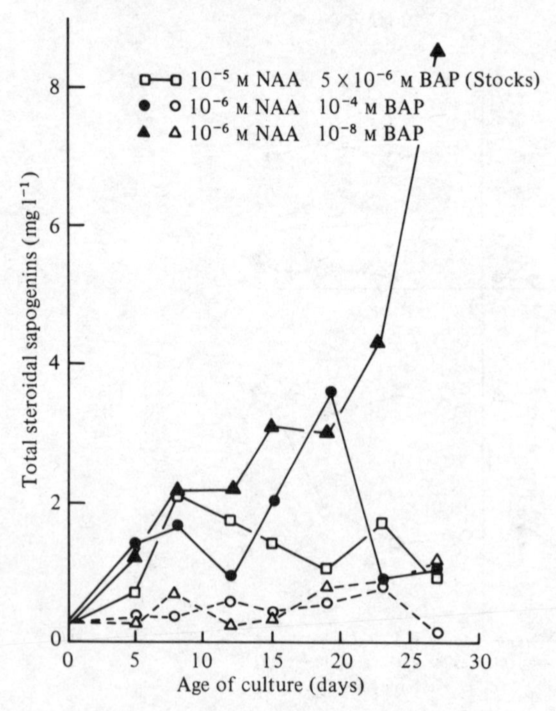

exogenous cytokinin, e.g. Fletcher (1969) has observed that leaf senescence in bean plants can be retarded by applications of BAP (1.5×10^{-4} M). The possibility of such a situation underlines the need for more attention to be paid to the role of growth regulator interactions in modulating secondary metabolism in vitro. Also, the phenomenon of growth regulator habituation should be more seriously investigated from the standpoint of its role in creating shifts in the metabolic potential of cell cultures held for prolonged periods of subculture. Loss of ability by *N. tabacum* and *S. aviculare* cells to produce nicotine and solasodine respectively has been attributed to the respective presence of auxin or cytokinin habituation in cultures maintained for extensive periods (Pearson, 1978; S. H. Mantell, R. W. Smith & H. Smith, unpublished observations).

Little study has been devoted to the effect of other growth regulators such as gibberellic acid (GA), abscisic acid (ABA) and ethylene or of growth retardants like β-chloroethyltrimethyl ammonium chloride (CCC) on secondary metabolism in cell cultures. Information is, however, available on the effects of some of these substances on embryogenesis. In general, ABA inhibits the growth of embryos whilst GA has an opposite effect. Interestingly, Fujimura & Komamine (1975) found that although neither of these regulators influenced the number of embryos produced by carrot cell cultures, ABA had an inhibitory effect on the progress of embryos to advanced stages of development. Not surprisingly, ABA inhibited both the growth and production of phenolic compounds by tobacco callus cultures (Li, Rice, Rohrbaugh & Wender, 1970). Marshall & Staba (1976) were unable to detect any effect of a single concentration (0.3 mg l^{-1}) of GA on diosgenin production in *Dioscorea* cultures. Ethylene, a known regulator of the physiology of whole plants (Bidwell, 1979) is known to be produced by tissue cultures, e.g. those of *Acer* (Mackenzie & Street, 1970), and its presence promotes the production of tannins and extractable PAL from these cultures (Westcott, 1976). Since all these regulators have a well-known and profound effect on the senescence of plant tissues in vivo (it has been suggested that GA induces juvenility in some plants (Schwabb, 1975)), it is likely that these substances could be used to modulate the aforementioned effects of auxins and cytokinins in vitro. The need for further investigations along these lines is made more compelling when it is considered that secondary metabolites themselves possess growth regulatory activity. For instance, nicotine (at concentrations of 50 mg l^{-1}) has been claimed by Peters, Wu, Sharp & Paddock (1974) to induce root formation in callus cultures of *N. rustica*, although this morphogenetic stimulus was negated in the presence of certain combinations of auxins and cytokinins. Roddick (1972) stated that, in general, alkaloids appear to inhibit the response of tissues to growth factors and even though some may not appear to possess auxin activity themselves they nevertheless interfere with auxin-dependent growth. Occurrence of such synergistic or antagonistic effects indicates that more detailed investigations into secondary metabolism in vitro using continuous culture systems, as advocated by Street (1977*b*), or using the immobilised systems described by Lindsey & Yeoman (this volume) (in which removal of medium might prevent a build-up of certain growth regulatory metabolites) are likely to prove particularly worthwhile. The removal of secondary metabolites in this way from productive cultures could also prevent their influence on the feedback regulation of biosynthetic enzyme pathways, e.g. that of nicotine on the regulation of ornithine decarboxylase, putrescine *N*-methyl transferase and *N*-methylputrescine oxidase as postulated by Waller & Nowaki (1978).

Macro- and micro-nutrients. The effect of nutrients employed in plant tissue culture media on both the primary and secondary metabolism of cells grown in these has been comprehensively reviewed by Dougall (1980). Generally, increased levels of nitrate, potassium, ammonium and phosphate tend to support rapid cell growth while depletion or deficiency of some of these

Fig. 9. Fresh weight and nicotine content of suspension cultures of *N. tabacum* cv. NC 2512, and the sugar, nitrate and phosphate content of the medium. (From Pearson, 1978.)

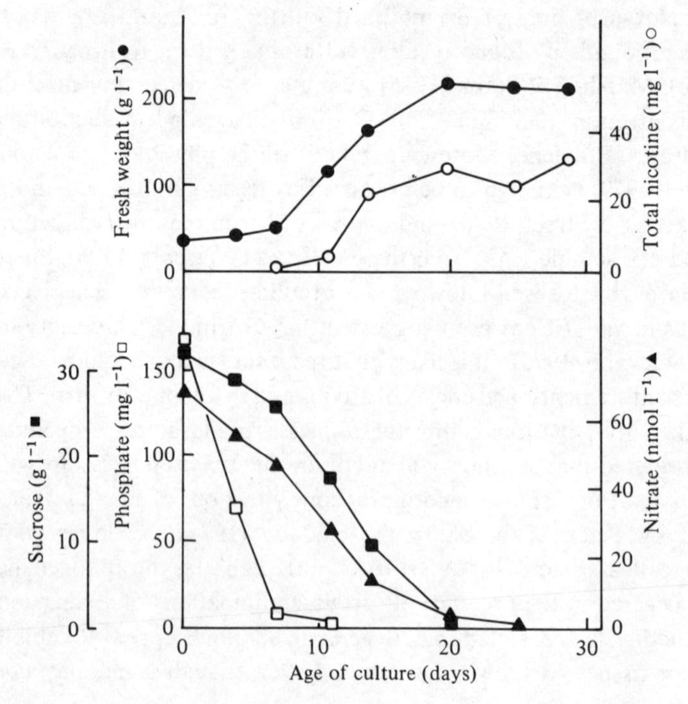

nutrients is associated with growth limitation and concomitant secondary metabolism. Thus, when increases in macro-nutrients like phosphates are made, log and exponential phases of cell growth are prolonged, e.g. as in *C. roseus* cultures (MacCarthy, Ratcliffe & Street, 1980), but in the presence of low initial phosphate, secondary metabolism is strongly stimulated, e.g. as in *Peganum* callus cultures (Nettleship & Slaytor, 1974). Significantly, lack of phosphate more than any other nutrient stimulates secondary metabolite biosynthesis.

Similar effects of reduced levels of nutrients have been observed on nicotine biosynthesis in our laboratory. Nicotine biosynthesis in tobacco cultures is associated with exhaustion of medium phosphate and depletion of other nutrients like nitrate and sucrose (Fig. 9). The effect of growing tobacco cells in media containing similar levels of nutrients to those found when nicotine biosynthesis is detected in cultures has been investigated. Results of experiments in which nicotine production was followed in 4-l batch cultures

Fig. 10. Growth and nicotine production by tobacco cells cultured in liquid MS medium containing (*a*) 150 mg l^{-1} phosphate, (*b*) 75 mg l^{-1} phosphate and (*c*) 15 mg l^{-1} phosphate during an auxin (from 10^{-5} M to 10^{-6} M NAA)/cytokinin (from 10^{-6} M to 10^{-7} M kinetin) stepdown procedure.

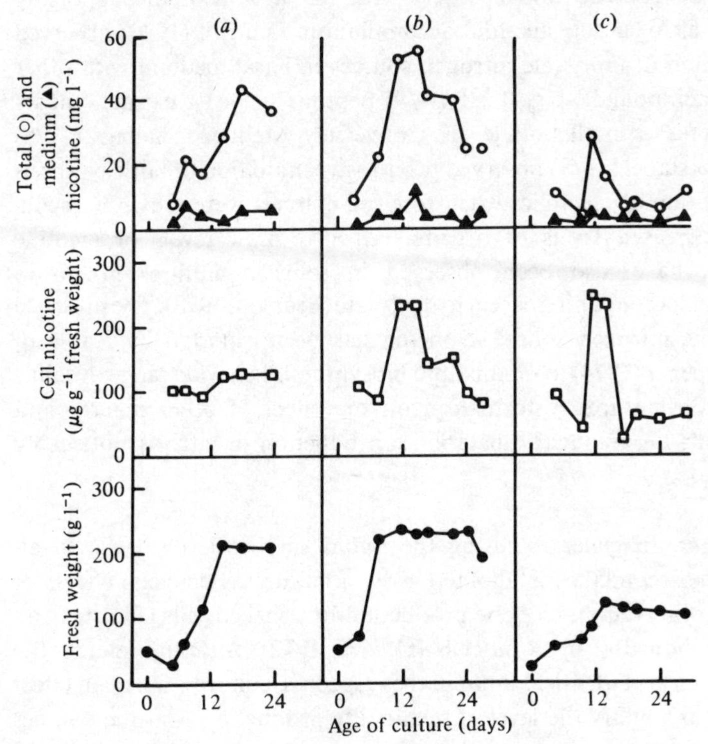

initiated from one stock of cells (Fig. 10) showed that reduction of initial inorganic phosphate levels to 50% and 10% of those normally employed (150 mg l^{-1}) induced an earlier peak of nicotine biosynthesis in cells. Nicotine yield was highest in the culture containing 50% of normal initial phosphate. Reductions to 10% of normal initial levels also markedly inhibited cell growth as measured by cell fresh weight. These results are not surprising in view of the fact that low intracellular phosphate levels are common in ageing plant tissues in vivo, and in microbial cells low medium phosphate leads to low energy charge levels within cells which results in derepression of secondary metabolite enzymes (Drew & Demain, 1977). In contrast to these beneficial effects of reducing initial phosphate levels in production media, there are reports of a few cases in which increased rather than decreased initial phosphate stimulated secondary metabolism. Carew & Krueger (1977) found that raised phosphate levels increased the yields of indole compounds in the medium of *C. roseus* and Zenk *et al.* (1975) obtained a 50% increase in anthraquinones in *M. citrifolia* cultures when phosphate was increased to 50 mM.

Zenk *et al.* (1977*a*) compared the growth of cell cultures of *C. roseus* on seven commonly used culture media. While cells grew well on all the media tested, alkaloid content varied widely with the MS medium supporting maximum levels of indole alkaloid accumulation. Dougall (1980) observed that substitution of inorganic nitrogen sources in basal medium with other nitrogenous compounds, e.g. 0.2% (w/v) peptone or yeast extract, causes significant increases in alkaloid levels. Conversely, Mehta & Shailaga (1978) in Kurz & Constabel (1979) observed progressive inhibition of both synthesis and accumulation of phenolics when rose cell cultures were grown in media containing increased levels of organic nitrogen. Low levels of nicotine accumulation have also been observed in tobacco cultures grown in production media containing casein hydrolysate (Pearson, 1978). The presence of this organic nitrogen source in media has been reported by Radwan, Mangold & Spener (1974) to inhibit lipid biosynthesis in a wide range of callus and suspension cultures. Information on the effect of other macro- and micro-nutrients on secondary metabolite production in vitro is noticeably lacking.

Carbon sources. In general, raising the initial sucrose levels leads to an increase in the secondary metabolite yields of cultures. Beneficial effects of sucrose were observed on nicotine production by tobacco cells (Fig. 11) and polyphenol production by *Rosa* cells (Davies, 1972). Although metabolite biosynthesis starts at identical times in the respective cultures, the main effect of sucrose is to amplify the level of metabolite production. Carbon sources other than sucrose and glucose have been tested for their suitability for supporting secondary metabolite accumulation in cultures. Zenk *et al.* (1975) tested 14 carbohydrates at 2% (w/v) levels and found that sucrose gave the highest yields of anthraquinones in *M. citrifolia* cultures. Little difference in steroid accumulations was found in *S. aviculare* cultures grown on media supplemented with either glucose or sucrose over a range of 15–30 g l^{-1} (S. H. Mantell, R. W. Smith & H. Smith, unpublished observations). However, solasodine production was completely inhibited in sucrose-supplemented media and at the highest glucose concentration tested. Few data are currently available on the effects of different osmotic conditions on primary let alone secondary metabolism in vitro. Perhaps the effect of high initial levels of sucrose is to raise the osmotic potential of media. In tobacco cultures containing 5% (w/v) sucrose, medium sucrose is not exhausted as it is in cultures with lower initial levels (Fig. 11). The presence of appreciable levels of sucrose (in the 2–3% range) late on in culture development could be significant. Certainly the presence of rapidly metabolised sugars such as glucose, which favour rapid growth and exert catabolite inhibition, have marked inhibitory effects on secondary metabolism in micro-organisms (Drew & Demain, 1977). Sucrose, on the other hand, is more slowly metabolised, particularly during the idiophase.

Fig. 11. Depletion of medium sucrose in nicotine production medium containing (*a*) 20 g l^{-1} or (*b*) 50 g l^{-1} initial sucrose. (L. P. Hazell, personal communication.)

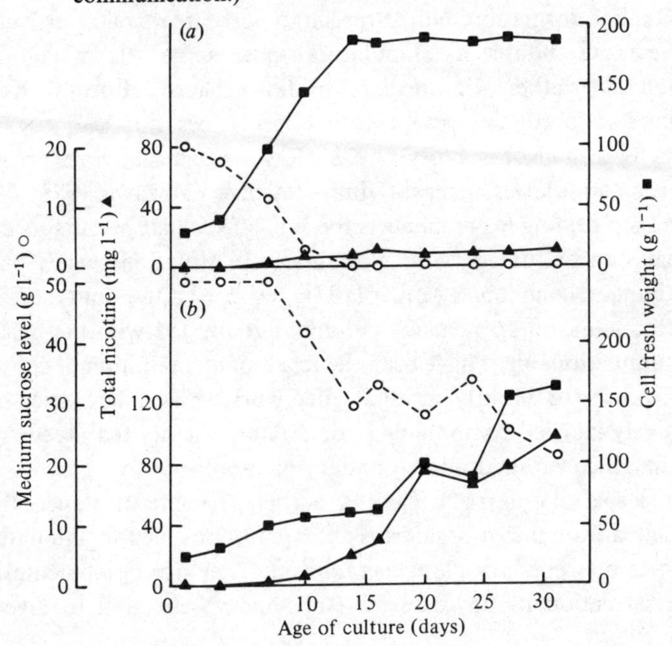

Precursors. Among attempts to increase the potential of plant cell cultures to produce particular metabolites have been those involving the feeding of cultures with known precursors and/or intermediates with the intention of stimulating particular enzyme pathways. Yeoman *et al.* (1980) have shown how encouraging this approach can be in their work on capsaicin production by callus cultures of *Capsicum frutescens* (see Lindsey & Yeoman, this volume, also). By growing cells on media containing radioactively-labelled phenylalanine and valine (amino acid precursors of capsaicin) and low levels of total nitrogen but no sucrose (treatments designed to limit growth and particularly to retard protein synthesis), label was incorporated into the product. Capsaicin yields could be further increased by growing cells on media containing more immediate precursors, particularly vanillylamine and isocapric acid, both supplied at 5 mM. Other notable examples of the stimulation of the production of secondary metabolites by their respective known precursors are those reported by Chowdhury & Chaturvedi (1979), who found that initial feeding of 100 mg l^{-1} cholesterol to *Dioscorea deltoidea* cultures increased diosgenin yields by *c.* 100%, and Zenk, El-Shagi & Ulbrich (1977*b*), who obtained a 100% stimulation of rosmarinic acid in *Coleus blumei* cultures initially fed with 500 mg l^{-1} phenylalanine. Timing of precursor feeding may be particularly important. Tryptamine (100 mg l^{-1}) added during the second or third week of culture stimulates cell growth and alkaloid metabolism (Krueger & Carew, 1978) but when it is initially fed (e.g. Alfermann & Reinhard, 1978; Döller, 1978) it causes inhibition of both cell growth and alkaloid formation. Not surprisingly, precursors of a particular product prove more suitable as stimulators under some sets of cultural conditions than under others. For instance, feeding tobacco cultures with the nicotine precursors putrescine and nicotinic acid does not lead to any measurable increase in nicotine biosynthesis. However, when cultures are fed with ornithine, nicotine levels increase (Ohta, Matsui & Yatazawa, 1978). One problem with such feeding experiments is the side-effects that precursors can themselves have on culture growth, particularly in situations where they inhibit it. Although Neümann & Muller (1971) recorded a two-fold increase in nicotine production in *N. tabacum* cultures initially fed with methylputrescine this stimulation might have been the result of an inhibition of culture growth by the precursor. Clearly, more detailed work needs to be carried out with radioactively-labelled compounds to determine whether fed precursors are in fact being incorporated into secondary metabolites.

Elicitors are a special group of triggering factors. They are substances that have been isolated from micro-organisms and are responsible for stimulating particular facets of secondary plant metabolism. They are instrumental in inducing the formation of phytoalexins (substances belonging to diverse groups of compounds like isoflavonoids, terpenoids, polyacetylenes and dihydrophenanthrenes) as part of plants' defence mechanisms against pathogens. For example, when the polysaccharide elicitor Pms, obtained from *Phytophthora megasperma* var. *sojae*, is added to soya bean cell suspension cultures growing in the dark, glyceollin (an isoflavonoid) appears 12–15 h later and accumulates thereafter (Ebel, Ayers & Albersheim, 1976). Prior to this development, a rapid increase of PAL activity starts, reaching its maximum level 20 h after the addition of Pms. Comparative density-labelling techniques have recently established that such increases in PAL activity associated with elicited responses are due in part to *de novo* synthesis of this enzyme (Lamb & Dixon, 1978). Current research on microbially produced elicitors in vitro is limited to cell cultures of the Leguminosae, yet it is an area of work which may grow in importance as such elicitors find applications in studies on the biosynthesis of diverse secondary metabolites in vitro.

Aeration and culture mixing

Both aeration and macromixing are integral components of the physicochemical nature of liquid cell suspensions. Culture aeration (i.e. exposure of constituents to the mechanical and/or chemical action of air) is dependent in these systems upon the agitation of culture media. One cannot therefore be considered in isolation from the other except in special cases where aeration is carefully controlled at a given flow rate employing various mixtures of gases (i.e. of oxygen and carbon dioxide). Such techniques have unfortunately been used little in plant cell culture work because they can only be carried out on the more recently developed large-scale cultivation systems (see Fowler, this volume). In the most widely used small-scale liquid culture systems, i.e. 40- to 200-ml rotary shake-flask cultures, upon which most current data on secondary metabolism in vitro are based, culture agitation and aeration are interdependent cultural components. These can influence the yield of secondary metabolites in vitro. In conical-flask cultures, culture volume (with its obvious influence on oxygen absorption coefficients, OAC, associated with the area of culture medium having an air–liquid interface) has a marked effect on nicotine biosynthesis in tobacco cultures. At low culture volumes (conditions which raise OAC values), nicotine production is enhanced (L. P. Hazell & D. W. Pearson, personal communication).

The importance of aeration and culture agitation has also been reported by Wagner & Vogelmann (1977) in studies on large-scale fermentors. Fermentor types examined were flat-blade turbine, draught tube (with turbine) and draught tube (with airlift) with respect to their performance in supporting growth and metabolite production by *M. citrifolia* and *C. roseus* cells. With the former cells, anthraquinone yields in airlift fermentors were *c.* 30% more than those in shake flasks and twice those in the other fermentors tested. Cell yields were relatively unaffected by the type of reactor used. In the case of indole alkaloid production by the latter cells, it was found that reactors with mechanical agitation were quite unsatisfactory whereas the airlift fermentor gave yields comparable to those in shake flasks. Some of the marked decreases in secondary metabolism observed during scale-up from small rotary shake-flask cultures to large multilitre fermentors could be due to differences in the aeration and mixing properties of cultures.

Culture medium pH

Optimal growth in plant cell cultures usually occurs in media with initial pH values in the ranges 5–6. Media containing undefined organic components like casein hydrolysate and yeast extract are usually well buffered so that the pH changes relatively little during the course of culture development. However, in media without these substances, shifts in pH during culture can be dramatic. Veliky (1977) found that *Ipomoea* cell cultures, which transform tryptophan into a variety of indole metabolites such as tryptophol, yield double the amount of this metabolite when cells are grown at pH 6.3 in a pH-stat compared with yields obtained when cells are grown in uncontrolled culture. When the pH drops to 4.8, tryptophol accumulation is completely inhibited.

Preculture treatment of cells

With the adoption of a two-stage batch culture procedure in which cells are propagated by rapid growth in a stock culture medium (SCM) and then transferred to a second medium, which has been modified to stimulate metabolite production, the type of pretreatment of cells in the stock culture stage appears to have marked effects on the subsequent accumulation of secondary metabolites in cultures held in the production medium (NPM). For instance, tobacco cells derived from stock media containing different levels of auxin accumulate different levels of nicotine when cultured in NPM (Fig. 12). Although significant carry-over of auxin from SCM to NPM media is a distinct possibility in the case of cells propagated in 10^{-4} M NAA, this is not likely in the case of cells grown in 10^{-6} M and 10^{-5} M NAA, since predicted carry-over levels at subculturing (i.e. 10^{-7} M and 2×10^{-6} M NAA respectively) would be well within the region of those normally found to be optimal in NPM for nicotine production (see Fig. 4). Hence, the preculture treatment on media containing 10^{-5} M NAA appears to predispose cells to accumulate nicotine. Stock culture age (14 or 21 days) also has a measurable effect on nicotine production by identical stocks of tobacco cells derived from SCM supplemented with NAA. Stocks of cells subcultured at 21 days produced higher levels of nicotine than did those subcultured at 14 days (L. P. Hazell & H. Smith, unpublished observations). In the same way, preculture conditions (stock culture age and stock culture auxin levels) also affect the amounts of diosgenin accumulating in *S. aviculare* cells grown on steroid production media (S. H. Mantell, R. W. Smith & H. Smith, unpublished observations).

Conclusions

The following cultural conditions influence the degree to which secondary metabolites accumulate in plant cell cultures.

(1) External cultural conditions
   Light
   Temperature
   Culture vessel agitation

Fig. 12. Effect of stock culture NAA concentration on the fresh weight and nicotine content of suspension cultures of *N. tabacum* cv. NC 2512. (From Pearson, 1978.)

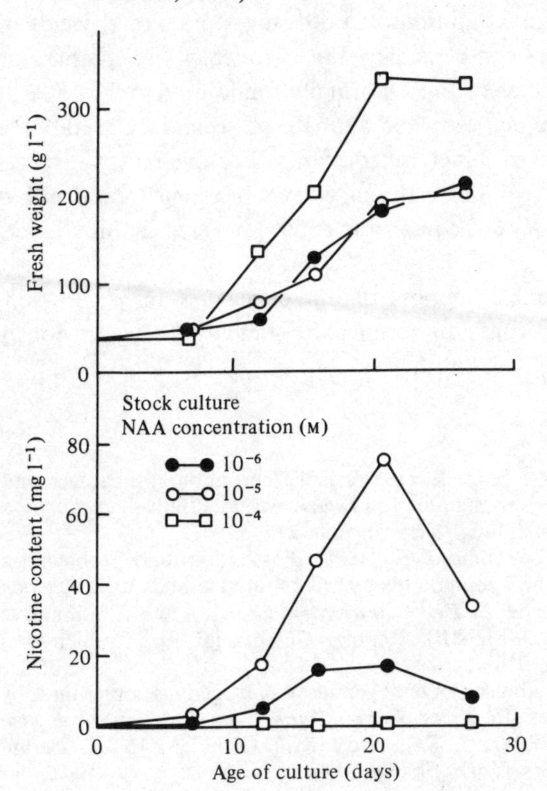

(2) Internal cultural conditions
 Medium components
 e.g. Growth regulators
 Macro- and micro-nutrients
 Carbon sources
 Precursors and elicitors
 Aeration and culture mixing
 Culture medium pH
(3) Preculture treatment of cells

As a general rule, cultural factors which promote accumulation of secondary plant metabolites in batch culture systems are those which support limited exponential growth and which perpetuate extended periods of decelerating and stationary phases of cell growth. The most satisfactory method currently available for achieving high accumulations of secondary plant metabolites per unit of biomass is to employ a two-stage culture strategy. A first stage, designed to produce high levels of biomass as rapidly as possible, is followed by a second that is designed to provide conditions which are conducive to the triggering and maintenance of active states of secondary metabolism. By creating appropriate cultural conditions at both stages of culture, yields of secondary plant metabolites can be achieved in vitro which are comparable to and in some cases in excess of those normally found in vivo. The use of plant cells immobilised in matrices like alginate as secondary metabolite production systems is still very much in its infancy. For this reason, aspects of these culture systems have not been discussed here but some treatment has been given to these by Lindsey & Yeoman and Alfermann *et al.* in this volume.

We acknowledge the valuable support of our colleagues D. W. Pearson, L. P. Hazell (who gave permission to include some of their previously unpublished results) and R. W. Smith.

References

Alfermann, W. & Reinhard, E. (1971). Isolierung anthocyanhaltizer und anthocyanfreier Gewebestamme von *Daucus carota*. Einfluss von Auxinen auf die Anthocyanbildung. *Experimentia*, **27**, 353.

Alfermann, A. W. & Reinhard, E. (1978). Possibilities and problems in production of natural compounds by cell culture methods.In *Production of Natural Compounds by Cell Culture Methods*, ed. A. W. Alfermann & E. Reinhard, pp. 3–15. BPT Report, Gesellschaft für Strahlen und Umweltforschung, Munich.

Barz, W. (1977). Catabolism of endogenous and exogenous compounds by plant cell cultures. In *Plant Tissue Culture and its Biotechnological Application*, ed. W. Barz, E. Reinhard & M. H. Zenk, pp. 153–71. Berlin, Heidelberg and New York: Springer-Verlag.

Barz, W., Kettner, M. & Husemann, W. (1978). On the degradation of nicotine in *Nicotiana* cell suspension cultures. *Planta Medica*, **34**, 73–8.
Bidwell, R. G. S. (1979). *Plant Physiology*, 2nd edn. New York: MacMillan.
Böhm, H. (1978). Regulation of alkaloid production in plant cell cultures. In *Frontiers of Plant Tissue Culture*, ed. T. A. Thorpe, pp. 201–11. Calgary: University of Calgary.
Böhm, H. (1980). The formation of secondary metabolites in plant tissue and cell cultures. *International Review of Cytology*, Supplement **11B**, 183–208.
Brain, K. R. (1976). Accumulation of L-DOPA in cultures from *Mucuna pruriens*. *Plant Science Letters*, **7**, 157.
Büchner, S. A. & Staba, E. J. (1964). Preliminary examination of *Digitalis* tissue cultures for cardenolides. *Journal of Pharmacy and Pharmacology*, **16**, 733.
Bu'Lock, J. D. (1975). The two-faced microbiologist: contributions of pure and applied microbiology to good research. *Developments in Industrial Microbiology*, **16**, 11–19.
Butcher, D. N. (1977). Secondary products in tissue cultures. In *Plant Cell, Tissue and Organ Culture*, ed. J. Reinert & Y. P. S. Bajaj, pp. 668–93. Berlin, Heidelberg and New York: Springer-Verlag.
Carew, D. P. & Krueger, R. J. (1977). *Catharanthus roseus* tissue culture: the effects of medium modifications on growth and alkaloid production. *Lloydia*, **40**, 326.
Chowdhury, A. R. & Chaturvedi, H. C. (1979). Cholesterol and biosynthesis of diosgenin by tuber callus of *Dioscorea deltoidea*. *Current Science*, **49**, 237–8.
Constabel, F., Shyluk, J. P. & Gamborg, O. L. (1971). The effect of hormones on anthocyanin accumulation in cell cultures of *Haplopappus gracilis*. *Planta*, **96**, 306–16.
Corduan, G. (1975). Produktion von Tropanalkaloiden in Gewebekulturen. *Planta Medica* Supplement, p. 22.
Corduan, G. & Reinhard, E. (1972). Synthesis of volatile oils in tissue cultures of *Ruta graveolens*. *Phytochemistry*, **11**, 917–22.
D'Amato, F. (1977). Cytogenetics of differentiation in tissue and cell cultures. In *Plant Cell, Tissue and Organ Culture*, ed. J. Reinert & Y. P. S. Bajaj, pp. 343–57. Berlin, Heidelberg and New York: Springer-Verlag.
Davies, M. E. (1972). Polyphenol synthesis in cell suspension cultures of Paul's Scarlet Rose. *Planta*, **104**, 50–65.
Davies, M. E. (1974). Effects of auxin on polyphenol accumulation and the development of phenylalanine-ammonia-lyase activity in dark-grown suspension cultures. *Biochemica et Biophysica Acta*, **362**, 417.
Döller, G. (1978). Influence of the medium on the production of serpentine by suspension cultures of *Catharanthus roseus* (L.) G. Don. In *Production of Natural Compounds by Cell Culture Methods*, ed. A. W. Alfermann & E. Reinhard, pp. 109–16. BPT Report, Gesellschaft für Strahlen und Umweltforschung, Munich.
Dougall, D. K. (1980). Nutrition and metabolism. In *Plant Tissue Culture as a Source of Biochemicals*, ed. E. J. Staba, pp. 21–58. Boca Raton, Florida: CRC Press.
Doumery, B. & Chouteau, J. (1975). Photodegradation of chlorophyll pigments and nicotine. *Annales du Tabac*, Section 2, 183–200.
Drew, S. W. & Demain, A. L. (1977). Effect of primary metabolites on secondary metabolism. *Annual Review of Microbiology*, **31**, 346–56.

Ebel, J., Ayers, A. R. & Albersheim, P. (1976). Host–pathogen interactions: XII Response of suspension-cultured soybean cells to the elicitor isolated from *Phytophthora megasperma* var. *sojae*, a fungal pathogen of soybeans. *Plant Physiology*, **57**, 775–9.

Everett, N. P., Wang, T. L., Gould, A. R. & Street, H. E. (1981). Studies on the control of the cell cycle in cultured plant cells. 2. Effects of 2,4-dichlorophenoxyacetic acid (2,4-D). *Protoplasma*, **106**, 15–22.

Everett, N. P., Wang, T. L. & Street, H. E. (1978). Hormone regulation of cell growth and development *in vitro*. In *Frontiers of Plant Tissue Culture*, ed. T. A. Thorpe, pp. 307–16. Calgary: University of Calgary.

Fletcher, R. A. (1969). Retardation of leaf senescence by benzyladenine in intact bean plants. *Planta*, **89**, 1–8.

Forrest, G. I. (1969). Studies on the polyphenol metabolism of tissue cultures derived from the tea plant (*Camellia sinensis* L.). *Biochemical Journal*, **113**, 765–72.

Fritsch, H., Hahlbrock, K. & Grisebach, H. (1971). Biosynthese von Cyanidin in Zellsuspensionkulturen von *Haplopappus gracilis*. *Zeitschrift für Naturforschung*, **266**, 581.

Fujimura, T. & Komamine, A. (1975). Effects of various growth regulators on embryogenesis in a carrot cell suspension culture. *Plant Science Letters*, **5**, 359–64.

Furuya, T., Kojima, H. & Syono, K. (1971). Regulation of nicotine biosynthesis by auxins in tobacco callus tissues. *Phytochemistry*, **10**, 1529–32.

Gamborg, O. L., Constabel, F., La Rue, T. A. G., Miller R. A. & Steck, W. (1971). The influence of hormones on secondary metabolite formation in plant cell cultures. In *Les Cultures de Tissus de Plantes*, pp. 335–44. Paris: Centre National de la Recherche Scientifique.

Goren, R. & Galston, A. W. (1966). Control by phytochrome of ^{14}C-sucrose incorporated into buds of etiolated pea seedlings. *Plant Physiology*, **41**, 1055–64.

Grisebach, H. & Hahlbrock, K. (1974). Enzymology and regulation of flavonoid and lignin biosynthesis in plants and plant cell suspension cultures. In *Metabolism and Regulation of Secondary Plant Products*, ed. V. C. Runeckles & E. E. Conn, pp. 21–52. New York: Academic Press.

Hahlbrock, K., Schröder, J. & Vieregge, J. (1980). Enzyme regulation in parsley and soybean cell cultures. *Advances in Biochemical Engineering*, **18**, 39–60.

Hiraoka, N. & Tabata, M. (1974). Alkaloid production by plants regenerated from cultured cells of *Datura innoxia*. *Phytochemistry*, **13**, 1671–5.

Hirotani, M. & Furuya, T. (1977). Restoration of cardenolide synthesis in redifferentiated shoots from callus cultures of *Digitalis purpurea*. *Phytochemistry*, **16**, 610–11.

Ikeda, T., Matsumoto, T. & Noguchi, M. (1976). Formation of ubiquinone by tobacco plant cells in suspension culture. *Phytochemistry*, **15**, 568–9.

Ikeda, T., Matsumoto, T. & Noguchi, M. (1977). Effects of inorganic nitrogen sources and physical factors on the formation of ubiquinone by tobacco plant cells in suspension culture. *Agricultural and Biological Chemistry*, **41**, 1197–201.

Jones, R. W. & Sheard, R. W. (1975). Phytochrome, nitrate movement, and induction of nitrate reductase in etiolated pea terminal buds. *Plant Physiology*, **55**, 954–9.

Kamimura, S. & Nishikawa, M. (1976). Growth and alkaloid production of the cultured cells of *Papaver bracteatum*. *Agricultural and Biological Chemistry*, **40**, 907–11.

Kartnig, T. (1977). Cardiac glycosides in cell cultures of *Digitalis*. In *Plant Tissue Culture and its Biotechnological Application*, ed. W. Barz, E. Reinhard, & M. H. Zenk, pp. 44–51. Berlin, Heidelberg and New York: Springer-Verlag.

Kaul, B. & Staba, E. J. (1968). *Dioscorea* tissue cultures. I. Biosynthesis and isolation of diosgenin from *Dioscorea deltoidea* callus and suspension cells. *Lloydia*, **31**, 171–9.

King, P. J. (1980). Cell proliferation and growth in suspension cultures. *International Review of Cytology, Supplement* **11A**, 25–54.

Kinnersley, A. M. & Dougall, D. K. (1980*a*). Increase in anthocyanin yield from wild-carrot cell cultures by a selection system based on cell-aggregate size. *Planta*, **149**, 200–4.

Kinnersley, A. M. & Dougall, D. K. (1980*b*). Correlation between the nicotine content of tobacco plants and callus cultures. *Planta*, **149**, 205–6.

Krikorian, A. D. & Steward, F. C. (1969). Biochemical differentiation: the biosynthetic potentialities of growing and quiescent tissue. In *Plant Physiology*, Vol. *VB*, ed. F. C. Steward, pp. 227–326. New York: Academic Press.

Krueger, R. J. & Carew, D. P. (1978). *Catharanthus roseus* tissue culture: the effects of precursors on growth and alkaloid production. *Lloydia*, **41**, 327–31.

Kurz, W. E. W. & Constabel, F. (1979). Plant cell cultures, a potential source of pharmaceuticals. *Advances in Applied Microbiology*, **25**, 209–40.

Lamb, C. J. & Dixon, R. A. (1978). Stimulation of *de novo* synthesis of L-phenylalanine ammonia-lyase during induction of phytoalexin biosynthesis in cell suspension cultures of *Phaseolus vulgaris*. *FEBS Letters*, **94**, 277–80.

Li, H. C., Rice, E. L., Rohrbaugh, L. M. & Wender, S. H. (1970). Effects of abscisic acid on phenolic content and lignin biosynthesis in tobacco tissue culture. *Physiologia Plantarum*, **23**, 928–36.

Luckner, M. (1980). Expression and control of secondary metabolism. In *Secondary Plant Products*, ed. E. A. Bell & B. V. Charlwood, pp. 23–63. Berlin, Heidelberg and New York: Springer-Verlag.

Luckner, M., Nover, L. & Böhm, H. (1977). *Secondary Metabolism and Cell Differentiation*. Berlin, Heidelberg and New York: Springer.

MacCarthy, J. J., Ratcliffe, D. & Street, H. E. (1980). The effect of nutrient medium composition on the growth cycle of *Catharanthus roseus* G. Don cells grown in batch culture. *Journal of Experimental Botany*, **31**, 1315–25.

MacCarthy, J. J. & Stumpf, P. K. (1980). Effect of different temperatures on fatty-acid synthesis and polyunsaturation in cell suspension cultures. *Planta*, **147**, 389–95.

Mackenzie, I. A. & Street, H. E. (1970). Studies on the growth in culture of plant cells. VIII. The production of ethylene by suspension culture of *Acer pseudoplatanus* L. *Journal of Experimental Botany*, **31**, 824–34.

Marshall, J. G. & Staba, E. J. (1976). Hormonal effects on diosgenin biosynthesis and growth in *Dioscorea deltoidea* tissue cultures. *Phytochemistry*, **15**, 53–5.

Martin, S. M. (1980). Environmental factors: B. Temperature, aeration and pH. In *Plant Tissue Culture as a Source of Biochemicals*, ed. E. J. Staba, pp. 143–8. Boca Raton, Florida: CRC Press.

Matsumoto, T., Ohunishi, K., Nishida, K. & Noguchi, M. (1972). Effect of physical factors and antibiotics on the growth of higher plant cells in suspension culture. *Agricultural and Biological Chemistry*, **36**, 2177.

Meins, F., Jr & Binns, A. N. (1978). Epigenetic clonal variation and the requirement of plant cells for cytokinins. In *The Clonal Basis for Development*, ed. S. Subtelny & I. M. Sussex, pp. 185–201. New York: Academic Press.

Murashige, T. & Skoog, F. (1962). A revised medium for rapid growth and bioassays with tobacco tissue cultures. *Physiologia Plantarum*, **15**, 473–97.

Nagel, M. & Reinhard, E. (1975). Das atherische Öl der Calluskulturen von *Ruta graveolens*. II. Physiologie zur Bildung des atherischen Oles. *Planta Medica*, **27**, 264.

Nettleship, L. & Slaytor, M. (1974). Adaptation of *Peganum harmala* callus to alkaloid production. *Journal of Experimental Botany*, **25**, 1114–23.

Neümann, D. & Muller, E. (1971). Beitrage zur Physiologie der Alkaloide. V. Alkaloidbildung in Kallus-und Suspensionkulturen von *Nicotiana tabacum*. *Biochemie und Physiologie der Pflanzen*, **162**, 503–13.

Ohta, S., Matsui, O. & Yatazawa, M. (1978). Culture conditions for nicotine production in tobacco tissue culture. *Agricultural and Biological Chemistry*, **42**, 1245–51.

Ohta, S. & Yatazawa, M. (1978). Effect of light on nicotine production in tobacco tissue culture. *Agricultural and Biological Chemistry*, **42**, 873–7.

Pearson, D. W. (1978). Nicotine production by tobacco tissue cultures. Ph.D. Thesis, Nottingham University.

Peters, J. E., Wu, P. H. L., Sharp, W. R. & Paddock, E. F. (1974). Rooting and the metabolism of nicotine in tobacco callus cultures. *Physiologia Plantarum*, **31**, 97–100.

Radwan, S. S., Mangold, H. K. & Spener, F. (1974). Lipids in plant tissue cultures. III. Very long-chain fatty acids in the lipids of callus cultures and suspension cultures. *Chemistry and Physics of Lipids*, **13**, 103–7.

Rajasekhar, E. W., Edwards, M., Wilson, S. B. & Street, H. E. (1971). Studies on the growth in culture of plant cells. XI. The influence of shaking rate on the growth of suspension cultures. *Journal of Experimental Botany*, **22**, 107–17.

Roddick, J. G. (1972). Effects of the steroidal alkaloid tomatine in auxin bioassays and its interaction with indole-3-acetic acid. *Planta*, **102**, 134–9.

Röller, U. (1978). Selection of plants and plant tissue cultures of *Catharanthus roseus* with high content of serpentine and ajmalicine. In *Production of Natural Compounds by Cell Culture Methods*, ed. A. W. Alfermann & E. Reinhard, pp. 95–108. Munich: Gesellschaft fur Strahlen und Umweltforschung MBH.

Rose, D. & Martin, S. M. (1975). Growth of suspension cultures of plant cells (*Ipomoea* sp.) at various temperatures. *Canadian Journal of Botany*, **53**, 315–20.

Schwabb, W. W. (1975). Applied aspects of juvenility and some theoretical considerations. In *Symposium on Juvenility in Woody Perennials*, ed. R. H. Zimmerman, p. 45. The Hague: International Society of Horticultural Science.

Seabrook, J. E. A. (1980). Laboratory culture. In *Plant Tissue Culture as a Source of Biochemicals*, ed. E. J. Staba, pp. 1–20. Boca Raton, Florida: CRC Press.

Seibert, M. & Kadkade, P. G. (1980). Environmental factors: A. Light. In *Plant Tissue Culture as a Source of Biochemicals*, ed. E. J. Staba, pp. 123–42. Boca Raton, Florida. CRC Press.

Sharma, O. P. & Khanna, P. (1980). Studies on steroidal sapogenins from tissue cultures of *Agave wightii*. *Lloydia*, **43**, 459–62.

Shiio, I. & Ohta, S. (1973). Nicotine production by tobacco callus tissues and effect of plant growth regulators. *Agricultural and Biological Chemistry*, **37**, 1857–64.

Staba, E. J. (Ed.) (1980). *Plant Tissue Culture as a Source of Biochemicals*. Boca Raton, Florida: CRC Press.

Stickland, R. G. & Sunderland, N. (1972). Production of anthocyanins, flavonols and chlorogenic acids by cultured callus tissues of *Haplopappus gracilis*. *Annals of Botany*, **36**, 443–57.

Stöckigt, J., Pfitzner, A. & Firl, J. (1981). Indole alkaloids from suspension cultures of *Rauwolfia serpentina* Benth. *Plant Cell Reports*, **1**, 36–9.

Street, H. E. (1977*a*). *Plant Tissue and Cell Culture*. Oxford: Blackwell Scientific.

Street, H. E. (1977*b*). Application of cell suspension cultures. In *Plant Cell, Tissue and Organ Culture*, ed. J. Reinert & Y. P. S. Bajaj, pp. 649–67. Berlin, Heidelberg and New York: Springer-Verlag.

Tabata, M., Ogino, T., Yoshioka, K., Yoshikawa, N. & Hiraoka, N. (1978). Selection of cell lines with higher yield of secondary products. In *Frontiers of Plant Tissue Culture*, ed. T. A. Thorpe, pp. 313–22. Calgary: University of Calgary.

Tabata, M., Yamamoto, H. & Hiraoka, N. (1971). Alkaloid production in the tissue cultures of some solanaceous plants. In *Les Cultures de Tissus de Plantes*, pp. 390–402. Paris: Centre National de la Recherche Scientifique.

Tabata, M., Yamamoto, H., Hiraoka, N. & Konoshima, M. (1972). Organization and alkaloid production in tissue cultures of *Scopolia parviflora*. *Phytochemistry*, **11**, 949–55.

Thomas, E. & Street, H. E. (1970). Organogenesis in cell suspension cultures of *Atropa belladonna* L. and *Atropa belladonna* cultivar *lactea* Döll. *Annals of Botany*, **34**, 657–69.

Veliky, I. A. (1977). Effect of pH on tryptophol formation by cultured *Ipomoea* sp. plant cells. *Lloydia*, **40**, 482.

Wagner, F. & Vogelmann, H. (1977). Cultivation of plant tissue cultures in bioreactors and formation of secondary metabolites. In *Plant Tissue Culture and its Biotechnological Applications*, ed. W. Barz, E. Reinhard & M. H. Zenk, pp. 245–52. Berlin, Heidelberg and New York: Springer-Verlag.

Waller, R. G. & Nowacki, E. K. (1978). *Alkaloid Biology and Metabolism in Plants*. New York and London: Plenum Press.

Weeks, W. W. (1970). Physiology of alkaloids in germinating seed of *Nicotiana tabacum*. PhD Thesis, University of Kentucky, Lexington, USA.

Wellmann, E. (1975). UV dose dependent induction of enzyme related to flavonoid biosynthesis in cell suspension cultures of parsley. *FEBS Letters*, **51**, 105–7.

Westcott, R. J. (1976). Changes in the phenolic metabolism of suspension cultures of *Acer pseudoplatanus* L. caused by the addition of 2-(Chloroethyl) phosphonic acid (CEPA). *Planta*, **131**, 209–10.

White, J. M. & Pike, C. S. (1974). Rapid phytochrome-mediated changes in adenosine 5′-triphosphate content of etiolated bean buds. *Plant Physiology*, **53**, 76.

Willits, C. O., Swain, M. L., Connelly, J. A. & Brice, B. A. (1950). Spectrophotometric determination of nicotine. *Analytic Chemistry*, **22**, 430–3.

Wilson, S. B., King, P. J. & Street, H. E. (1971). Studies on the growth in culture of plant cells. XII. A versatile system for the large scale batch or continuous culture of plant cells. *Journal of Experimental Botany*, **21**, 177–207.

Yamada, Y., Sato, F. & Hagimori, M. (1978). Photoautotrophism in green cultured cells. In *Frontiers of Plant Tissue Culture*, ed. T. A. Thorpe, pp. 453–62. Calgary: University of Calgary.

Yeoman, M. M. & Forche, E. (1980). Cell proliferation and growth in callus cultures. *International Review of Cytology*, Supplement, **11A**, 1–24.

Yeoman, M. M., Miedzybrodzka, M. B., Lindsey, K. & McLaughlan, W. R. (1980). The synthetic potential of cultured plant cells. In *Plant Cell Cultures: Results and Perspectives*, ed. F. Sala, B. Parisi, R. Cella & O. Ciferri, pp. 327–44. Amsterdam, New York and Oxford: Elsevier North-Holland Biomedical Press.

Zenk, M. H. (1978). The impact of plant cell culture on industry. In *Frontiers of Plant Tissue Culture*, ed. T. A. Thorpe, pp. 1–13. Calgary: University of Calgary.

Zenk, M. H., El-Shagi, H., Arens, H., Stöckigt, J., Weiler, E. W. & Deus, B. (1977*a*). Formation of the indole alkaloids, serpentine and ajmalicine in cell suspension cultures of *Catharanthus roseus*. In *Plant Tissue Culture and its Biotechnological Application*, ed. W. Barz, E. Reinhard & M. H. Zenk, pp. 27–43. Berlin, Heidelberg and New York: Springer-Verlag.

Zenk, M. H., El-Shagi, H. & Schulte, U. (1975). Anthraquinone production by cell suspension cultures of *Morinda citrifolia*. *Planta Medica Supplement*, pp. 79–101.

Zenk, M. H., El-Shagi, H. & Ulbrich, B. (1977*b*). Production of rosmarinic acid by cell suspension cultures of *Coleus blumei*. *Naturwissenschaften*, **64**, 585–6.

PART II

Plant propagation by tissue culture

G. HUSSEY

In vitro propagation of horticultural and agricultural crops

Introduction

The vegetative propagation of plants has been practised for centuries and although many improvements in conventional methods have been introduced over the years, the more recent application of tissue culture techniques (i.e. 'micropropagation') has considerably expanded both their scope and potential. Vegetative cloning is an integral part of the breeding and exploitation of heterozygous species and depends on the expression of various asexual reproductive phenomena in suitable types of cuttings or grafts. Tissue culture provides a powerful tool for extending these regenerative processes to a wider range of cells and tissues in diverse species, possibly in all species. Because the vectors of systemic disease are completely excluded, in vitro methods are eminently suited to the rapid propagation and storage of disease-tested plants. Cloned individual plants of heterozygous crops that are usually propagated by seed may be used as parents for F_1 hybrid seed production, as desirable cultivars in their own right, for uniform experimental material or for the production of protoplasts.

This chapter will examine the basic principles involved in micropropagation and make some attempt to assess its present standing and limitations. Some possible future developments will also be discussed. The range of in vitro techniques that are available for cloning plants will be illustrated by reference to typical examples selected from the numerous crops being researched worldwide. More comprehensive lists of references to culture in vitro have been given by Murashige (1974), Pierik (1979) and Conger (1981).

In vitro procedures used in propagation

The procedures used for in vitro propagation comprise: (1) the selection of suitable explants, their sterilisation and transfer to nutrient medium, (2) the proliferation of shoots on multiplication medium and (3) the transfer of shoots to a rooting (or storage) medium and planting out. These have been formally designated 'Stages I, II and III' by Murashige (1974) and a number of workers have adopted this terminology (e.g. Debergh & Maene, 1981,

Hughes, 1981). Although generally useful, these clear-cut distinctions do not necessarily apply to all plants.

The basic principles of tissue culture techniques have been described in detail in standard texts and reviews (Street, 1977; Thorpe, 1981; and those particularly concerned with propagation: Murashige, 1974; de Fossard, 1977; Hussey, 1980 and Conger, 1981).

Explants: origin and sterilisation

The choice of explant will depend on the species being used and on which of the various methods of shoot proliferation described below is considered the most appropriate. The season and stage of growth of the parent plant may be important in determining the reaction of the explant depending on which organ is used. Although many plants culture well only when actively growing, explants from others may produce shoots only during the dormant phase. Bulb scale explants of *Lilium speciosum* regenerated bulbils freely when taken from plants during spring and autumn periods of growth but not when taken during summer or winter (Robb, 1957). Flower stem explants of *Tulipa*, on the other hand, gave rise to shoots only during dry storage; once stem elongation had commenced, the capacity of this organ to regenerate was lost (Wright & Alderson, 1980).

Plants used for starting tissue cultures are generally grown in as clean an environment as possible, overhead watering or splashing of foliage being strictly avoided. Shoot or meristem tips and the centres of bulbs, corms and rhizomes usually have sufficient protection by way of mature leaves or scales to be dissected out sterile. The covering structures are carefully removed one by one after wiping with 70% ethanol. Explants from more exposed organs are sterilised with dilute hypochlorite solution (or other of the decontaminants listed by Street, 1977) and washed with sterile distilled water. An alternative technique for obtaining aseptic organs from young plants makes use of sterile plantlets raised from seed in suitably transparent and sterilisable containers.

Nutrient media

Plant organs, tissues and cells are mostly able to grow on media containing only the principal basic nutrients that are transported in the vascular system such as the salts of nitrogen and other essential elements, sucrose and certain amino acids, vitamins and growth factors (two useful reviews of basic nutrient media are Gamborg, Murashige, Thorpe & Vasil, 1976; Huang & Murashige, 1977). The exact composition of the medium, however, has to be adjusted according to the requirements of the different groups of plants and some species and cultivars require additional supplements to sustain adequate growth. The nutrient medium is generally solidified with agar (0·7–1·0% w/v) but liquid cultures are preferable for some plants (such as bromeliads) and at certain stages in the culture of others. An alternative method utilises a filter paper bridge or wick to support the explant above liquid medium with the free ends of the paper immersed in the medium.

The hormonal control of organogenesis in plant tissue cultures stems mainly from the work of Skoog & Miller (1957) with tobacco callus. According to what has become known as the 'Skoog-Miller model' the formation of shoots is promoted by high levels of cytokinin relative to auxin while the reverse favours the development of roots. The model holds for a large number of plants but it is by no means universal. Not only do endogenous levels of cytokinin and auxin apparently differ according to the species, organ and growth stage, but other factors influence the ability of organs and tissues to regenerate shoots or roots. The levels of hormones required for organogenesis vary widely and are determined empirically for each species.

A number of cytokinin compounds are available but those mostly used are 6-benzylaminopurine (BAP), 6-furfurylaminopurine (kinetin) and 6-(γ, γ, dimethylallyamino)-purine (2iP). Many different auxins are used. Indole-3-acetic acid (IAA), the naturally occurring auxin, is the least active and is readily broken down by plant tissues. More useful are the stronger and more stable analogues such as indole-3-butyric acid (IBA) and naphthaleneacetic acid (NAA). The most active auxins are 2,4-dichlorophenoxyacetic acid (2,4-D), 2,4,5-trichlorophenoxyacetic acid (2,4,5-T), 4-chlorophenoxyacetic acid (CPA) and 4-amino-3,5,6-trichloropicolinic acid (Pichloram or TCP) and these induce rapid callus proliferation. High levels of these substances, especially 2,4-D, strongly suppress organogenesis.

Light and temperature conditions

Little or no photosynthesis need occur in organs or shoots cultured on media containing sucrose but some light is necessary for morphogenesis and chlorophyll formation. Fluorescent tubes are generally used as a light source, with intensities falling in the range 1000–5000 lx. There has been comparatively little work on investigating optimal conditions of light and temperature but various reports show that these can be important in some species. Murashige (1977) has found that the optimum light intensity for shoot formation in *Gerbera* and a large number of herbaceous genera is in the region of 1000 lx, intensities as low as 300 lx and as high as 3000–10000 lx being strongly inhibitory. Gautheret (1969) has shown that adventitious root formation in *Helianthus tuberosus* tuber sections is severely reduced by intensities of greater than 7000 lx.

As far as daylength is concerned, Murashige (1977) considered the optimum to be 16 h for a wide range of plants although there was nothing to signify a photoperiodic effect. Hussey & Falavigna (1980) found that more adventitious shoots were formed in 16-h than in 8-h days in cultures of *Allium cepa*. With shoot cultures of *Solanum tuberosum* maximum node formation took place in continuous light with fewer nodes and weaker shoots being formed in 8-h days (Hussey & Stacey, 1981). Plants known to be photoperiodically sensitive may well show an equivalent in vitro response. In *Vitis* not only are short days necessary for shoot and root formation (Chée & Pool, 1982), but in vitro behaviour is also influenced by short-day pretreatment of plants that are used to make cultures (Alleweldt & Radler, 1962).

Tissue cultures are normally grown successfully at temperatures of 20–25 °C and this may reflect the optimum temperature for growth of shoots in the majority of temperate species. The usual environmental temperature requirements, however, may also be required in vitro and these should be taken into account during in vitro propagation. Work in this laboratory with bulbous species (e.g. *Allium* and *Narcissus*) has indicated that optimum temperatures are about 18 °C and Fonnesbech & Fonnesbech (1980) have shown that proliferation of *Monstera deliciosa* was greater at 30 °C than at either 21 or 24 °C.

Although the effects of light and temperature may not always be large, there is clearly a case for systematic investigation with different environments when culturing new species. The same is probably true for some well-established culture systems where the conditions used by the original investigators are unquestioningly followed. Light and temperature studies are often omitted because of the widespread use of controlled environment shelves in large rooms that are run at only one or two standard daylength and temperature regimes.

Methods of shoot induction and proliferation

The mechanisms of vegetative propagation centre around the formation and multiplication of shoot meristems, each meristem being a potential plant. In vitro cultures used for propagation may be started either from existing meristems – the embryo, main shoot or subsequently formed axillary shoots – or from organ explants that are suitable for the induction of adventitious meristems in the form of shoot apices or embryos. When organ explants are used, adventitious structures may be formed either directly within the parent tissue or from intermediate callus or cell suspension cultures. These facts are summarised in Scheme 1.

The distinction between these various types of multiplication is important because the genetic uniformity of the plants produced may be affected. The factors concerned are discussed in a subsequent section.

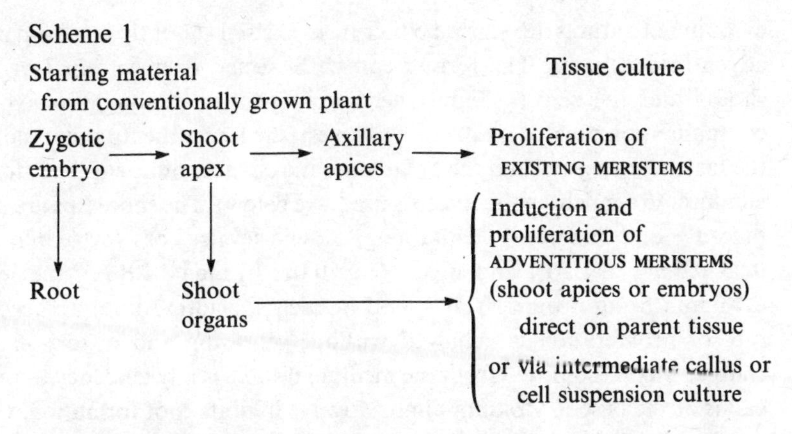

Multiplication by axillary shoots

Most higher plants have an indeterminate mode of growth in which the leaf axils contain subsidiary meristems each of which is capable of growing into a shoot that is identical to the main axis. According to the degree of branching that is charactristic of the species (and influenced to some extent by the environment), only a limited number of axillary meristems develop, the majority being inhibited by apical dominance. Although the mechanism of apical dominance appears to involve a number of different hormone systems (see review by Phillips, 1975), in many plants the outgrowth of axillary shoots seems to depend ultimately on the supply of cytokinin to the meristem. Shoot tips cultured on basic medium which is free of hormones typically develop into single seedling-like shoots with strong apical dominance. When cultured on medium containing cytokinin, axillary shoots often develop prematurely and are followed by secondary, then tertiary etc. shoots in a proliferating cluster (Fig. 1). Once such a cluster has developed sufficiently it may be divided up into smaller clumps of shoots or separate shoots which will form similar clusters when subcultured on fresh medium. Provided the basic nutrient formulation is adequate for normal growth this process may apparently be continued indefinitely. Proliferating shoots of *Gladiolus* have been maintained in this laboratory for 11 years (see Hussey, 1977) with no apparent physiological deterioration.

The rate at which precocious axillary shoots can be produced depends on the rate of leaf formation that is attainable in vitro and this varies with the species and the suitability of the nutrient medium. Within limits the rate of proliferation may be controlled by the concentration of cytokinin and the particular compound used (see Fig. 1). Multiplication rates vary widely but the range $\times 5$ to $\times 10$ per 4–8 weeks is typical in plants that respond well.

The technique of proliferation by axillary shoots should be applicable to any plant that produces regular axillary shoots and responds to an available cytokinin. Cultures are started either from excised shoot tips or from in vitro adventitious shoots. The former can be dissected from terminal or lateral shoots and the size (1–5 mm) need only be small enough to avoid the contamination present on the outer leaves. The larger the tip that is cultured, the faster it will grow and the greater are the chances of its survival (for virus elimination a much smaller piece is used, see below). The shoot tips are usually placed on medium containing low levels of cytokinin (e.g. 0·05–0.5 mg l^{-1} BAP) and auxin (e.g. 0.01–0.1 mg l^{-1} I B A), the level of cytokinin being progressively raised at each subculture until an acceptable rate of proliferation is achieved without yellowing and distortion of the shoots. Auxin is omitted from the medium if there is any tendency to produce callus at the base. Cytokinin almost always inhibits root formation, thereby making the shoot clusters more easily separable at each subculture. Rooting often occurs spontaneously on transfer to medium lacking cytokinin, especially in monocotyledons.

Reports of effective multiplication by enhanced axillary shoot production date from the mid-1970s and include both woody and herbaceous dicotyledons and monocotyledons. Examples are *Gerbera* (Murashige, Serpa & Jones, 1974), strawberry (Boxus, 1974), apple (Jones, Hopgood & O'Farrell, 1977), *Gladiolus* (Hussey, 1977), sugar beet (Hussey & Hepher, 1978), pineapple (Matthews & Rangan, 1979) and teak (Gupta, Nadgir, Mascarentias & Jagannathan, 1980).

Fig. 1. Shoot cultures of *Sparaxis tricolour*. Left, on medium without cytokinin; middle and right respectively, on medium with 0.5 and 2.0 mg l^{-1} BAP. Each culture vessel is 46 mm in diameter.

Virus elimination. Plants multiplied by axillary shoot cultures that have been derived from shoot tips 1–5 mm long will almost always carry any virus infection present in the mother plants. In order to eliminate viruses, the technique known as meristem tip culture is used. In this technique a much smaller piece of apical tissue is cultured, generally the apical dome and one or two leaf primordia comprising an explant 0.3–0.5 mm high. Meristem tip culture is often preceded by heat treatment (for 6–12 weeks at 30–40 °C) of the mother plant in order to attenuate the viruses. It seems that viruses either do not penetrate easily into the shoot apex or their replication there is inhibited by the trauma of excision or by other factors; the result is that some of the plants grown up from meristem tips provc to be completely free of detectable viruses. Generally the smaller the tips that are cultured, the greater are the chances of recovering a virus-free plant though the rates of growth and survival of the explants are correspondingly reduced.

Meristem tip culture is now routinely carried out on many horticultural and agricultural crops. The nutrient media requirements resemble those for the establishment of cultures of shoot tips, close attention being given to the hormone levels in the early stages in order to reduce callus development to a minimum. Excess callus can swamp the growth of the meristem or give rise to adventitious shoot apices which may differ genetically from the parent plant (see e.g. Hackett & Anderson, 1967).

Comprehensive accounts of the production of virus-free plants have been given by Quak (1977) and Walkey (1978, 1980).

Multiplication by adventitious shoots

Numerous observations are recorded in the botanical literature of shoots arising adventitiously on mature organs, particularly on leaves, stems and roots. Such phenomena are frequently exploited in horticultural practice by the use of cuttings for propagation (for references see Priestley & Swingle, 1929; Sinnott, 1960; Broertjes, Haccius & Weidlich, 1968; Peterson, 1975; Hartmann & Kester, 1975). Almost every type of organ can be used as a cutting. For instance, the formation of shoots occurs on the leaves of *Begonia* and other ornamentals, on the scales of various bulbs and on stem and root pieces of trees such as apple. The fact that this kind of reproductive regeneration occurs so widely throughout the plant kingdom led early in this century to the concept of the 'totipotency' of plant cells and tissues (Sinnott, 1950, 1960; Krikorian & Berquam, 1969). The more recent use of tissue culture techniques has increasingly reinforced this concept by the induction of shoots in vitro in more and more species.

Shoots formed directly on organ tissue. Adventitious regeneration direct on organ explants can be obtained in a wide variety of plants but in many of these adequate multiplication may be obtained more readily by the use of axillary shoots, which are likely to be genetically more stable (see below). There are many plants, however, that form axillary shoots too slowly for rapid propagation. These include conifers and some of the palms (see Jones, this volume) as well as various ornamentals and bulbous monocotyledons.

The tissue culture method is essentially a scaled-down and refined version of the technique used with conventional cuttings; aseptic conditions and adequate nutrient supplies allow the size of the explant to be reduced to only a few mm if necessary. In many cases the most suitable organ for adventitious regeneration is that normally used in conventional propagation though much younger, more meristematic tissue will frequently show more vigorous regeneration. With other plants, tissues not normally associated with vegetative reproduction (e.g. anther filaments) can prove to be totipotent.

Saintpaulia ionantha (African violet) is conventionally propagated from leaf cuttings, one or more shoots arising from the base of each petiole (Hartmann & Kester, 1975). Bilkey, McCown & Hildebrandt (1978) have described an in vitro method using 2-mm thick petiole sections which can yield up to 20000 plantlets from each petiole. *Petunia hybrida* (Daykin, Langhans & Earle, 1976) and *Kalanchöe blossfeldiana* (Smith & Nightingale, 1979) may similarly be propagated to large numbers from small leaf or stem pieces.

Many bulbous species have very slow natural propagation rates and various conventional means such as chipping, scaling (including 'twin-scaling') and scooping (Hartmann & Kester, 1975; Alkema & van Leeuwn, 1977) are employed to facilitate multiplication. The tissue culture technique enables this type of adventive shoot or bulbil production to be greatly increased not only by facilitating more prolific regeneration from smaller explants but also by making possible the continued recycling of in vitro material for continuous production (see references cited by Hussey, 1980*b*). Although shoots can be obtained in vitro from a number of different organs, the most regenerative regions are the basal tissues of leaves and scales immediately above the basal plate (Hussey, 1980*a*). In vitro shoots can be induced on scale or leaf piece explants on media containing auxin or auxin plus cytokinin (Pierik & Steegmans, 1975; Seabrook, Cumming & Dionne, 1976). Developing in vitro shoots from a number of bulbous plants can be split up and used as secondary explants for further adventitious shoot production (Hussey & Falavigna, 1980; Hussey, 1982).

Recent work in this laboratory has established what appears to be an effective method of continuous propagation by adventitious shoots applicable to many species that form bulbs or corms. Shoots induced on suitable explants in vitro are trimmed to approximately 5 mm high, this fragment including 2–3 mm of basal plate tissue. In some bulbs such as *Iris*, a mixture of axillary and adventitious shoots arise from between the young leaves and scales. In others such as *Narcissus*, *Nerine*, *Hyacinthus* and *Leucojum*, apical dominance is very pronounced and to obtain a good response the main apex has to be destroyed after trimming. This is achieved by making two vertical cuts at right angles to the level of the apex without penetrating low enough to split the basal plate. The same procedure is later repeated on each of the clusters of adventitious shoots that arise from between the truncated scales.

Shoots from callus. An in vitro callus is an unorganised mass of proliferating cells and may be obtained from almost any type of plant. Callus formation from explants is occasionally spontaneous, representing an extension of the wound reaction, but generally requires an auxin in the medium, often in combination with a cytokinin. The concentrations of the hormones most effective in producing callus vary with the species and the organ used, but tend to be higher than those needed to induce shoots directly from the explant where this is possible. Fig. 2 shows (left) shoots formed directly on a swollen leaf base explant of *Narcissus* cultured on medium containing 16 mg l^{-1} BAP and 0.5 mg l^{-1} NAA, and (right) callus with a few shoots induced on a similar explant cultured on medium containing 16 mg l^{-1} BAP and 4 mg l^{-1} NAA.

Fig. 2. Leaf base explants of *Narcissus* which give rise to (left) adventitious shoots and (right) callus with shoots. Each culture vessel is 46 mm in diameter.

In most plants callus can be detached from the explant, subcultured and serially propagated on medium containing the hormone levels that were used to induce it. On *lowering* the hormone levels, or appropriately adjusting the auxin:cytokinin ratio, some calluses will regenerate shoots or embryos. Mass production of callus followed by shoot regeneration would seem to be the ideal method of large-scale propagation but two serious drawbacks at present limit its use to only a few species. With repeated subculture (as would be necessary to obtain large quantities) the capacity of many calluses to regenerate shoots is diminished or even lost; at the same time the proportion of polyploid, aneuploid and other genetically aberrant cells progressively increases, making it likely that the callus will regenerate plants that differ from the parent type. The problem of the genetic stability of tissue cultures will be considered in detail below.

Rooting and planting out. Adventitious embryos develop directly into seedling-like plantlets but in vitro shoots proliferating on medium containing cytokinin are almost always devoid of roots. Root formation often occurs spontaneously on transfer to medium lacking cytokinin, especially in certain monocotyledons that normally form adventitous roots readily in vivo. A rooting hormone in the medium is more generally required for the final stage, IBA being commonly used at concentrations of between 0.5 and 10 mg l^{-1}.

Plantlets produced in vitro are equivalent to small rooted cuttings but both root and shoot systems need to adapt to normal conditions on planting out. The best time for transplanting is when the roots are beginning to grow out and the developing leaves are becoming photosynthetically self-supporting. Before planting out some photosynthesis can be encouraged by increasing the light intensity and allowing carbon dioxide to enter the culture vessels by slackening the screw tops or other covering. It needs to be emphasised, however, that because different species have become adapted to diverse environmental conditions, only very general guidelines for planting out can be formulated. Each type of plant produced in vitro is likely to have its own optimum requirements for successful establishment in soil or compost, and these have to be ascertained empirically.

Debergh & Maene (1981) have proposed that shoots be rooted *after* removal from culture whenever possible by using suitably modified conventional rooting procedures. They point out that rooting in vitro is often the most labour-intensive stage of in vitro propagation because the shoots have to be manipulated singly rather than in proliferating clusters, and that in vitro roots rarely adapt to normal conditions without some damage and consequent delay in growth.

Plants that normally form resting organs can be encouraged under the right conditions to form them in vitro as an alternative to direct planting. Tubers (Fig. 3, right), cormels, bulbils and rhizomes make convenient and easily transportable planting material and in some cases could be adaptable to mechanisation. Cold treatment or hormone application may be necessary with some temperate plants to break dormancy.

Limitations and problems

The effectiveness of in vitro methods in improving vegetative propagation coupled with the advantages of built-in disease protection makes them an attractive practical alternative to conventional techniques with many types of plant. Shoots may be multiplied rapidly and considerable savings made in the time and space required by conventional procedures. Disease-tested material can be bulked in large quantities without the costly precautions that

Fig. 3. Potato (*Solanum tuberosum*). Left, shoot cultures; right, in vitro tubers formed directly from subcultured nodes. Each culture vessel is 46 mm in diameter.

are otherwise needed to prevent reinfection during propagation. Once a nucleus of disease-free material is established in culture it will remain safe from reinfection until planted out. With some crops an in vitro method provides, or could provide, the only practicable means of producing clonal material.

These obvious advantages, however, may very easily be offset by one or more of the problems and disadvantages that are inherent in the tissue culture method, and these factors on the debit side will now be considered. Promising and attractive laboratory results can be beguiling and many workers have become over-enthusiastic about the possibilities of micropropagation without seriously considering its limitations. On the other hand there are some who take a pessimistic view, dismissing its practical role as hazardous and over-expensive. One of the purposes of the following discussion is to get the perspective right by treating the problems in some detail. They will be dealt with under four headings.

Expertise and equipment

Compared with a conventional propagating bench, a tissue culture laboratory requires a large capital investment with correspondingly higher running costs. The cost of labour constitutes the largest single item of expenditure. Furthermore, personnel should be conversant with standard microbiological techniques and have sufficient knowledge of the botany of the material and its in vitro behaviour to exercise the care and consistency necessary for good results. Without reliability and attention to detail disaster is never far away. Careful organisation can reduce all labour-intensive operations to a minimum; nevertheless, the end product with all its advantages is bound to be an expensive one.

Range of applicability

Although in theory it seems that any plant can be cultured by supplying the appropriate metabolites in suitable in vitro conditions, so far only a small proportion of plants can be cultured well enough to be effectively propagated. There are still many important species in which shoot multiplication or regeneration is difficult or impossible, the lack of reaction usually being associated with poor in vitro growth. A more systematic approach to devising culture systems is now increasingly employed in which the structure, life-cycle and biochemistry of the plant are taken into account, but the procedures are still predominantly empirical. The time and expense involved should be worthwhile for the major crop plants but not necessarily so for the minor crops. For the general run of desirable ornamentals and miscellaneous plants, in vitro propagation may simply not be worth the effort unless they are readily amenable to it.

Technical problems

Phenolics and other inhibiting factors. A number of plants produce excessive amounts of phenolic substances whose oxidation products not only darken both the tissues and the medium but often strongly inhibit growth. Where the problem is confined to the reaction of the initial explant it may be prevented by treatment with ascorbic or citric acid. In the case of teak, *Tectona grandis*, polyvinylpolypyrrolidone proved effective (Gupta *et al.*, 1980).

Growth and regeneration have sometimes been improved by the addition of activated charcoal to the medium and there is evidence that inhibiting substances can be removed (Fridborg, Pedersen, Landström & Eriksson, 1978). Unfortunately, activated charcoal also binds hormones and other metabolites (Constantin, Henke & Mansur, 1977; Weatherhead, Burdon & Henshaw, 1978) and this lack of selectivity limits its use in tissue culture media.

Vitrification. An occasional problem with cultured plants is the development of swollen, distorted leaves which become irreversibly translucent and eventually necrotic, a condition that may lead to the death of the shoot cluster. This has been referred to as 'watersoaking' by Hussey (1977, 1978) and more recently described as 'vitrification' by Debergh, Harbaoui & Lemeur (1981). The phenomenon occurs in shoots cultured on medium containing cytokinin and can generally be prevented by making sure that the concentrations are no higher than necessary. In the globe artichoke, *Cynara scolymus*, Debergh *et al.* were able to overcome vitrification only by raising the agar concentration to 1.1% (w/v), but this necessitated the use of a mixture of the cytokinins kinetin and 2iP to obtain acceptable multiplication rates.

Bacterial contamination. A persistent difficulty in tissue culture laboratories is the occurrence of contamination by slowly growing and often macroscopically invisible micro-organisms, particularly bacteria. These may have no obvious effect on proliferating shoots but vigour is often depressed and subsequent rooting may be partly inhibited. Most are slow-growing saprophytic organisms that have survived the original decontamination procedures and their incidence can be lessened by using only clean starting material that has been grown without overhead watering. It is important that the original cultures be examined microscopically and tested with microbe detecting media so that any infected cultures can be rejected. In some laboratories cultures have become infected with *Bacillus subtilis*. This demonstrates how easily tissue cultures can become contaminated in spite of the use of standard aseptic methods.

A more serious problem concerns bacteria that are carried within the tissues of the plant and cannot be removed by the normal sterilisation techniques. Knauss & Miller (1978) have reported the contamination of commercial plant tissue cultures of several ornamental plants with *Erwinia carotovora*, an important pathogen of many horticultural plants. The presence of the bacterium caused a reduction in both vigour and chlorosis in cultures of plants not known to be susceptible in conventional comercial production (*Nephrolepis exaltata*, *Saxifraga sarmentosa* and *Pteris* spp.) Such cultures may serve as carriers of *Erwinia* in culture operations.

Quality of plants propagated by in vitro techniques

There are a number of reasons why plants propagated by tissue culture may not consistently resemble the original. These may be considered under the categories of (a) short-term developmental changes, so-called epigenetic effects, and (b) genetic differences arising from gene or chromosomal mutations.

Epigenetic changes. Plantlets formed in vitro may carry over the effects of the culture conditions to subsequent growth after planting out. The first few leaves formed may be abnormal in shape or senesce prematurely. These effects will be important only when the plantlet produced in vitro is being used directly for a saleable product, e.g. ferns or foliage plants, and can be avoided by careful choice of media and time of transfer to compost.

Where the apical organisation is affected the changes may be more persistent, e.g. increased branching or leaf arrangements. Anderson, Abbott & Wiltshire (1982) have described an abnormality in micropropagated strawberry plants, persisting for at least two years after planting out, in which an excessive number of small branch crowns are accompanied by disturbed phyllotaxis, giving the plants a stunted, bushy appearance. Dissection of the growing point revealed multiple shoot apices in a linear or occasionally circular arrangement. The condition was exacerbated by higher BAP concentrations but alleviated by gibberellin (GA_3). The authors concluded that it could be avoided by using appropriate levels of hormones while still obtaining acceptable multiplication rates.

Genetic changes. Spontaneous mutations occur occasionally in all crops. The majority of these changes are deleterious and during conventional propagation any affected plants can be routinely discarded. In the culture tube, however, mutant shoots may not be recognisable; changes in characters like leaf shape, flower colour or fruit quality will only be evident after the propagules have been planted out and grown up. Spontaneous mutations occurring in vitro are therefore likely to go undetected during multiplication. The earlier that a mutation occurs during the bulking-up process the greater will be the proportion of mutant propagules produced.

The risks of incurring a significant proportion of mutant plants during micropropagation will clearly be reduced by starting with as many shoot tips or explants as possible in order to broaden the base of the initial material. The recycling of small samples of in vitro material (which might contain a mutant shoot) is also best avoided, the safest policy being to start any multiplication with plants that have been carefully checked for trueness-to-type.

The incidence of mutation, however, is not simply a question of the numbers of individuals being propagated; much more important is the method of in vitro multiplication being used. Experience with both experimental and practical work shows that organised meristems are genetically the most stable parts of the plant and that multiplication by precocious axillary shoots is the method least likely to incur mutation. Evidence from both propagation and mutation breeding is consistent on this point. Plant breeders, who are trying to *produce* mutant plants, as opposed to propagators, who are trying to avoid them, find that shoot systems are fairly resistant to mutation. Axillary meristems arise on the flanks of the shoot apex and include several discrete tissue layers: one or more tunica layers plus the corpus. Mutation is a single-celled event. A spontaneous or induced mutant cell can divide to form only a limited area of tissue within one layer, resulting in a chimera that usually proves unstable and short-lived.

The adventitious meristem, by contrast with the axillary meristem, is much more susceptible to mutation as it is usually derived from either a single cell or a small group of cells. The plant breeder is thus able to produce wholly mutant shoots (rather than chimeras) much more easily by using regenerating organ tissue for the application of mutagens or even by relying on spontaneous mutation (for an extended treatment of this topic see Broertjes & van Harten, 1978). The propagator should therefore view with suspicion any in vitro system involving adventitious meristems.

Mutation in totipotent cells may be enhanced as a result of the possible mutagenic action of media constituents such as NAA, 2,4-D or BAP but there is no convincing evidence of any direct effects of these compounds. The most likely source of genetic variability is the changes that occur naturally during differentiation, particularly polyploidy, which occurs in 90% of plant species (D'Amato, 1975). The variability in plants regenerated in vitro from somatic cells is often considerable (Skirvin, 1978) and although the underlying mechanisms are very imperfectly understood, it would appear that genetic changes frequently occur in plant tissues and come to light only after regeneration by in vitro procedures. Naturally occurring adventitious regeneration, and to a large extent that exploited in conventional cuttings, comes from tissue that is typically non-polysomatic, such as epidermis, or from certain parts of the phloem. But with the use of hormones, other tissues containing polyploid cells are stimulated to divide and either form organised meristems directly within the tissue or form varying amounts of intermediate callus growth according to the type of hormones used and their concentrations. The genetic uniformity of adventitious shoots will therefore depend on both the nature of the explant and the hormones used. The chances of mutation will be reduced by using the youngest, least differentiated organ tissue with the hormone levels adjusted to produce the minimum of unwanted callus.

When callus growth is induced, endopolyploid cells will probably be included from the beginning. Callus, however, even if deliberately started from a strictly non-polysomatic explant, typically develops increasing proportions of polyploid and eventually aneuploid cells, especially if 2,4-D or NAA is used (Sunderland, 1977). As well as chromosomal mutations, apparent gene mutations affecting media requirements or pigment formation occur. Callus cultures may develop considerable clonal variation and lines may be isolated that show quite striking differences in appearance, growth rate, response to growth hormones and degree of totipotency.

Although there are a number of reports of apparently normal diploid plants being regenerated from callus cultures, many examples of polyploid and aneuploid regenerants have been recorded. Examples are listed by Murashige (1974) and by D'Amato (1977). There are also clear-cut cases of gene mutations: *Chrysanthemum* plants from callus cultures showed changes in flower colour (Ben-Jaacov & Langhans, 1972) and selection of useful mutants has been possible in sugar cane (Heinz, Krishnamurthi, Nickell & Maretzki, 1977; Nickell, 1977).

Thus, with present techniques, the method of choice for in vitro propagation is proliferation by axillary shoots in order to ensure genetically uniform propagules. Propagation laboratories select this method whenever possible and uniform material is now routinely produced. In a few species comparisons have been possible between plants propagated by axillary shoots and those regenerated from callus. *Asparagus* multiplied by axillary shoot enhancement by Murashige *et al.* (1972) showed no genetic aberrations while those regenerated from callus by Malnassy & Ellison (1970) included up to 70% polyploids. Davis, Baker & Hanan (1977) found no mutants out of approximately 1500 carnations propagated by axillary shoots, while in the earlier work of Hackett & Anderson (1967) involving callus, carnations were produced with deviations in flower colour.

Multiplication by adventitious shoots is appropriate where axillary shoots are infrequent or unresponsive. With this method of propagation careful choice of explant and attention to hormone levels is clearly necessary. In some species, abnormal plants are readily formed adventitiously as, for example, in marrowstem kale (*Brassica oleracea*), in which plants regenerated direct from leaf and stem explants included over 70% tetraploids and 15% octaploids (Horák *et al.*, 1975). Not all adventitious shoots derive from one tissue layer or from single cells. In a number of bulbous plants it has been shown that adventitious meristems are formed at the base of scales or leaves from at least two layers of cells (Hussey & Falavigna, 1980; Hussey, 1982).

Current uses of micropropagation

It follows from the considerations in the previous section that the effort and expense of propagating plants by tissue culture will be worthwhile only for special purposes connected with breeding or research or where there are problems with conventional methods. The current uses of micropropagation may be categorised as follows.

(1) In breeding work, micropropagation is particularly useful for the maintenance and multiplication of modest numbers of special genotypes or potential new cultivars including any products of genetic engineering involving in vitro procedures.

(2) Newly selected cultivars can be bulked up rapidly to the level of several thousands in months rather than years; much time and effort is thus saved in the initial stages of introducing them into commercial production. Orchids were the first plants to benefit in this way and many ornamentals, including bulb crops, are now being multiplied rapidly for this purpose (Holdgate, 1977).

(3) Many plants, while not necessarily difficult to propagate conventionally, are very susceptible to virus and these include lilies, strawberries and potatoes. The advantageous use of micropropagation here lies in the rapid production of large quantities of guaranteed disease-free material either for direct planting or as a basis for further conventional increase. The large-scale production of disease-free potato plants may become commercially important. It normally takes six years of multiplication in the field to produce foundation seed stock starting from disease-tested cuttings. Inevitably during this time, in spite of locations selected for low aphid populations, some reinfection occurs. The equivalent of at least four years of field multiplication can be produced in vitro within a few months (Hussey & Stacey, 1981 and references cited, see Fig. 3).

(4) Some heterozygous horticultural crops such as *Gerbera* and *Anthurium* that are conventionally propagated only by seed can now be cloned rapidly from individuals selected for colour and vigour.

(5) F_1 hybrid seed crops such as those of cauliflower (Crisp & Walkey, 1974) can be produced from cloned selected parents. In the onion the

disease-free maintenance of male sterile lines by tissue culture provides an alternative to the difficult back-crossing method (Hussey & Falavigna, 1980).

(6) In woody species, breeding and selection by sexual hybridisation has been slow because of the long generation times involved, and many are difficult to propagate vegetatively. The in vitro cloning of elite specimens has enormous potential application as discussed fully by Jones (this volume).

Future prospects

Range of species amenable to culture

The number of species that can be successfully propagated by tissue culture may be expected to increase as media and techniques are improved. It is not clear at present how far the inadequate growth obtained with so many plants is due to deficiencies in the nutrient medium or to other physical and chemical factors connected with the method of culture. In vitro behaviour must largely reflect in vivo behaviour and it would be naive to assume that all species can be made to react in a more or less standard way in the culture vessel. It is possible that to some extent the successful tissue culture systems happen to be those which can easily be adapted to the in vitro techniques originally developed for plants like carrot and tobacco and that major modifications are necessary in order to obtain optimum performance from plants genetically programmed to behave differently. A less optimistic view is that only certain species happen to be adaptable to the in vitro method and that for various reasons others are inherently unable to show normal growth and development in isolated tissues in a closed aseptic environment. The disqualifying factors are largely speculative but possibilities such as wound reactions that affect nutrient transport, the formation of inhibitors and the adverse effects of volatile substances are easily suggested and should form the subject of future research.

Towards mechanisation

The development of in vitro propagation in the immediate future will be limited mainly by the labour-intensive nature of the procedures which confine practical application to the relatively high-value material described in the previous section. The large-scale production of horticultural plants that sell easily can be profitable if there are problems with conventional propagation, and the same will be true for many woody crops. The mass propagation of selected plants of heterozygous field crops, however, cannot be economically justified with current procedures and new methods which are amenable to some form of mechanisation will have to be devised. Present tissue culture methods represent a sophisticated horticultural practice that is almost impossible to automate. The first step towards mechanisation would be the removal of the need for precise division and orientation of the proliferating material so that it could be cut up at random or shaken or vibrated in liquid to make it fall apart. Three possible systems will be considered as examples of the lines along which research might be directed.

Proliferating adventitious shoots on young leaves. Adventitious shoots are most likely to be genetically uniform if regenerated from young meristematic tissue such as developing leaves. A system in which young adventitious shoots are regularly cut up to provide explant material for further adventitious shoot production could give a high proliferation rate while avoiding the risks of polyploidy and other genetic aberration that might arise from the use of older leaves. *Ornithogalum thyrsoides* provides a model system for this type of culture (Hussey, 1976). Grass-like shoots derived from mature leaves or scales can be chopped up into small pieces a few mm long which will regenerate similar shoots in liquid medium (Fig. 4). If such a system could be adapted to field crops, mass propagation might be feasible (with dicotyledonous plants axillary shoots would be involved as well but the principle would be the same). An example might be sugar beet, one of the most difficult arable crops to breed. Great benefits would accrue from massed clonal material for use either

Fig. 4. Proliferating adventitious shoot cultures of *Ornithogalum thyrsoides*. Each culture vessel is 30 mm in diameter.

Fig. 5. *Freesia* callus regenerating shoots. The culture vessel is 46 mm in diameter.

in breeding programmes or for eventual field planting. Plants may be multiplied by axillary shoot proliferation (Hussey & Hepher, 1978; Fig. 6) and it is also possible to derive clusters of adventitious shoots from the petioles of such plantlets (Fig. 7). So far this reaction has been obtainable regularly in one genotype but also occurs sporadically in others tested. Control of this phenomenon is clearly much more difficult in sugar beet than in *Ornithogalum* but the establishment of the appropriate conditions could pave the way to mechanised shoot production.

Stable regenerating callus. An ideal callus would be one in which (a) genetically stable, diploid, meristematic cells similar to those found in the region of the shoot apex could be multiplied continuously and (b) plantlets could be recovered in large numbers after making appropriate changes in the medium (Fig. 8). Meristematic cells, however, typically divide slowly, are almost impossible to culture alone and since they are genetically programmed to form the organised structures of the shoot apex, generally continue to do so in vitro. At present the only practicable method of frustrating this organised development is to disrupt it to varying degrees by the use of hormones thereby producing a mass of cells which is loosely termed callus. Most calluses consist of mixtures of irregularly differentiating cells interspersed with areas of smaller more meristematic cells. In fast-growing callus polyploid

Fig. 6. Proliferating axillary shoot cultures of sugar beet (*Beta vulgaris*). Each culture vessel is 46 mm in diameter.

Fig. 7. Adventitious shoots arising on petioles from shoot cultures of sugar beet.

and other genetically aberrant cells increasingly predominate and some may prove to be totipotent (Fig. 9). In order to ensure that a callus remains totipotent and regenerates only true-to-type plants it is important that the diploid meristematic cells continue to hold their own against the enlarging irregularly differentiating cells which should not produce plants. Callus that regenerates uniform plants has been described for a number of species, e.g. *Lilium* (Sheridan, 1968), *Freesia* (Davies, 1972), *Chrysanthemum* (Earle & Langhans, 1974), tomato (De Langhe & De Bruijne, 1976) and *Hemerocallis* (Krikorian, Staicu & Kann, 1981). The *Chrysanthemum* callus continuously produced plantlets over a four-year period and histological examination revealed superficial meristematic areas around masses of typical highly vacuolated callus-like cells. In this laboratory, various lines of *Freesia* callus have continued to produce plantlets for up to 10 years (Fig. 5) and histological examination shows superficial meristematic areas around more vacuolated inner cells (Fig. 10). Walkey & Woolfitt (1968) described a regenerating callus of *Nicotiana rustica* which went on producing plantlets for 10 years. When samples were cultured and examined in this laboratory, the callus was found

Fig. 8. Ideal callus propagation (flow diagram).

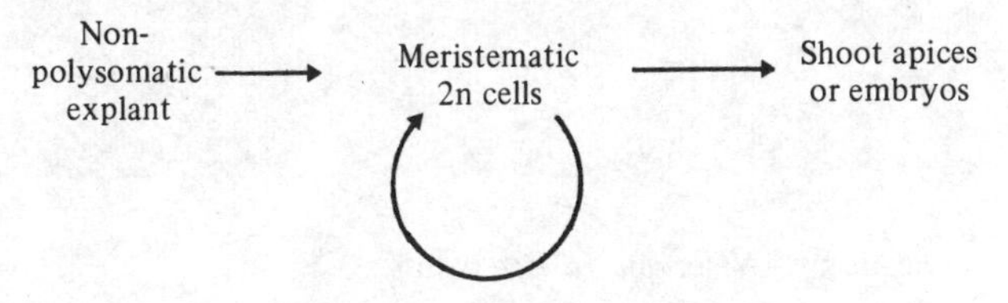

Fig. 9. Propagation by mixaploid callus (flow diagram).

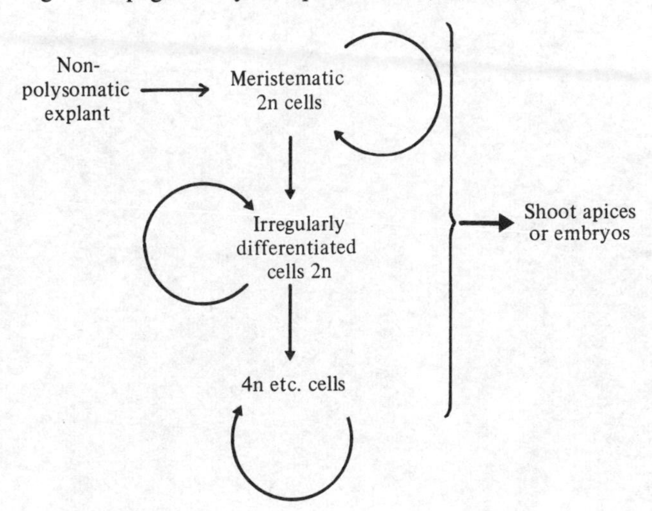

to have a structure similar to that of both the *Freesia* and *Chrysanthemum* calluses (Fig. 11). All these calluses can be proliferated by cutting them up at random.

It would appear that in calluses of the type just described the shoot apices forming on the surface are inhibiting (possibly by some kind of apical

Fig. 10. Section of *Freesia* callus.

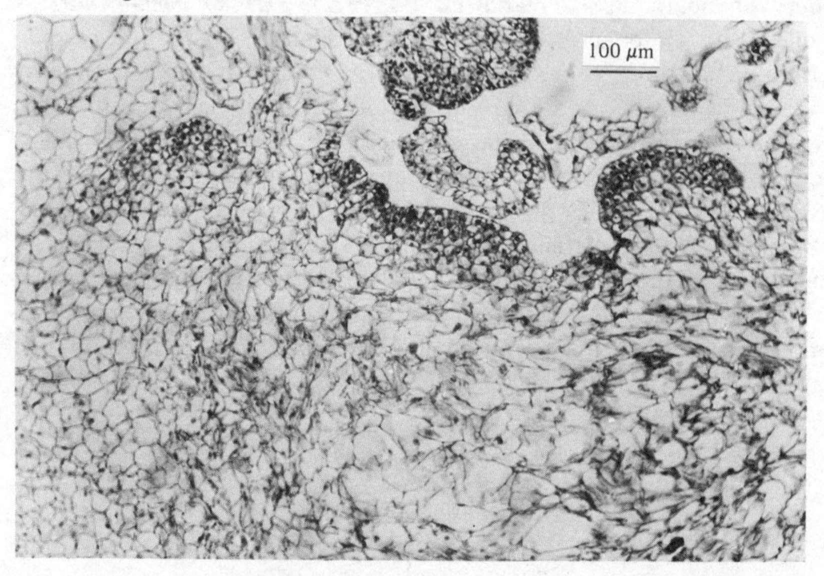

Fig. 11. Section of *Nicotiana rustica* callus.

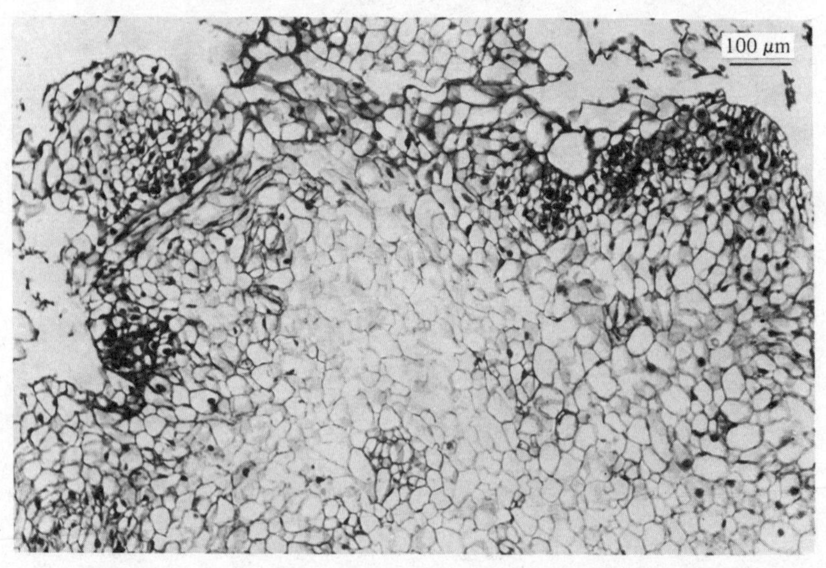

dominance) regeneration in the inner more vacuolated cells which act only as mechanical and nutritional support (Fig. 12). The presence of the meristematic tissue and regenerating shoot apices may be crucial to the formation and maintenance of such calluses as well as to the use of hormone levels which do not disturb the balance of growth of meristematic and internal tissues. Although callus of this type is clearly more easily obtained in certain plants it could be worthwhile deliberately trying to derive it from others. Careful attention would have to be given to the structure, as revealed histologically, of the callus during its development and establishment as well as to the hormone levels necessary to achieve this. All too often the initial callus is obtained from rapidly proliferating cell masses that are selected on a purely macroscopic basis without regard to their origin and structure.

Mass embryogenesis. Mass somatic embryogenesis from cell suspension cultures has long been regarded as the ultimate goal in plant propagation and improvement, and a number of workers have adopted this philosophy in their approach to micropropagation (e.g. Krikorian & Kann, 1979). Most of the research so far has been based on the carrot from the laboratories of Steward (1958) and Reinert (1959) but embryo production has since been reported in many other plants including cotton (Price & Smith, 1979), date palm (Tisserat & DeMason, 1980), grape (Srinivasan & Mullins, 1980), coffee (Staritsky, 1970) and pearl millet (Vasil & Vasil, 1981). Very few systems, however, show any immediate promise as far as practical application is concerned and research has a long way to go before this approach can play an important role in propagation. Tisserat, Esan & Murashige (1978) have pointed out that asexual embryo formation in tissue cultures is relatively widespread among angiosperms and although occurring more frequently in certain families (e.g. Ranunculaceae, Solanaceae and Umbelliferae) it has been observed in nearly 150 species in more than 30 families of flowering plants.

Once efficient mass embryogenesis has been obtained in an economically

Fig. 12. Propagation by stable callus (flow diagram).

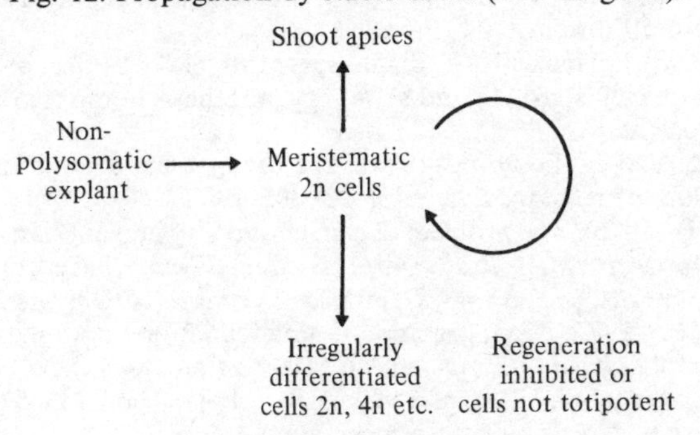

important plant there remains the problem of how to plant out the propagules in the field. Adventitious embryos have no food reserves, as in a seed, and some means is required to get the naked embryos into the soil before they have developed too far to be handled mechanically. All kinds of devices are being considered and these include plastic strips or pellets of various kinds in which the developing embryos together with the necessary nutrients could be trapped. An alternative approach is to harness the principle of fluid drilling in order to inject the developing plantlets directly into the field. Such ideas are not altogether fanciful but serve to emphasise the difficulties inherent in the use of adventitious embryos for large-scale planting.

References

Alkema, H. Y. & van Leeuwn, C. J. M. (1977). Propagation of a number of miscellaneous bulb crops by means of the twin scale method. *Bloembollencultuur*, **88**, 32–3.

Alleweldt, G. & Radler, F. (1962). Interrelationship between photoperiod behaviour of grapes and growth of plant tissue cultures. *Plant Physiology*, **37**, 376–9.

Anderson, H. M., Abbott, A. J. & Wiltshire, S. (1982). Micro-propagation of strawberry plants *in vitro* – effect of growth regulators on incidence of multi-apex abnormality. *Scientia Horticulturae*, **16**, 331–41.

Ben-Jaacov, J. & Langhans, R. W. (1972). Rapid multiplication of *Chrysanthemum* plants by stem-tip proliferation. *HortScience*, **7**, 289–90.

Bilkey, P. C., McCown, B. H. & Hildebrandt, A. C. (1978). Micropropagation of African Violet from petiole cross sections. *HortScience*, **13**, 37–8.

Boxus, P. (1974). The production of strawberry plants by *in vitro* micropropagation. *Journal of Horticultural Science*, **49**, 209–10.

Broertjes, C., Haccius, B. & Weidlich, S. (1968). Adventitious bud formation on isolated leaves and its significance for mutation breeding. *Euphytica*, **17**, 321–44.

Broertjes, C. & van Harten, A. M. (1978). *Application of Mutation Breeding Methods in the Improvement of Vegetatively Propagated Crops*. Amsterdam: Elsevier Scientific Publishing Co.

Chée, R. & Pool, R. M. (1982). The effects of growth substances and photoperiod on the development of shoot apices of *Vitis* cultured *in vitro*. *Scientia Horticulturae*, **16**, 17–27.

Conger, B. V. (Ed.) (1981). *Cloning Agricultural Plants via In Vitro Techniques*. Boca Raton, Florida: CRC Press.

Constantin, M. J., Henke, R. R. & Mansur, M. A. (1977). Effect of activated charcoal on callus growth and shoot organogenesis in tobacco. *In Vitro*, **13**, 293–6.

Crisp, P. & Walkey, D. G. A. (1974). The use of aseptic meristem culture in cauliflower breeding. *Euphytica*, **23**, 305–13.

D'Amato, F. (1975). The problem of genetic stability in plant tissue and cell cultures. *In Crop Genetic Resources for Today and Tomorrow*, ed. O. Frankel & J. G. Hawkes, pp. 333–48. Cambridge: Cambridge University Press.

D'Amato, F. (1977). Cytogenetics of differentiation in tissue and cell cultures. In *Applied and Fundamental Aspects of Plant Cell, Tissue and Organ Culture*, ed. J. Reinert & Y. P. S. Bajaj, pp. 343–57. Berlin, Heidelberg and New York: Springer-Verlag.

Davies, D. R. (1972). Speeding up the commercial propagation of freesias. *Grower*, **77**, 711.

Davis, M. J., Baker, R. & Hanan, J. J. (1977). Clonal multiplication of carnation by micropropagation. *Journal of the American Society for Horticultural Science*, **102**, 48–53.

Daykin, M., Langhans, R. W. & Earle, E. D. (1976). Tissue culture of the double petunia. *HortScience*, **11**, 35.

Debergh, P., Harbaoui, Y. & Lemeur, R. (1981). Mass propagation of globe artichoke (*Cynara scolymus*): Evaluation of different hypotheses to overcome vitrification with special reference to water potential. *Physiologia Plantarum*, **53**, 181–7.

Debergh, P. C. & Maene, L. J. (1981). A scheme for commercial propagation of ornamental plants by tissue culture. *Scientia Horticulturae*, **14**, 335–45.

De Langhe, E. & De Bruijne, E. (1976). Continuous propagation of tomato plants by means of callus cultures. *Scientia Horticulturae*, **4**, 221–7.

Earle, E. D. & Langhans, R. W. (1974). Propagation of *Chrysanthemum in vitro*. II. Production, growth and flowering of plantlets from tissue cultures. *Journal of the American Society for Horticultural Science*, **99**, 352–8.

Fonnesbech, A. & Fonnesbech, M. (1980). *In vitro* propagation of *Monstera deliciosa*. *HortScience*, **15**, 740–1.

de Fossard, R. S. (1977). *Tissue Culture for Plant Propagators*. Armidale: University of New England Printing.

Fridborg, G., Pedersen, M., Landström, L. & Eriksson, T. (1978). The effect of activated charcoal on tissue cultures: adsorption of metabolites inhibiting morphogenesis. *Physiologia Plantarum*, **43**, 104–6.

Gamborg, O. L., Murashige, T., Thorpe, T. A. & Vasil, I. K. (1976). Plant tissue culture media. *In Vitro*, **12**, 473–8.

Gautheret, R. J. (1969). Investigations on the root formation in the tissues of *Helianthus tuberosus* cultured *in vitro*. *American Journal of Botany*, **56**, 702–17.

Gupta, P. K., Nadgir, A. L., Mascarentias, A. F. & Jagannathan, V. (1980). Tissue culture of forest trees: clonal multiplication of *Tectona grandis* L. (teak) by tissue culture. *Plant Science Letters*, **17**, 259–68.

Hackett, W. P. & Anderson, J. M. (1967). Aseptic multiplication and maintenance of differentiated carnation shoots derived from shoot apices. *Proceedings of the American Society for Horticultural Science*, **90**, 365–9.

Hartmann, H. T. & Kester, D. E. (1975). *Plant Propagation, Principles and Practices*. Prentice Hall.

Heinz, D. J., Krishnamurthi, M., Nickell, L. G. & Maretzki, A. (1977). Cell, tissue and organ culture in sugar cane improvement. *In Applied and Fundamental Aspects of Plant Cell, Tissue and Organ Culture*, ed. J. Reinert & Y. P. S. Bajaj, pp. 3–17. Berlin, Heidelberg and New York: Springer-Verlag.

Holdgate, D. P. (1977). Propagation of ornamentals by tissue culture. In *Applied and Fundamental Aspects of Plant Cell, Tissue and Organ Culture*, ed. J. Reinert & Y. P. S. Bajaj, pp. 18–43. Berlin, Heidelberg and New York: Springer-Verlag.

Horák, J., Luštinec, J., Měsíček, J., Kamínek, M. & Poláčková, D. (1975). Regeneration of diploid and polyploid plants from the stem pith explants of diploid marrow stem kale (*Brassica oleracea* L.) *Annals of Botany*, **39**, 571–7.

Huang, L. & Murashige, T. (1977). Plant tissue culture media: major constituents; their preparation and some applications. *Tissue Culture Association Manual*, **3**, 539–48.

Hughes, K. W. (1981). Ornamental species. In *Cloning Agricultural Crops via In Vitro Techniques*, ed. B. V. Conger, pp. 5–50. Boca Raton, Florida: CRC Press.

Hussey, G. (1976). Plantlet regeneration from callus and parent tissue in *Ornithogalum thyrsoides*. *Journal of Experimental Botany*, **27**, 375–82.

Hussey, G. (1977). *In vitro* propagation of *Gladiolus* by precocious axillary shoot formation. *Scientia Horticulturae*, **6**, 287–96.

Hussey, G. (1978). *In vitro* propagation of the onion *Allium cepa* by axillary and adventitious shoot proliferation. *Scientia Horticulturae*, **9**, 227–36.

Hussey, G. (1980*a*). Propagation of some members of the Liliaceae, Iridaceae and Amaryllidaceae by tissue culture. In *Petaloid Monocotyledons*, ed. C. D. Brickell, D. F. Cutler & M. Gregory, pp. 33–42. London and New York: Academic Press.

Hussey, G. (1980*b*). *In vitro* propagation. In *Tissue Culture for Plant Pathologists*, ed. D. S. Ingram & J. P. Helgeson, pp. 51–61. London: Blackwell Scientific Publications.

Hussey, G. (1982). *In vitro* propagation of *Narcissus*. *Annals of Botany*, **49**, 707–19.

Hussey, G. & Falavigna, A. (1980). Origin and production of *in vitro* adventitious shoots in the onion. *Allium cepa* L. *Journal of Experimental Botany*, **31**, 1675–86.

Hussey, G. & Hepher, A. (1978). Clonal propagation of sugar beet plants and the formation of polyploids by tissue culture. *Annals of Botany*, **42**, 477–9.

Hussey, G. & Stacey, N. J. (1981). *In vitro* propagation of potato (*Solanum tuberosum* L.) *Annals of Botany*, **48**, 787–96.

Jones, O. P., Hopgood, M. E. & O'Farrell, D. (1977). Propagation *in vitro* of M26 apple rootstocks. *Journal of Horticultural Science*, **52**, 235–8.

Knauss, J. F. & Miller, J. W. (1978). A contaminant, *Erwinia carotovora*, affecting commercial plant tissue cultures. *In vitro*, **14**, 754–6.

Krikorian, A. D. & Berquam, D. L. (1969). Plant cell and tissue cultures: the role of Haberlandt. *Botanical Reviews*, **35**, 59–88.

Krikorian, A. D. & Kann, R. P. (1979). Micropropagation of daylilies through aseptic culture techniques. *Hemerocallis Journal*, **33**, 44–61.

Krikorian, A. D., Staicu, S. A. & Kann, R. P. (1981). Karyotype analysis of a daylily clone reared from aseptically cultured tissues. *Annals of Botany*, **47**, 121–31.

Malnassy, P. & Ellison, J. H. (1970). Asparagus tetraploids from callus tissue. *HortScience*, **5**, 444–5.

Matthews, V. H. & Rangan, T. S. (1979). Multiple plantlets in lateral bud and leaf explant *in vitro* cultures of pineapple. *Scientia Horticulturae*, **11**, 319–28.

Murashige, T. (1974). Plant propagation through tissue cultures. *Annual Review of Plant Physiology*, **25**, 135–66.

Murashige, T. (1977). Manipulation of organ culture in plant tissue cultures. *Botanical Bulletin Academia Sinica*, **18**, 1–24.

Murashige, T., Serpa, M. & Jones, J. B. (1974). Clonal multiplication of *Gerbera* through tissue culture. *HortScience*, **9**, 175–80.

Murashige, T., Shabde, M. N., Hasegawa, P. M., Takatori, F. H. & Jones, J. B. (1972). Propagation of asparagus through shoot apex culture. I. Nutrient medium for formation of plantlets. *Journal of the American Society for Horticultural Science*, **97**, 158–61.

Nickell, L. G. (1977). Crop improvement in sugar cane: studies using *in vitro* methods. *Crop Science*, **17**, 717–19.

Peterson, R. L. (1975). The initiation and development of root buds. In *The Development and Function of Roots*, ed. J. G. Torrey & D. T. Clarkson, pp. 125–61. New York and London: Academic Press Ltd.

Phillips, I. D. J. (1975). Apical dominance. *Annual Review of Plant Physiology*, **26**, 341–67.

Pierik, R. L. M. (1979). *In Vitro Culture of Higher Plants.* Wageningen, Ponsen en Looijen, Kniphorst: Scientific Bookshop PO Box 67.

Pierik, R. L. M. & Steegmans, H. H. M. (1975). Effect of auxins, cytokinins, gibberellins, abscisic acid and ethephon on regeneration and growth of bulbets on excised bulb scale segments of hyacinth. *Physiologia Plantarum*, **34**, 14–17.

Price, H. J. & Smith, R. H. (1979). Somatic embryogenesis in suspension cultures of *Gossypium klotzschianum* Anderss. *Planta*, **145**, 305–7.

Priestley, J. H. & Swingle, C. F. (1929). Vegetative propagation from the standpoint of plant anatomy. *Technical Bulletin, 151.* Washington, DC: US Department of Agriculture.

Quak, F. (1977). Meristem culture and virus free plants. In *Applied and Fundamental Aspects of Plant Cell, Tissue and Organ Culture*, ed. J. Reinert & Y. P. S. Bajaj, pp. 598–615. Berlin, Heidelberg and New York: Springer-Verlag.

Reinert, J. (1959). Uber die Kontrolle der Morphogenese und die Induktion von Adventureembryonen an Gewebakulturen aus Karotten. *Planta*, **53**, 318–33.

Robb, S. H. (1957). The culture of excised tissue from bulb scales of *Lilium speciosum. Journal of Experimental Botany*, **8**, 348–52.

Seabrook, J. E. A., Cumming, B. G. & Dionne, L. A. (1976). The *in vitro* induction of adventitious shoot and root apices on *Narcissus* (daffodil and narcissus) cultivar tissue. *Canadian Journal of Botany*, **54**, 814–19.

Sheridan, W. F. (1968). Tissue culture of the monocot *Lilium. Planta*, **82**, 189–92.

Sinnott, E. W. (1950). *Cell and Psyche. The Biology of Purpose.* Chapel Hill: University of North Carolina Press.

Sinnott, E. W. (1960). *Plant Morphogenesis.* New York: McGraw Hill.

Skirvin, R. M. (1978). Natural and induced variation in tissue culture. *Euphytica*, **27**, 241–66.

Skoog, F. & Miller, C. O. (1957). Chemical regulation of growth and organ formation in plant tissues cultured *in vitro*. In *The Biological Action of Growth Substances.* Symposia of the Society for Experimental Biology No. 11, ed. H. K. Porter, pp. 118–31. Cambridge: Cambridge University Press.

Smith, R. H. & Nightingale, A. E. (1979). *In vitro* propagation of *Kalanchöe*. *HortScience*, **14**, 20.

Srinivasan, C. & Mullins, M. G. (1980). High frequency somatic embryo production from unfertilised ovule of grapes. *Scientia Horticulturae*, **13**, 245–52.

Staritsky, G. (1970). Embryoid formation in callus of coffee. *Acta Botanica Neerlandica*, **19**, 509–14.

Steward, F. C. (1958). Interpretations of the growth from free cells to carrot plants. *American Journal of Botany*, **45**, 709–13.

Street, H. E. (Ed.) (1977). *Plant Tissue and Cell Culture.* London: Blackwell Scientific Publications.

Sunderland, N. (1977). Nuclear cytology. In *Plant Tissue and Cell Culture*, ed. H. E. Street, pp. 177–205. London: Blackwell Scientific Publications.

Thorpe, T. A. (1981). *Plant Tissue Culture; Methods and Applications in Agriculture*. New York: Academic Press.
Tisserat, B. & DeMason, D. A. (1980). A histological study of development of adventive embryos in organ cultures of *Phoenix dactylifera* L. *Annals of Botany*, **46**, 465–72.
Tisserat, B., Esan, E. B. & Murashige, T. (1978). Somatic embryogenesis in angiosperms. *Horticultural Reviews*, **1**, 1–78.
Vasil, V. & Vasil, I. K. (1981). Somatic embryogenesis and plant regeneration from suspension cultures of pearl millet (*Pennisteum americanum*). *Annals of Botany*, **47**, 669–78.
Walkey, D. G. A. (1978). In vitro methods for virus elimination. In *Frontiers of Plant Tissue Culture*, ed. T. A. Thorpe, pp. 245–54. Calgary: Calgary University.
Walkey, D. G. A. (1980). Production of virus-free plants by tissue culture. In *Tissue Culture Methods for Plant Pathologists*, ed. D. S. Ingram & J. P. Helgeson, pp. 109–17. London: Blackwell Scientific Publications.
Walkey, D. G. A. & Woolfitt, J. (1968). Clonal multiplication of *Nicotiana rustica* L. from shoot meristem culture. *Nature*, **220**, 1346.
Weatherhead, M. A., Burdon, J. & Henshaw, G. G. (1978). Some effects of activated charcoal as an additive to plant tissue culture media. *Zeitschrift für Pflanzenphysiologie*, **89**, 141–7.
Wright, N. A. & Alderson, P. G. (1980). The growth of tulip tissues *in vitro*. *Acta Horticulturae*, **109**, 263–70.

O. P. JONES

In vitro propagation of tree crops

Introduction

A wide range of tree species is of very significant economic value in habitats ranging from tropical to northern temperate zones. Progress in crop improvement is usually very slow. This is related to the long-term nature of the tree life-cycle: pine forests are normally managed on a 20- to 25-year rotation, apple and other 'top fruit' orchards have a commercial life of some 20 years or more and that of an oil palm plantation is even longer.

Multiplication of adult trees selected for proven performance is achieved by vegetative propagation. This is often difficult and time consuming and in several cases, such as monocotyledonous palms and some forest species, is virtually impossible by conventional propagation procedures.

In recent years, methods of tissue culture in vitro that provide rapid vegetative propagation of herbaceous plants have been applied successfully to woody species. Initially these methods met with little success with trees, particularly in the adult form, and up to five years ago the majority of reports on such propagation in vitro revealed only low or sporadic production of plants and were usually related to seed or seedling material (Winton, 1978). These pioneering studies greatly encouraged increased effort and in recent years methods of culture in vitro have been applied successfully to a wide range of species, including many in the adult form.

This review is concerned with the methods and applications of the in vitro propagation of tree crop species. The emphasis is on the principles involved as illustrated by examples, with no attempt to present a comprehensive survey of the many tree species presently being investigated in laboratories throughout the world.

General methods

Methods of plant propagation in vitro include shoot culture with proliferation of axillary or adventitious shoots and callus culture with regeneration of shoots or embryoids (Hussey, 1978).

The shoot culture method has been applied to a wide range of tree species including fruit, forest and ornamental trees. Callus culture has been used less frequently but, nevertheless, has been successful with important tree crops such as *Citrus* and oil palm. Thus far, the more sophisticated variations of propagation in vitro using cell suspensions, pollen and protoplast culture have seldom been completely successful with trees and a detailed description of these aspects will not be attempted in this review.

Juvenility is of major significance in relation to the vegetative propagation of trees by conventional and in vitro methods. The stage of growth of the seedling following germination is described as the juvenile phase. This is a phase of very active growth, frequently characterised by morphological features such as unique leaf shape and the presence of spines, when the tree does not initiate flowers and vegetative propagation is usually achieved readily. This phase continues for a number of years that varies according to species, but eventually the tree attains the capacity to produce flowers, fruits and seeds. The tree now enters the adult phase, which is characterised by a decreased capacity for vegetative propagation, a feature that usually becomes progressively more marked as the tree ages further. However, the transition from the juvenile phase appears to be reversible to some extent and it is common for some tissues of the adult tree to have some of the physiological characteristics of seedlings such as a high capacity for adventitious root production. Such tissues are described as 'rejuvenated'. Knowledge of the physiological and biochemical basis of rejuvenation is very incomplete (Kester, 1976).

Tree propagation in vitro by shoot culture

Methods

Shoot culture is often sufficiently reliable with herbaceous plants to be used commercially (Holdgate, 1977). The method involves culturing explants such as shoot tips and buds in tubes or other suitable culture vessels containing a nutrient medium. The cultures are maintained in illuminated growth rooms where they grow readily and their axillary buds extend prematurely to produce new shoots. These are excised at intervals of about one month and transferred to fresh medium where further axillary shoot production occurs; shoot culture lines can be multiplied indefinitely by this method of sequential subculture. With some cultures, the new shoots are produced from adventitious buds that arise directly from explants and sometimes the shoot proliferation occurs from both axillary and adventitious buds. The production of plantlets is completed by excising the axillary or adventitious shoots and transferring them singly to another type of medium that promotes adventitious root formation. It is essential that all these in vitro operations are carried out under sterile conditions with the use of airflow cabinets to minimise contamination with micro-organisms which would quickly kill the cultures.

Application of shoot culture to tree species is now described.

Initial explant. Axillary shoot cultures have usually been initiated from shoot tip or nodal bud explants. Adventitious shoot cultures have also been initiated from such explants and also from portions of internodes, leaves, cotyledons, hypocotyls and embryos.

The physiological status of the explant has been of prime importance with adult trees. Extensive studies at the Association Forêt-Cellulose (AFOCEL) research institute in France indicated that explants from adult forest trees were frequently slow to commence growth in vitro, or failed completely, unless selected from rejuvenated tissues. Such tissues may occur naturally, for example as the vigorous shoots, commonly known as 'suckers', that arise directly from the roots of various species, or as the masses of adventitious buds that are a feature of the 'lignotubers' of *Eucalyptus* spp. Where rejuvenation does not occur naturally, it has been induced by various treatments such as the grafting of shoots onto seedlings, shoot pruning, the maintenance of high fertiliser levels, vegetative propagation or spraying with cytokinins (Franclet, 1979).

With more highly cultivated crops such as fruit trees, the necessity of taking explants from rejuvenated tissues has seldom been emphasised, but this is probably because most of the treatments that have been used to rejuvenate forest trees feature in the routine management of cultivated trees. However, even with highly cultivated apple trees, propagation in vitro has been achieved most readily when the initial explants have been taken from the shoots of young nursery trees within about one month following bud-break, with cold storage being used to maintain a year-round supply of trees at this early stage of development (Jones, 1973; Jones, Hopgood & O'Farrell, 1977).

Surface sterilisation. Explants must be free from micro-organisms when placed on nutrient media and this has usually been achieved by surface sterilisation with solutions of sodium or calcium hypochlorite. Such treatments have failed with some adult trees, presumably because of penetration of tissues by micro-organisms (Franclet, 1979; Jones, Pontikis & Hopgood, 1979). In such circumstances antibiotics and systemic fungicides have been used to sterilise the shoot tips of ornamental trees. However, the effects of such treatment on the subsequent development of the shoots has yet to be assessed (Phillips, Arnott & Kaplan, 1981; A. J. Abbott & D. Constantine; personal communication, 1981).

Culture media. Defined media consisting of mineral nutrients, sugar, vitamins and growth regulators have been used to culture shoots. The macro- and micro-nutrients have usually been based on the medium of Murashige & Skoog (1962), with sucrose at a level of 2 or 3% (w/v) as the most common carbohydrate source. The presence of a cytokinin, usually 6-benzylamino-purine (BAP) at a concentration of about 1 mg l^{-1}, has been essential for the proliferation of both axillary and adventitious shoots (Jones, 1967), whilst addition of an auxin such as indole-3-butyric acid (IBA) and gibberellic acid A_3 has sometimes enhanced shoot elongation. Furthermore, the presence of the phenolic compound phloroglucinol (1,3,5-trihydroxybenzene) at a concentration of about 150 mg l^{-1} has enhanced the shoot proliferation of some apple cultivars and also a limited number of other tree species. This compound is a breakdown product of phloridzin, a major phenolic constituent of apple trees and its action may take place via effects on the metabolism of auxin (Jones, 1976). Activated carbon, at about 2%, has often been an essential medium constituent for the elongation of shoots of gymnosperms. Possibly, this carbon absorbs growth inhibitors exuded by the plant tissues or contains impurities such as monophenylamines which promoted growth (Carbanne, Martin-Tanquy & Martin, 1977; Boulay, 1979). The pH of media has been adjusted to 5.2–5.8 before the latter have been solidified with agar. Media have also been used in liquid form with filter-paper supports or the use of shaking to improve the aeration of the cultures. Sterilisation has usually been achieved by autoclaving but some thermolabile substances have been filter-sterilised into media.

Similar media have been used for rooting the shoots except that the cytokinins have been omitted or reduced and auxin levels have sometimes been raised. It has often been beneficial at this stage to reduce the concentrations of macro-nutrients and sucrose by half.

Culture conditions. Relatively little systematic research relating to trees has been published on this topic and the effects of the type of culture vessel, temperature and light have not been adequately studied. Cultures have been generally grown in tubes, flasks, jars or boxes, with various closures such as cotton wool, aluminium foil or sealed lids. In the absence of specific data relating to variations in growth-room conditions, temperatures of 20–30 °C and light of about 2000 lx for 18 h daily have been generally used.

Results – shoot proliferation

Juvenile material. Explants of seeds and seedlings have usually responded readily to culture in vitro and rapid proliferation of axillary or adventitious shoots has been achieved with just such juvenile material of species of *Betula* (McCown & Amos, 1979), *Cinchona* (Hunter, 1979), *Coffea* (Dublin, 1980), *Eucalyptus* (Durand & Boudet, 1979), *Liquidambar* (Sommer, 1982), *Pinus* (Sommer & Brown, 1974; Rancillac, 1979), *Picea* (Campbell & Durzan, 1976; Von Arnold & Eriksson, 1978) and *Pseudotsuga* (Cheng, 1975).

Adult trees – angiosperms. Rapid shoot proliferation has been achieved with a wide range of species with initial explants being taken from normal aerial shoots of nursery or field-grown trees. These have included shoots from (a) fruit trees, for example rootstocks of apple (Jones *et al.*, 1977; Cheng, 1979), scion cultivars of apple (Abbott & Whiteley, 1976; Lane, 1978; Jones *et al.*, 1979; Zimmerman & Broome, 1980; Sriskandarajah & Mullins, 1981), rootstocks and scions of plum and cherry (Quoirin & Lepoivre, 1977; Jones, 1979; Jones & Hopgood, 1979), scions of pear (Lane, 1979) and rootstocks of peach (Zuccherelli, 1979) and (b) forest and ornamental trees such as species of *Fagus*, *Quercus*, *Ulmus* (Chalupa, 1979), *Tectona* (Gupta, Nadgir, Mascarenhas & Jagannathan, 1980), *Ilex*, *Magnolia* (A. J. Abbott & D. Constantine; personal communication, 1981), and *Salix* (Beauchesne, 1978).

The use of explants from naturally rejuvenated tissues has been advantageous with some angiosperm forest trees. Explants from suckers of *Populus* spp. began to grow and proliferate on the nutrient medium immediately, whereas those from the aerial shoots frequently did not begin to grow for several months (Christie, 1978). A similar more rapid establishment was achieved with *Juglans* and *Eucalyptus* spp. by use of explants from suckers and lignotubers respectively (Boulay, 1980).

Most of the shoot proliferation of angiosperm trees has appeared to be from axillary shoots but the origins of these shoots have seldom been examined critically and sometimes adventitious shoots could also have been produced. However, the rate of proliferation has usually been consistently rapid and sufficient to give the potential for the production of at least one million shoots annually.

Adult trees – gymnosperms. In the case of adult trees, the use of explants from rejuvenated tissue has usually been essential for the achievement of rapid shoot proliferation. Thus explants were taken from suckers of *Sequoia sempervirens* (Boulay, 1977; Ball, 1978) or from trees of *Pinus pinaster* that had been subjected to various treatments to induce rejuvenation (David, David, Faye & Isemukali, 1979; Franclet, David, David & Boulay, 1980). Moreover, for several species the explants were from shoots that had been grafted onto seedlings (Boulay, 1980).

Shoot proliferation from explants of gymnosperms has often occurred by means of adventitious shoots and has generally been less rapid than from explants of angiosperms. Nevertheless, there has usually been the potential for producing about 200000 shoots annually.

Results – rooting

Angiosperms. It has been common, even with cultures originating from adult trees, for more than 70% of the shoots to be rooted within 4–6 weeks of transfer to rooting medium. Furthermore, the shoots of some species have been dipped in auxin and rooted directly into compost in humid greenhouses. This high degree of success with rooting allied to the rapid shoot proliferation already described suggests that mass propagation of a wide range of angiosperm trees should now be feasible. However, it is emphasised that many species remain recalcitrant. Notable examples of this are *Citrus* spp. where all attempts to develop plantlets from shoot tips have failed (Button & Kochba, 1977).

Gymnosperms. Rooting in vitro, even with shoots originating from seed or seedling explants, has been generally much more difficult than with angiosperms and there has been only a small amount of success. Nevertheless, shoots of *S. sempervirens*, *P. pinaster* and several other species have been rooted directly into compost in a greenhouse as was done with some of the angiosperms. However, a further problem with gymnosperms has been that a proportion of the rooted plantlets have been plagiotropic and therefore unsuitable as forest trees. This has been especially serious with *Pseudotsuga menziesii* (Boulay, 1980).

As a result of difficulties with the establishment of cultures, rooting and the plagiotropic habit, the range of gymnosperm forest trees that can be mass produced at present appears to be limited. At the AFOCEL laboratories in France, where propagation in vitro of gymnosperms appears to be most advanced, the only adult trees that have been propagated up to 100000 times are of *S. sempervirens*. However, it is considered that similar propagation is feasible with adult trees of *P. pinaster*, *Sequoiadendron giganteum*, *Taxodium distichum*, *Cryptomeria japonica* and *Cunninghamia lanceolata* (Franclet, 1979; Boulay, 1980).

Rejuvenation in vitro

The rejuvenation of adult tree tissues that may occur in vivo is enhanced by the process of culture in vitro (Franclet, 1979). Most information on such rejuvenation in vitro relates to apple trees and it has become clear that this process is an essential factor for the successful propagation in vitro of adult trees. Thus with the apple cultivar Northern Spy the rate of shoot proliferation increased about four fold over the initial four months of sequential subculture (J. Hutchinson; personal communication, 1981) and there have been similar results in the author's laboratory with the rootstocks M.7 and M.25 and the scion cultivar Greensleeves. Furthermore, these latter investigations indicated that the capacity of the shoots to initiate adventitious roots also increased with the number of subcultures; shoots of M.7 required two months of subculture before more than 70% of the shoots could be rooted, whilst the corresponding period for the scion cultivar Greensleeves was five months (Jones, Marks & Waller, 1982). Moreover, this period appeared to be even longer for other cultivars; shoots of the cultars Jonathan and Red Delicious required six and 36 months of subculture respectively before rooting could be achieved (Sriskandarajah, Mullins & Nair, 1982).

Effects of phloroglucinol on rejuvenation in vitro of apple trees. Phloroglucinol (PG) in the nutrient medium produced a two to three fold increase in the proliferation and rooting of shoots of the rootstock M.7 (Jones, 1976; Jones & Hatfield, 1976). Recent investigations with M.7 have indicated that these effects of the phenolic were sustained only during the initial three months of culture when the capacity for proliferation and rooting was low (Jones, Hopgood & Marks, 1981) and there have been similar observations with the cultivar Northern Spy (J. Hutchinson; personal communication, 1981). By contrast, the rootstock M.9 had a more sustained requirement for PG. Only 38% of M.9 shoots of a culture line that had been established for 12–18 months were rooted on a medium with IBA but this increased to 90% when PG was present in both the medium for shoot proliferation and that for rooting. However, following the maintenance of the culture line for a further 18 months, 85% rooting was achieved in the absence of PG and about 90% in its presence (James & Thurbon, 1981 and personal communication). Thus the results with M.9 are similar to those with M.7 and cv. Northern Spy except that a much longer period of subculture was needed to achieve high levels of rooting in the absence of PG.

There are other reports of the effects of PG on apple shoots. With the apple rootstock A2, the phenolic improved the initiation and growth of roots and also the survival of plants in soil more markedly when the shoot cultures were derived from adult plants than when they were derived from seedlings (Welander & Huntrieser, 1981). Furthermore, investigations with shoot culture lines of eight apple cultivars that had been subcultured for more than 12 months indicated that PG had little or no effect with seven cultivars but with one, the cultivar Spartan, it consistently increased the percentage of rooted shoots from 20 to more than 80. This effect with cv. Spartan was apparent only with shoot cultures that were initiated from twelve-year-old trees;

cultures from two-year-old trees had no requirement for the phenolic (Zimmerman & Broome, 1981). Thus it has become clear that the effect of PG is dependent both on the cultivar tested and on the physiological state of the tissues. The effect of the phenolic could be regarded as hastening the process of rejuvenation in vitro and from the results with the M.9 rootstock in particular, it appears that a sustained requirement for this substance may reflect a slow rate of rejuvenation. Moreover, PG may hasten rejuvenation with a range of tree species since it has markedly enhanced shoot proliferation with species of *Prunus* (Jones & Hopgood, 1979; Garland & Stolz, 1981) and *Cinchona* (Hunter, 1979). Other phenolic compounds may have similar effects to those of PG. Thus rutin and quercetin were more effective than PG for the rooting in vitro of peach (Mosella-Chancel, Macheix & Jonard, 1980).

Tree propagation by callus culture

Rapidly proliferating masses of cells known as calluses can be initiated from explants from any part of a plant and cultured in vitro indefinitely by sequential subculture. With a limited range of species, such calluses can be induced to regenerate plants via the production of adventitious shoots or somatic embryos commonly known as embryoids. These methods of plant culture have the potential for even more rapid propagation than shoot culture.

Methods and results

The methods are similar to those for shoot culture except that the nutrient medium usually contains high concentrations of auxins such as 2,4-dichlorophenoxyacetic acid (2,4-D) or naphthaleneacetic acid (NAA). Withdrawal of auxin or varying the ratio of auxin to cytokinin may induce adventitious shoot or embryoid regeneration. The regenerated shoots can be excised and rooted or the embryoids grown into complete plants on a nutrient medium without growth regulators.

As with shoot culture, the physiological status of the initial explant has been of prime importance, and calluses from seeds and seedlings have usually had the greatest capacity for differentiation. Roots, shoots, embryoids and sometimes plantlets have been obtained from just such juvenile material of a limited range of angiosperm and gymnosperm trees (Abbott, 1978). Furthermore, in the case of *Ulmus americana* and *Liquidambar styraciflua*, cell suspensions produced by shaking calluses from seedlings have produced embryoids and plantlets (Durzan & Lopushanski, 1975; Sommer & Brown, 1980).

Plants have been regenerated from calluses from adult trees of a very limited range of species. Shoots were regenerated from cambial explants of trees of *U. campestris* that were approximately 180 years of age (Jacquiot, 1966) and there has been similar regeneration in the case of calluses from shoots of forest clones of *Populus* spp. (Lester & Berbee, 1977). With *Prunus*, shoots have been regenerated from several species, but only when the callus was initiated from plants that had been rejuvenated through propagation *in vitro* by shoot culture (Druart, 1980). However, these callus systems from adult trees are not as effective as the corresponding shoot culture systems for the sustained and rapid production of plants. Notable exceptions to this are several species of *Coffea* which have been multiplied with high frequency through embryoids produced in calluses from leaf explants (Sondahl & Sharp, 1977).

Nucellar tissue. This can be regarded as highly rejuvenated tissue of the adult tree and with *Citrus* spp. embryoid production in callus from such tissue has led to the propagation of proven clones (Button, 1977). Embryoids have also been produced from nucellar callus of apple but at a much lower frequency than with *Citrus* (Eichholtz & Robitaille, 1980).

Propagation of palms by callus culture. Oil palm (*Elaeis guineensis*) has been studied extensively since vegetative propagation is not normally possible with this species. Initial studies were carried out on seedlings, and shoots developed in calluses from apical shoot tissue following transfer to a series of media containing varying concentrations of sucrose and mineral salts (Rabechault, Ahée & Guénin, 1972). Embryoids, from which plants developed, were also reported in calluses evolved from root and leaf base tissue of seedlings. Details of the media were not described but embryogenesis was reported as being 'more a function of the origin of the initial explant than of subsequent cultural conditions' (Jones, 1974). Subsequently, embryoids were induced in calluses evolved from adult palms following incubation in media with cytokinins and auxins for between five months and two years. However, once the capacity for embryogenesis was initiated, it was retained for at least three years with about 50 plantlets being produced per gram of callus, which increased in weight between ten and thirty fold every two months (Rabechault & Martin, 1976). More recently, these techniques have been improved to produce embryoids much more rapidly (Ahée *et al.*, 1981). Similar results have been obtained at the Unilever Laboratories in England following more than 12 years of research. Experimental details have not been published but clones of palms from embryoids are presently under field trial in Malaysia (Hawkes, 1980).

Seedling and adult palms of date (*Phoenix dactylifera*) have also been propagated from embryoids produced in calluses evolved from shoot and leaf explants. The methods used were similar to those described for oil palm and the survival and differentiation of the callus was enhanced by the presence of 0.3% activated carbon in the medium (Tisserat, 1979).

The formation of roots and embryoid-like structures has been reported in calluses evolved from seedling and adult palms of coconut (*Cocos nucifera*) but plants have not yet been obtained (Eeuwens, 1976; Guzman, Rosario & Ubalde, 1978; Fulford, Passey & Butler, 1980).

Establishment of plantlets in soil

Plants propagated in vitro often do not transfer readily to an open soil environment. A weaning stage has usually been used in which the plants have been transplanted from in vitro conditions into humid conditions in greenhouses and the humidity has then been reduced gradually over three to four weeks to that of the open soil environment. During this stage, great care is necessary with trees, particularly with the water regime, or large losses can occur. Antitranspirants have been used to reduce the vulnerability of apple trees at this stage with promising results (J. Hutchinson; personal communication, 1981). Some fruit tree plantlets established readily in soil but did not grow subsequently unless first chilled or treated with gibberellins (Howard & Oehl, 1981).

Genetic variation of trees propagated in vitro

Thus far, information on this vital topic is limited because much of the in vitro propagation of trees remains at the laboratory stage of development.

Organised shoot meristems appear to be genetically stable and able to resist possible mutagenic effects of the culture system (Hussey, 1978). The limited results available so far for trees support this view. Apple rootstocks produced by axillary shoot proliferation appeared to be true to the parental material as assessed by morphological characteristics (Jones *et al.*, 1977) and plum rootstocks produced by similar methods have also been assessed as true-to-type under a commercial certification scheme (M. Waddy & R. Knight; personal communication, 1981).

The chances of producing genetically abnormal plants appears to be greater when the propagation is carried out by culturing adventitious shoots that originate directly from explants since polyploid cells may be present in the original explant and could result in polyploid shoots being produced (Hussey, 1978). However, there appears to be no published information concerning the genetic uniformity of trees produced by this method.

With callus culture, there have been numerous reports of cytological irregularities leading to the regenerated plants being mutant or polyploid and this has been the finding with plants from calluses of *Populus* spp. (Lester & Berbee, 1977). Such instability appears to be characteristic of calluses with many large parenchymatous cells growing under the influence of growth substances, especially 2,4-D or NAA. A callus from the nucellus of a *Citrus* sp. was composed almost entirely of small meristematic cells and the regenerated plants were genetically uniform. Furthermore, this callus became habituated and did not require growth regulators, and this could have also contributed to the genetic stability (Button, 1977).

Calluses of oil palm have appeared to be genetically stable since the plants from the regenerated embryoids have been uniform within clones. The anatomical structure of these calluses has not been described (Corley, Wong & Wooi, 1981).

These assessments of the variability of trees propagated in vitro indicate that genetic variation has not been a major problem with at least some of the shoot culture and callus systems in use. Nevertheless, it is highly desirable to assess this variability at the propagation stage before expensive field planting has been completed. Techniques for 'biochemical fingerprinting' may eventually provide this early assessment and are already being developed for fruit trees (Vanstee, 1980).

Applications of in vitro tree propagation

Potential applications are well documented, especially for *Citrus* (Button & Kochba, 1977), *Coffea* (Sondahl & Sharp, 1977) and forest trees (Sommer & Brown, 1977), and include the mass propagation, transport and storage of disease-free material and breeding programmes. Although much of tree propagation in vitro remains at the laboratory stage of development, some of these applications are now being realised and examples with specified tree crops are now described.

Mass propagation

Propagation in vitro of trees has been regarded as commercially viable only for the initial rapid multiplication of new varieties or in other situations where mother tree stock is limited (Jones, 1979). However, with the increasing success in these areas, nurserymen are becoming more aware of possible financial benefits of the use of in vitro methods for the routine production of plants (Anon. 1981).

Fruit tree rootstocks. Fruit tree rootstocks are now being mass-produced by axillary shoot culture in commercial laboratories in Europe, Canada and the USA. Experience of such commercial production is probably greatest in Italy where production from a single laboratory has risen from 38000 trees in 1978 to more than two million trees in both 1980 and 1981. These trees have been of high quality, most being rootstocks of peach that were needed to satisfy local demand. The transfer of plantlets to soil accounted for 40–80% of the production cost but, nevertheless, the enterprise has been sufficiently profitable to support substantial research into increasing the efficiency of the transfer to soil and also the propagation of fruit tree scion cultivars (Zimmerman, 1979; S. Barducci; personal communication, 1981).

Experiences in other commercial laboratories have not been so favourable. There have been technical problems such as inconsistent rooting with some apple rootstocks and shoot cultures becoming slow-growing with tightly-rolled translucent leaves. This latter condition has been described as 'vitrified' (Natavel, 1980) and has recently been alleviated in the case of shoot cultures of *Cynara scolymus* by raising the concentration of agar in the medium (Debergh, Harbaoui & Lemeur, 1981). Such technical problems together with the high cost of establishment in soil and the low price presently commanded by some of the rootstocks have resulted in propagation in vitro being commercially unattractive in some circumstances (J. Aynsley; personal communication, 1981). Nevertheless, the general experience with the initial ventures into commercial production appears to have been sufficiently favourable to encourage further investment, as illustrated by the increasing number of laboratories that are now being planned for fruit and ornamental tree production (Anon., 1981).

In addition to commercial production, in vitro propagation of fruit trees has been applied on a somewhat smaller scale to breeding programmes through the rapid production of promising new lines for field trial (Longbottom, Tobutt & Jones, 1980), and has also proved useful for rapid year-round production of uniform trees for research projects that would otherwise depend on single batches of trees produced annually by conventional methods (Atkinson & Wilson, 1980).

Self-rooted apple trees. The traditional method of propagating apple trees consists of budding or grafting the scion cultivar onto rootstocks which are themselves raised by stooling or layering. This process takes three years and demands expensive nursery facilities and skills. Propagation of self-rooted trees would be expected to be much more rapid, which could be of major economic advantage, especially in relation to modern high-density orchards. However, most scion cultivars have proved very difficult to root from cuttings and it is only recently that some have been propagated from hardwood cuttings under carefully controlled conditions (Child & Hughes, 1978). Furthermore, dwarf cultivars have the greatest potential for success as self-rooted trees, but have very slow-growing shoots and could not be expected to produce enough shoot cuttings for rapid propagation.

By contrast, a wide range of apple scion cultivars have now been rapidly propagated in vitro in various laboratories throughout the world. Furthermore, at East Malling Research Station in England, the methods have been applied by the fruit tree breeders to their new semi-dwarf cultivar Malling Greensleeves and also to dwarf types derived from cv. Wijick McIntosh, a dwarf mutant of the cultivar McIntosh. In the field, these Wijick types produce single leading shoots and few laterals, whereas in vitro they produce axillary shoots as prolifically as much more vigorous types such as cv. Bramley or cv. Golden Delicious (Longbottom *et al.*, 1980).

Samples of trees of five apple scion cultivars which had been propagated in vitro were planted in the orchard at East Malling in 1978 and 1979 for observation and comparison with conventional grafted trees. This is a feasibility study of establishing an orchard with self-rooted trees, involving assessment of performance in the absence of control over size and flowering that is presently provided by the appropriate choice of rootstock. Thus far, all the self-rooted trees have continued to grow vigorously and came into cropping in 1980 or 1981. Thus the rejuvenation in vitro that has been referred to in relation to the propagation of such trees did not result in delayed flowering as would be expected if the plants had all the characteristics of seedlings (Gayner, Hopgood, Jones & Watkins, 1981). Planting of self-rooted apple trees has also been in progress at the United States Department of Agriculture (USDA) research centre, Beltsville, since 1979, and results thus far are similar to those at East Malling (Zimmerman & Broome, 1980).

Self-rooted trees, through propagation in vitro, could have a similar application to the wide range of fruit and other tree crops that are traditionally propagated by grafting onto rootstocks. Thus self-rooted trees of cherry, plum and pear have also been produced for orchard trial at East Malling (Jones, 1979; Oehl, 1979; Gayner, Hopgood, Jones & Watkins, 1981).

Forest trees. Forest trees have been propagated traditionally by seeds. Propagation in vitro from seeds and seedlings has some application, mainly to increase output from types that are low in seed production (Rancillac, 1979). However, the most important application is in the mass production of elite adult trees and this is expected to produce dramatic improvements in forest productivity (Sommer & Brown, 1977). Such mass production may be feasible with at least some species in the near future. Already more than 2000 trees have been propagated in vitro from selected adult gymnosperms at the AFOCEL institute in France for assessment in nursery and forest trials. It is envisaged that such trees will soon be used to establish nurseries of high quality rejuvenated trees for subsequent forest planting (Franclet, 1979; Boulay, 1980).

Oil palm. As in the case of forest trees, rapid propagation of elite material is the most important application and this is already in progress on a limited scale. The first clonal palms from the Unilever Laboratory were planted in the field in Malaysia in 1977 and these plantings have now been extended to 12000 palms from 30 clones. At the present early development stage of the propagation methods, the cost of this clonal material is about five times that of seedlings. However, it is estimated that this extra cost will be recovered in five years through the 30% higher yield which is expected from the best clones and that thereafter there will be an additional annual profit of about 30% (Corley, Wong & Wooi, 1981; L. H. Jones; personal communication, 1981).

Disease-free trees

Many economically important trees are infected throughout by viruses as well as being susceptible to attacks by systemic bacteria and fungi. Disease-free trees may be obtained in small numbers by meristem shoot-tip culture and heat treatment. Propagation in vitro provides an ideal method for rapidly multiplying such material. Recently, virus-free rootstocks have been propagated in vitro in England for commercial distribution overseas. Since this material was maintained under the sterile conditions of culture in vitro, it was possible, following the appropriate testing, to guarantee its disease-free status and this simplified greatly the quarantine procedures involved in the distribution (M. Waddy & R. Knight, personal communication, 1981). Furthermore, methods have been described for the storage at 3 °C of apple shoot cultures (Lundergan & Janick, 1979). This would enable disease-free material to be maintained for many months with a minimum of storage area and labour, as is in progress already with strawberry (Boxus, Quorin & Laine, 1977). Such storage gives added convenience in relation to the distribution of disease-free material.

Plants regenerated from nucellar callus are usually virus free since such pathogens are rarely seed-transmitted. Viruses have been eliminated by this method from *Citrus* clones that do not produce nucellar seedlings naturally (Button, 1977). Unfortunately, nucellar plants of *Citrus*, whether produced in vivo or in vitro, exhibit undesirable juvenile characteristics such as the presence of thorns and delayed sexual maturity. For this reason attention has been transferred to meristem shoot-tip grafting onto seedlings as a solution to the problems of virus elimination in *Citrus* (Navarro, Roistacher & Murashige, 1975).

Conclusions and future developments

Recent progress has been excellent and already the success of in vitro propagation with adult trees and palms is bringing important practical benefit; the future of propagation in vitro as an invaluable aid to the cropping of trees appears to be assured.

Thus far, shoot culture methods are outstanding in their applicability to the widest range of tree species. Nevertheless, there remains much scope for improving these methods on the commercial scale and also for applying them to subjects that presently remain recalcitrant. Thus there is a particular need to improve knowledge of juvenility and maturation and also of the control of shoot proliferation, adventitious root formation and the reactions of plantlets to transfer to soil. Some of the recalcitrant types may yield to the application of methods for inducing rejuvenation and also to modification of existing culture methods such as the changes in the inorganic and cytokinin components of the culture medium that led to the successful propagation of *Liquidambar styraciflua* (Sommer, 1982).

However, shoot culture methods are relatively crude and labour intensive with the need for continued division of shoots, for rooting and for establishment in soil. Thus embryogenesis from callus or cell suspension would appear to have more attractive possibilities in the long term. Such methods have the potential for the most rapid propagation of all and experience with *Citrus* spp. and oil palm indicates that genetic instability need not be a major problem. Moreover, embryoids may prove to be more amenable than cultured shoots or plantlets to mechanised planting techniques such as 'fluid drilling' for germinated seeds (Currah, Gray & Thomas, 1972). Clearly, there is a need to improve methods for achieving embryoid regeneration with trees. Study of biochemical markers of morphogenesis such as IAA–peroxidase relations (Johnson & Carlson, 1978) and the biochemical basis for rejuvenation phenomena would appear to require some priority as research topics that might lead to such improvement.

Application of propagation in vitro to the production of desirable new tree varieties could be of great value since breeding by normal sexual hybridisation is very labour intensive and can take many years. Furthermore, there are many commercially important trees that are sterile and improved variants cannot be obtained by conventional breeding. In the near future, it is conceivable that some of these limitations will be overcome by the production of mutants and polyploid types through regeneration from callus and single cells that are genetically unstable or have been subjected to mutagenic agents (Speigel-Roy & Kochba, 1973; Gayner, Jones, Hopgood & Watkins, 1981). In the longer term, application of pollen culture and protoplast culture techniques would be expected to transform possibilities for the breeding of trees as for other crops (Winton, Parnham, Johnson & Einspahr, 1974). Already there have been some significant achievements with trees in respect of such methods. Haploid calluses have been evolved from the pollen of several trees species (Winton & Stettler, 1974) and isolated protoplasts have given rise to callus colonies of *P. pinaster* (David & David, 1979) and complete plants with *Citrus* spp. (Vardi, Speigel-Roy & Galun, 1975).

References

Abbott, A. J. (1978). Practice and promise of micropropagation of woody species. *Acta Horticulturae*, **79**, 113–28.

Abbott, A. J. & Whiteley, E. (1976). Culture of *Malus* tissues *in vitro*. I. Multiplication of apple plants from isolated apple apices. *Scientia Horticulturae*, **4**, 183–9.

Aheé, J., Arthuis, P., Cas, G., Duval, Y., Guenin, G., Hanower, J., Lievoux, D., Loiret, C., Malaurie, B., Pannekier, C., Raillur, D., Varechon, C. & Zuckermann, L. (1981). La multiplication végétative *in vitro* du palmier à huile par embryogénese somatique. *Oleagineux*, **36**, 113–16.

Anon. (1981). Raisers neglect micropropagation. *Grower*, *London*, **96**, No. 23, 2.

Atkinson, D. & Wilson, S. A. (1980). The growth and distribution of fruit tree roots and some consequences for nutrient uptake. In *Mineral Nutrition of Fruit Trees*, ed. D. Atkinson, J. E. Jackson, R. O. Sharples & W. M. Waller, pp. 137–50. London: Butterworths.

Ball, E. (1978). *Round Table Conference. In vitro Multiplication of Woody Species*, ed. P. L. Boxus, pp. 181–93, Gembloux, Belgium: CRA.

Beauchesne, G. (1978). *Round Table Conference. In vitro Multiplication of Woody Species*, ed. P. L. Boxus, p. 65. Gembloux, Belgium: CRA.

Boulay, M. (1977). Multiplication rapide du *Sequoia sempervirens* en culture *in vitro*. In *Annales de Recherches Sylvicoles*, 1977, AFOCEL, 37–65.

Boulay, M. (1979). Multiplication et clonage rapide du *Sequoia sempervirens* par la culture *in vitro*. In *Annales de Recherches Sylvicoles*, AFOCEL. Etudes et Recherches. No. 12, 6/79. *Micropropagation d'Arbres Forestiers*, 49–56.

Boulay, M. (1980). La micropropagation des arbres forestiers. *Comptes Rendus des Séances de l'Academie d'Agriculture de France*. **66**, Numero Special No. 8, 697–708.

Boxus, P. L., Quoirin, M. & Laine, J. M. (1977). Large scale propagation of strawberry plants from tissue culture. In *Applied and Fundamental Aspects of Plant Cell, Tissue and Organ Culture*, ed. J. Reinert & Y. P. S. Bajaj, pp. 130–43. Berlin: Springer-Verlag.

Button, J. (1977). International exchange of disease-free citrus clones by means of tissue culture. *Outlook in Agriculture*, **8**, 155–9.

Button, J. & Kochba, J. (1977). Tissue culture in the citrus industry. In *Applied and Fundamental Aspects of Plant Cell Tissue and Organ Culture*, ed. J. Reinert & Y. P. S. Bajaj, pp. 70–92. Berlin: Springer-Verlag.

Button, J., Kochba, J. & Bornman, C. H. (1974). Fine structure of and embryoid development from embryonic ovular callus of Shamouti Orange (*Citrus sinensis* Osb.). *Journal of Experimental Botany*, **25**, 446–57.

Campbell, R. A. & Durzan, D. J. (1976). Induction of multiple buds and needles in tissue cultures of *Picea glauca*. *Canadian Journal of Botany*, **53**, 1652–7.

Carbanne, F., Martin-Tanquy, J. & Martin, C. (1977). Phénolamines associées à l'induction florale et à l'état reproducteur. *Physiologie Végétale*, **15**, 429–43.

Chalupa, V. (1979). *In vitro* propagation of some broadleaved forest trees. *Communicationes Instituti Forestalis Cechosloveniae*, **11**, 159–70.

Cheng, T. Y. (1975). Adventitious bud formation in cultures of Douglas fir (*Pseudotsuga menziesii*). *Plant Science Letters*, **5**, 97.

Cheng, T. Y. (1979). Micropropagation of clonal fruit tree rootstocks. *Compact Fruit Tree*, **12**, 127–37.

Child, R. D. & Hughes, R. (1978). Factors influencing rooting in hardwood cuttings of apple cultivars. *Acta Horticulturae*, **79**, 43–8.

Christie, C. B. (1978). Rapid propagation of aspens and silver poplars using culture techniques. *The International Plant Propagators Society, Combined Proceedings*, **28**, 255–60.

Corley, R. H. V., Wong, C. Y. & Wooi, K. C. (1981). Early results from the first oil palm clone trials. *Oil Palm in Agriculture in the Eighties*. PORIM Conference, Kuala Lumpur, June 1981, 1–27.

Currah, I. E., Gray, D. & Thomas, T. H. (1972). Establishment of drilled crops. *Report of the National Vegetable Research Station, 1972*, 67–8.

David, A. & David, H. (1979). Isolation and callus formation from cotyledon protoplasts of Pine (*Pinus pinaster*). *Zeitschrift für Pflanzenphysiologie*, **94**, 173–9.

David, A., David, H., Faye, M. & Isemukali, K. (1979). Culture *in vitro* et micropropagation du pin maritime. In *Annales de Recherches Sylvicoles*, AFOCEL. Etudes et Recherches. No. 12, 6/79. *Micropropagation dArbres Forestiers*, 33–40.

Debergh, P., Harbaoui, Y. & Lemeur, R. (1981). Mass propagation of globe artichoke (*Cynara scolymus*): Evaluation of different hypotheses to overcome vitrification with special reference to water potential. *Physiologia Plantarum*, **53**, 181–7.

Druart, P. L. (1980). Plantlet regeneration from root callus of different *Prunus* species. *Scientia Horticulturae*, **12**, 339–42.

Dublin, P. (1980). Induction de bourgeons néoformes et embryogénese somatique. Deux voies de multiplication *in vitro* des caféiers cultivés. *Café, Cacao, Thé* (Paris), **24**, 121–30.

Durand, R. & Boudet, A. M. (1979). Le bouturage *in vitro* de l'eucalyptus. In *Annales de Recherches Sylvicoles*, AFOCEL, Etude et Recherches. No. 12, 6/79. *Micropropagation d'Arbres Forestiers*, 57–66.

Durzan, D. J. & Lopushanski, S. M. (1975). Propagation of American Elm via cell suspension cultures. *Canadian Journal of Forest Research*, **5**, 273–7.

Eeuwens, C. J. (1976). Mineral requirement of coconut tissues. *Physiologia Plantarum*, **36**, 23–8.

Eichholtz, D. A. & Robitaille, H. A. (1980). Asexual apple embryos *in vitro*. *Compact Fruit Tree*, **13**, 142–4.

Franclet, A. (1979). Rejeunissement des arbres adultes en vue de leur propagation végétative. In *Annales de Recherches Sylvicoles*, AFOCEL. Etudes et Recherches. No. 12, 6/79. *Micropropagation d'Arbres Forestiers*, 3–18.

Franclet, A., David, A., David, H. & Boulay, M. (1980). Premier mise en évidence morphologique d'un rejeunissement de méristèmes primaires caulinaires de Pin maritime âgé (*Pinus pinaster* Sol.). *Compte Rendus Academie Science, Paris*, **290**, serie D, 927–30.

Fulford, R. M., Passey, A. J. & Butler, M. (1980). Vegetative propagation of coconuts by tissue culture. *Report of East Malling Research Station for 1979*, 184–5.

Garland, P. & Stolz, L. P. (1981). Micropropagation of Pissardi Plum. *Annals of Botany*, **48**, 387–90.

Gayner, J. A., Hopgood, M. E., Jones, O. P. & Watkins, R. (1981). Propagation *in vitro* of fruit plants. *Report of East Malling Research Station for 1980*, 145.

Gayner, J. A., Jones, O. P., Hopgood, M. E. & Watkins, R. (1981). Mutant production and regeneration from callus *in vitro*. *Report of East Malling Research Station for 1980*, 144.

Gupta, P. K., Nadgir, A. L., Mascarenhas, A. F. & Jagannathan, V. (1980). Tissue culture of forest trees: clonal multiplication of *Tectona grandis* L. (Teak) by tissue culture. *Plant Science Letters*, **17**, 259–68.

Guzman, E. V., Rosario, A. G. & Ubalde, E. (1978). Proliferation, growth and organogenesis in coconut embryo and tissue culture. *Philippine Journal of Coconut Studies*, **3**, 1–10.

Hawkes, N. (1980). Test tube palms. Unilever Magazine, **38**, 43–5.

Holdgate, D. P. (1977). Propagation of ornamentals by tissue culture. In *Applied and Fundamental Aspects of Plant Cell, Tissue and Organ Culture*, ed. J. Reinert and Y. P. S. Bajaj, pp. 18–43. Berlin: Springer-Verlag.

Howard, B. H. & Oehl, V. H. (1981). Improved establishment of *in vitro* propagated plum micropropagules following treatment with GA_3 or prior chilling. *Journal of Horticultural Science*, **56**, 1–7.

Hunter, C. S. (1979). *In vitro* culture of *Cinchona ledgeriana* L. *Journal of Horticultural Science*, **54**, 111–14.

Hussey, G. (1978). The application of tissue culture to the vegetative propagation of plants. *Science Progress, Oxford*, **65**, 185–208.

Jacquiot, C. (1966). Plant tissue and excised organ cultures and their significance in forest research. *Journal of the Institute of Wood Science*, **16**, 22–34.

James, D. J. & Thurbon, I. J. (1981). Phenolic compounds and other factors controlling rhizogenesis *in vitro* of the apple rootstocks M.9 and M.26. *Zeitschrift für Pflanzenphysiologie*, **105**, 11–20.

Johnson, M. A. & Carlson, J. A. (1978). IAA–oxidase in Douglas Fir development. (Abstracts) Fourth International Congress of Plant Tissue Culture, p. 166. University of Calgary. In *Frontiers of Plant Tissue Culture*, ed. T. A. Thorpe. Calgary: Calgary University.

Jones, L. H. (1974). Propagation of clonal oil palms by tissue culture. *Oil Palm News*, **17**, 1–8.

Jones, O. P. (1967). Effect of benzyl adenine on apple shoots. *Nature*, **215**, 1514–15.

Jones, O. P. (1973). Effects of cytokinins in xylem sap from apple trees on apple shoot growth. *Journal of Horticultural Science*, **48**, 181–8.

Jones, O. P. (1976). Effect of phloridzin and phloroglucinol on apple shoots. *Nature*, **262**, 392–3. Erratum, **262**, 724.

Jones, O. P. (1979). Propagation *in vitro* of apple trees and other woody fruit plants: methods and applications. *Scientific Horticulture*, **30**, 44–8.

Jones, O. P. & Hatfield, S. G. S. (1976). Root initiation in apple shoots cultured *in vitro* with auxins and phenolic compounds. *Journal of Horticultural Science*, **51**, 495–9.

Jones, O. P. & Hopgood, M. E. (1979). The successful propagation *in vitro* of two rootstocks of *Prunus*: the plum rootstock Pixy (*P. insititia*) and the cherry rootstock F12/1 (*P. avium*). *Journal of Horticultural Science*, **54**, 63–6.

Jones, O. P., Hopgood, M. E. & Marks, T. R. (1981). Morphogenetic substances in xylem sap. *Report of East Malling Research Station for 1980*, 142–3.

Jones, O. P., Hopgood, M. E. & O'Farrell, D. (1977). Propagation *in vitro* of M.26 apple rootstocks. *Journal of Horticultural Science*, **52**, 235–8.

Jones, O. P., Marks, T. R. & Waller, B. J. (1982). Propagation *in vitro*. *Report of East Malling Research Station for 1981*, 159.

Jones, O. P., Pontikis, C. A. & Hopgood, M. E. (1979). Propagation *in vitro* of five apple scion cultivars. *Journal of Horticultural Science*, **54**, 155–8.

Kester, D. (1976). The relationship of juvenility to plant propagation. *The International Plant Propagators' Society, Combined Proceedings*, **26**, 71–84.

Lane, D. W. (1978). Regeneration of apple plants from shoot meristem-tips. *Plant Science Letters*, **13**, 281–5.

Lane, D. W. (1979). Regeneration of pear plants from shoot meristem-tips. *Plant Science Letters*, **16**, 337–42.

Lester, D. T. & Berbee, K. G. (1977). Within-clone variation among black poplar trees derived from callus culture. *Forest Science*, **23**, 122–31.

Longbottom, H., Tobutt, K. R. & Jones, O. P. (1980). Propagation *in vitro* of fruit trees. *Report of East Malling Research Station for 1979*, 187.

Lundergan, C. A. & Janick, J. (1979). Low temperature storage of *in vitro* apple shoots. *HortScience*, **14**, 514.

McCown, B. & Amos, R. (1979). Initial trials with commercial micropropagation of birch selections. *The International Plant Propagators' Society, Combined Proceedings*, **29**, 387–93.

Mosella-Chancel, L., Macheix, J. J. & Jonard, R. (1980). Les conditions du microbouturage *in vitro* du Pecher (*Prunus persica* Batsch): influence combinées des substances de croissance et de divers composés phénoliques. *Physiologie Végétale*, **18**, 597–608.

Murashige, T. & Skoog, F. (1962). A revised medium for rapid growth and bioassays with tobacco tissue cultures. *Physiologia Plantarum*, **15**, 473–97.

Natavel, L. M. (1980). L'utilisation des culture *in vitro* pour la multiplication de quelques espèces légumières et fruitières. *Comptes Rendus des Séances de l'Académie d'Agriculture de France*, **66**, numero special no. 8, 681–96.

Navarro, L., Roistacher, C. N. & Murashige, T. (1975). Improvement of shoot-tip grafting *in vitro* for virus-free *Citrus*. *Journal of the American Society for Horticultural Science*, **100**, 471–97.

Oehl, V. H. (1979). Own-root varieties. *Report of East Malling Research Station for 1978*, 60.

Phillips, R., Arnott, S. M. & Kaplan, S. E. (1981). Antibiotics in plant tissue culture: Rifampicin effectively controlled bacterial contaminants without affecting the growth of short-term explant cultures of *Helianthus tuberosum*. *Plant Science Letters*, **21**, 235–42.

Quoirin, M. & Lepoivre, P. (1977). Etude de milieux adaptés aux cultures *in vitro* de *Prunus*. *Acta Horticulturae*, **78**, 437–42.

Rabechault, H., Ahée, J. & Guénin, G. (1972). Recherches pour la culture *in vitro* des embryos de palmier à huile (*Elaeis guineensis* jacq.). *Oleagineux*, **27**.

Rabechault, H. & Martin, J. (1976). Multiplication végétative du palmier à huile (*Elaeis guineensis* Jacq.) à l'aide de culture de tissus foliaires. *Comptes Rendus des Séances de l'Académie de Paris*, t, **283**, serie D, 1735–7.

Rancillac, M. (1979). Mise au point d'une méthode de multiplication végétative *in vitro* du Pin maritime (*Pinus pinaster*). In *Annales de Recherches Sylvicoles*, AFOCEL. Etudes et Recherches. No. 12, 6/79. *Micropropagation d'Arbres Forestiers*, 41–8.

Sommer, H. E. (1982). Organogenesis in woody angiosperms: applications to vegetative propagation. *Bulletin Société Botanique* (*France*), *Actualities Botanique*, in press.

Sommer, H. E. & Brown, C. L. (1974). Plantlet formation in pine tissue cultures. *American Journal of Botany*, **61**, supplement, page 11.

Sommer, H. E. & Brown, C. L. (1977). Application of tissue culture to forest tree improvement. In *Plant Cell and Tissue Culture*, ed. W. R. Sharp, P. O. Larsen, E. F. Paddock & V. Raghavan, pp. 461–91. Ohio: Ohio State University Press.

Sommer, H. E. & Brown, C. L. (1980). Embryogenesis in tissue cultures of sweetgum. *Forest Science*, **26**, 257–60.

Sondahl, M. R. & Sharp, W. R. (1977). Research in *Coffea* spp. and application of tissue culture methods. In *Plant Cell and Tissue Culture*, ed. W. R. Sharp, P. O. Larsen, E. F. Paddock & V. Rahavan, pp. 527–84. Ohio: Ohio State University Press.

Speigel-Roy, P. & Kochba, J. (1973). Mutation breeding in citrus. In *Induced Mutations in Vegetatively Propagated Plants*, pp. 91–103. Vienna: International Atomic Energy Agency.

Sriskandarajah, S. & Mullins, M. G. (1981). Micropropagation of Granny Smith apple and factors affecting root formation *in vitro*. *Journal of Horticultural Science*, **56**, 71–6.

Sriskandarajah, S., Mullins, M. G. & Nair, Y. (1982). Induction of adventitious rooting *in vitro* in difficult-to-propagate cultivars of apple. *Plant Science Letters*. **24**, 1–9.

Tisserat, B. (1979). Propagation of date palm (*Phoenix dactylifera* L.) *in vitro*. *Journal of Experimental Botany*, **30**, 1275–83.

Vanstee, J. (1980). Finger printing fruit trees. *American Fruit Grower*, 100, N10, October 1980. p. 9.

Vardi, A., Speigel-Roy, P. & Galun, E. (1975). Citrus cell culture – isolation of protoplasts. Plating densities, effects of mutagens and regeneration of embryos. *Plant Science Letters*, **4**, 231–6.

Von Arnold, S. & Eriksson, T. (1978). Induction of adventitious buds on embryos of Norway spruce grown *in vitro*. *Physiologia Plantarum*, **44**, 283–7.

Welander, M. & Huntrieser, I. (1981). The rooting ability of shoots raised *in vitro* from the apple rootstock A2 in juvenile and in adult growth phase. *Physiologia Plantarum*, **53**, 301–6.

Winton, L. L. (1978). Morphogenesis in clonal propagation of woody plants. In *Frontiers of Plant Tissue Culture*, ed. T. A. Thorpe, pp. 419–26. Calgary: Calgary University.

Winton, L. L., Parnham, R. A., Johnson, M. A. & Einspahr, D. W. (1974). Tree improvement by callus, cell and protoplast culture. *Technical Association of the Pulp and Paper Industry*, (*TAPPI*), **57**, 151–2.

Winton, L. L. & Stettler, R. F. (1974). Utilisation of haploidy in tree breeding. In *Haploids in Higher Plants, Advances and Potential*, ed. K. J. Kasha, pp. 259–73. Guelph: University of Guelph.

Zimmerman, R. H. (1979). The laboratory of micropropagation at Cesana, Italy. *The International Plant Propagators' Society, Combined Proceedings*, **29**, 398–400.

Zimmerman, R. H. & Broome, O. C. (1980). Apple cultivar micropropagation. In *Proceedings of the Conference on Nursery Production of Fruit Plants through Tissue Culture. Applications and feasibility*, pp. 54–8. Beltsville, Maryland: USDA.

Zimmerman, R. H. & Broome, O. C. (1981). Phloroglucinol and *in vitro* rooting of apple cultivar cuttings. *Journal of the American Society for Horticultural Science*, **106**, 648–52.

Zuccherelli, G. (1979). Metodologie nella moltiplicazione industriale *in vitro* dei portainnesti clonali del pesco: Pescomardorlo GF 677, Susino GF 43, Damasco 1869, S. Givliano 655/2. In *Techniche di Colture in vitro per la Propagazione su Vasta Scala Della Specie Ortoflorafrutticole*, ed. E. Bellini, pp. 147–54. Pistoia: Consiglio Nazionale delle Ricerche.

PART III

Germplasm maintenance and storage

E. JAMES

Low-temperature preservation of living cells

Introduction

The science of cryobiology really began towards the end of the nineteenth century with the advent of techniques for liquefying air and its constituent gases. This enabled very low temperatures to be attained and the state of true suspended animation to be achieved. The next significant contribution to cryobiology came in 1949 when Polge, Smith & Parkes demonstrated that glycerol could protect living cells against freezing damage.

Many of the physical and chemical processes that occur during freezing and thawing were described during the pre-cryoprotectant era and have been recorded in the book *Life and Death at Low Temperatures* by Luyet & Gehenio (1940). The advent of cryoprotectants generated a new interest in low-temperature preservation which has precipitated the present widespread use of cryopreservation techniques in many diverse fields. Much of the impetus for these developments was stimulated by the pioneering work of Audrey Smith and her co-workers and has been outlined in her book *Biological Effects of Freezing and Supercooling* (Smith, 1961).

Cryopreservation (Greek, kryos = frost) is literally 'preservation in the frozen state'. However, in practice this is generally taken to mean storage at very low temperatures: above solid carbon dioxide (−79 °C), in low-temperature deep freezers (−80 °C or below), in vapour phase (approximately −140 °C) or in liquid nitrogen (−196 °C).

The chief advantage of storage at very low temperatures is obviously the ability to slow considerably, or even to halt, metabolic processes and biological deterioration. In addition, cryopreserved material remains genetically stable and the genetic drift which is a phenomenon of organisms maintained by standard techniques can be avoided. The term 'stabilate' has been coined for that 'discrete population or clone of organisms with specific biological characteristics preserved on a unique occasion' (Lumsden & Hardy, 1965). In the frozen state, however, there is no chance to escape from, nor to repair damage produced by, background ionising radiation; Ashwood-Smith (1980) though, has pointed out that it would take approximately 32000 years for an equivalent of 600 rad of damage to be accumulated. Cryopreservation also brings considerable savings in the costs of equipment and personnel and reduces the risk of losing valuable material through contamination, human error or equipment failure.

Responses of cells to low temperature

Cold shock

Most cell types can at least be cooled to 0 °C without being damaged, but some gram-negative bacteria (Farrell & Rose, 1968), many species of mammalian spermatozoa (Blackshaw, 1954), most cell types from the pig (Polge, Wilmut & Royson, 1974), *Chlorella* (Morris, 1976*a*), rapidly dividing cells of *Cyanophyta* (Jancz & Maclean, 1973) and some higher plant cells (Levitt, 1972) cannot.

The term 'cold shock' (chilling injury or thermal shock) has been used to describe this sensitivity to a reduction in temperature. It is independent of, or can occur in the absence of, freezing. Cytoplasmic enzymes and solutes have been shown to leak from cells following cold shock injury suggesting that damage has occurred to the cell membrane. The differential contraction of membrane components has been proposed to account for this injury (Lovelock, 1955) while Brandts (1967) suggested that cold shock was due to an irreversible protein denaturation.

Cold shock is often related to the rate of cooling, rapid cooling usually being more damaging than slow cooling (Macleod & Calcott, 1976; Morris, 1976*a*). It may, as for erythrocytes, be mediated by exposure to high concentrations of sodium chloride (Lovelock, 1955; Morris & Farrant, 1973). This injury can also be reduced or avoided by incorporating cryoprotectants, specific phospholipids (Butler & Roberts, 1975) or various other compounds (Polge, 1980) into the suspending medium or by a change in the concentration of specific anions (Morris, 1975). Farrant & Morris (1973) have suggested that since electrolyte concentrations increase following freezing (see Solute concentration), some of the mechanisms involved in cold shock may also be acting on cells in the frozen state.

Morris (1975) describes the damage produced in many cells by long periods of in vitro culture at lowered temperatures as 'indirect chilling injury'. However, this form of injury appears to develop as a result of metabolic processes becoming out of step, for instance when active ionic pumps cannot keep pace with passive diffusion (Farrant, 1980).

Cold shock is not a factor in the cryopreservation of most cell types, but it is as well to bear it in mind as a possible explanation for damage.

Cold tolerance and frost hardiness

In many instances the exposure of cells to low temperatures for a period of time can lead to the development of tolerance to cold (cold acclimatisation) and to sub-zero temperatures (frost hardiness). These processes occur naturally in many plants and animals when the ambient temperature decreases as winter approaches and they are an important consideration in plant breeding programmes.

There are several mechanisms by which organisms become cold tolerant or frost hardened. These include an increased ability for the tissues to remain supercooled, the manufacture of cryoprotective compounds to lower the freezing-point of the tissues, occasionally the ability to withstand or to encourage the formation of intracellular ice and certain behavioural changes in animals.

An effect which appears to be found in many plants and animals exposed to low temperatures for prolonged periods is the modification of the lipid constituents of the cell membranes. Upon a reduction in temperature, membrane lipids undergo a phase change, which takes place over a temperature range of approximately 10–20 °C due to the heterogeneity of the lipids. Enzyme activity and the transmembrane transport of water and solutes depends on the membranes remaining sufficiently fluid. For cells that are sensitive to cold shock, the solidification of membrane lipids is apparently particularly damaging. The temperature at which this phase change occurs can frequently be lowered by increasing the degree of unsaturation of the membrane phospholipids. This has been reported for many cell types including *Escherichia coli* (Marr & Ingram, 1962), *Chlorella* (Patterson, 1970), alfalfa (Gerloff, Richardson & Stahlmann, 1966), *Prototheca* (Morris, 1976*b*) and *Paramoecium* (Polyansky, 1963) grown at artificially lowered temperatures. Similar changes have also been reported to occur in many plant species during autumn (Levitt, 1972), in hibernating squirrels (Aloia, 1978; Rotermund & Veltman, 1981) and in fish (Cossins, 1983). However, this may not be a generalised phenomenon since other workers have suggested that an increase in the total membrane lipid content during hardening of some species (Siminovitch, Singh & De La Roche, 1975; Morris, 1976*b*) or a change in the rate of turnover of phospholipid (Willemot, Hope, Williams & Michand, 1977) is important.

Factors affecting the freezing process

Cultivation at a lowered temperature obviously affects a cell's metabolic rate, and while cultivation at temperatures below 20 °C can significantly increase the tolerance to cryopreservation (Fig. 1), Morris and Clarke (1978) have shown that similar effects can be produced simply by reducing the nutrient concentration or by including metabolic inhibitors in cell cultures at 20 °C. The susceptibility of plant cells to freezing is often related to their degree of vacuolation (Morris, 1980) and this appears to be a function of growth temperature and age of culture, stationary-phase cells generally being more susceptible than exponential-phase cells (Withers & Street, 1977; Morris, 1980) although the reverse may occur as in the case of some unicellular green algae (Morris & Canning, 1978) where vacuolation and freezing sensitivity decrease in the stationary phase.

Supercooling

Water becomes supercooled (subcooled or undercooled) if it is cooled to a temperature below its freezing-point without the formation of ice. The freezing-point of pure water is by definition 0 °C but it is possible to supercool water down to − 38 °C (Angell, Shuppert & Tucker, 1973), i.e. its homogeneous nucleation temperature.

The addition of solutes, e.g. sodium chloride and cryoprotectants, to water depresses both the freezing-point and the homogeneous nucleation temperature, i.e. the temperature down to which the solution can be supercooled (MacKenzie, 1977) (Fig. 2). An increase in barometric pressure has the same effect (MacKenzie, 1977).

Cells which can withstand cooling to 0 °C can also normally survive supercooling. Several species of plants (George, Burke, Pellett & Johnson, 1974) and animals, in particular many species of insects (Ring, 1980), employ

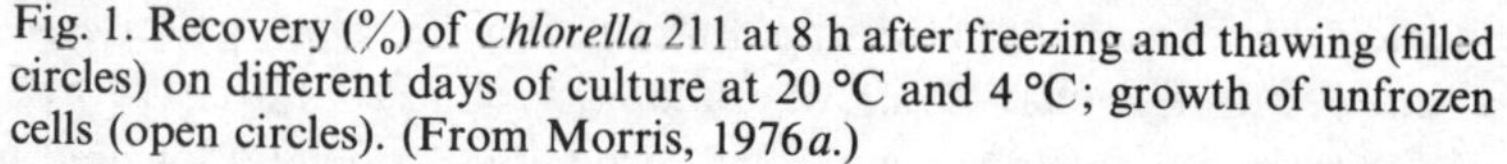

Fig. 1. Recovery (%) of *Chlorella* 211 at 8 h after freezing and thawing (filled circles) on different days of culture at 20 °C and 4 °C; growth of unfrozen cells (open circles). (From Morris, 1976*a*.)

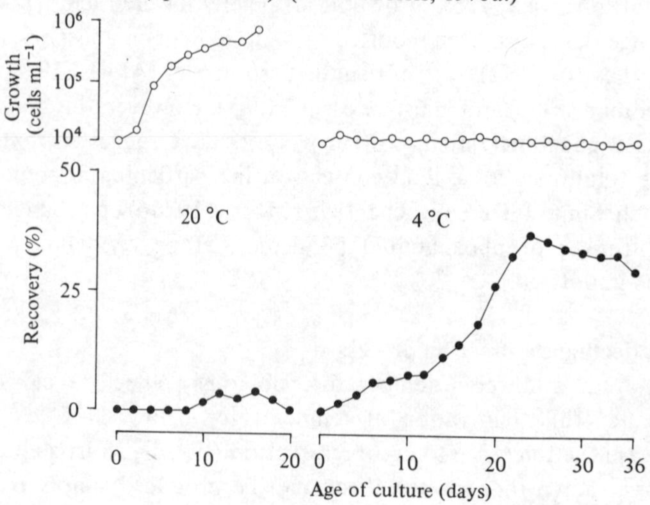

various methods to enhance the degree of supercooling. This can involve the synthesis of a range of fatty acids, proteins or carbohydrates which have the ability to 'bind' water and to depress the freezing- and supercooling points. Supercooling points as low as −55 °C have been recorded for larvae of the gall fly *Eurostra solidagensis* (Salt, 1956).

However, water in the supercooled state is relatively unstable and the probability of spontaneous nucleation occurring increases with time as well as with further lowering of temperature. Thus enhanced supercooling may not be sufficient to ensure survival of a particularly long and cold winter or long periods of low-temperature storage.

Ice nucleation

In living cells there are many molecules or particles which can act as ice nucleation sites. Homogeneous nucleation thus becomes a less likely event than the seeding of ice around these particles (heterogeneous nucleators).

Fig. 2. The effect of various solutes on the depression of the equilibrium freezing-point of water (upper curves) and the temperature of the homogeneous nucleation of ice (lower curves). PVP, polyvinylpyrrolidone. (From MacKenzie, 1977.)

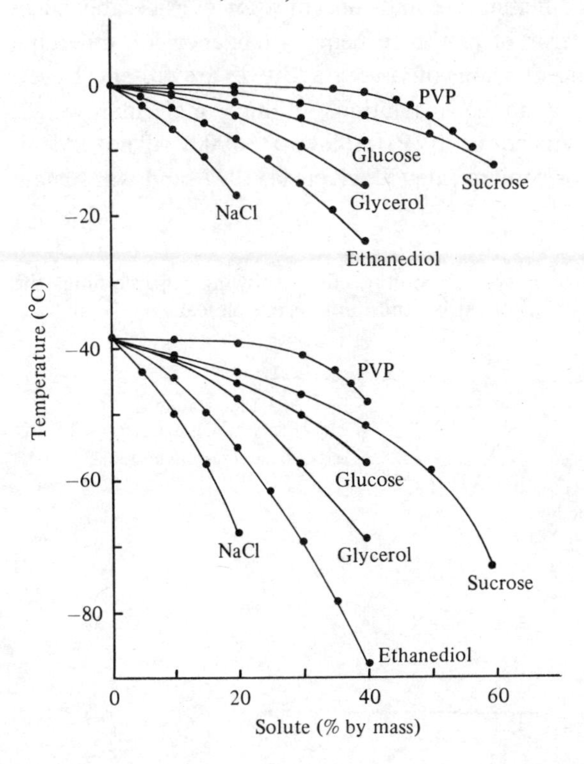

The nucleation of ice involves the conversion of randomly orientated molecules with a high level of free energy to an ordered structure with a low level of free energy, the excess of energy being liberated as heat of fusion. If account is not taken of this sudden evolution of latent heat during the controlled cooling of biological material, considerable damage can be done to the cells. This damage is most severe if the sample is allowed to supercool to any extent. As soon as ice begins to form, the temperature will rise rapidly towards the actual freezing-point of the solution and then will fall rapidly back to the temperature of the cooling chamber or bath (Fig. 3).

Nucleation and crystal growth require time to develop and the rates at which these processes proceed determine the morphological structure of the ice. The extent of supercooling, the rate of cooling, the number of potential nucleation sites present, the viscosity of the medium, the types of solutes present and the temperature all interact and help to determine the rate of nucleation and the type of crystals formed.

If nucleation occurs at the freezing-point and the sample is then cooled slowly, regular hexagonal ice is formed. At faster cooling rates the ice will form into irregular dendrites. At very fast cooling rates a large number of small ice spears grow around each centre of nucleation producing spherulites (Luyet, 1967). Solutes of different type and concentration considerably affect these crystallisation patterns of ice, there being a tendency for spherulite formation when high concentrations of viscous additives are present (Luyet, 1967). Some concentrated aqueous solutions of high molecular weight polymers, e.g. polyvinylpyrrolidone (PVP) (mol. wt > 30000) will not crystallise at any temperature or cooling rate (MacKenzie, 1977) and will remain amorphous.

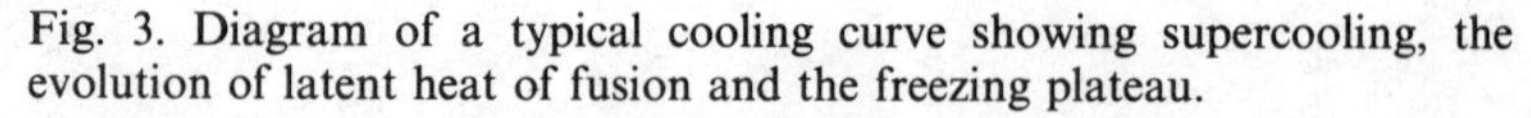
Fig. 3. Diagram of a typical cooling curve showing supercooling, the evolution of latent heat of fusion and the freezing plateau.

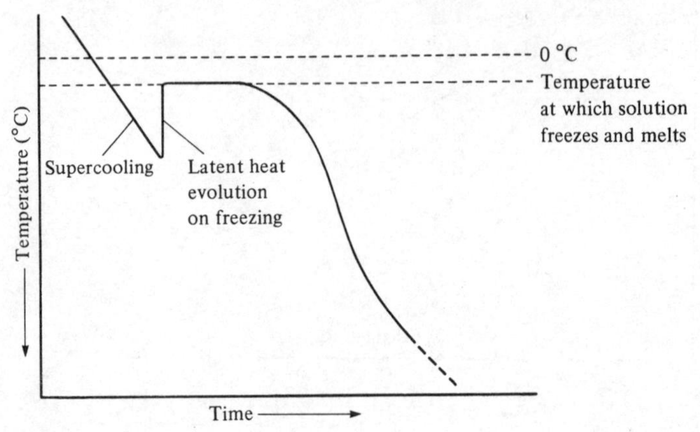

Solute concentration

The first liquid–solid phase change to occur during the cooling of an aqueous solution is the formation of pure (H_2O) ice. Any solutes present are thus concentrated by the gradual removal of the water of solution (Fig. 4). The freezing-point of the remaining solution is effectively depressed by the presence of the dissolved solutes. Further cooling will continue to remove water from solution until finally the eutectic temperature is reached. A second phase change then occurs and the concentrated solution itself solidifies.

Both the cell wall and the plasma membrane act as a barrier to ice growth. Since the cytoplasm of the cell is initially hypertonic relative to the extracellular medium, ice generally forms outside the cell first. The extracellular solutes thus become concentrated and the osmotic gradient across the cell membrane is reversed. The cell will thus tend to lose water, becoming progressively dehydrated, and the intracellular solutes will in consequence also become more concentrated. This is one of the two major causes of cell injury during cooling (Mazur, Leibo & Chu, 1972), the other being the formation of intracellular ice.

It is also possible that abnormally high external solute concentrations may cause solute to be driven into the cell (Lovelock, 1953). Solute effects are, however, dependent on both the length of time of exposure and the temperature (being reduced at lower temperatures and with shorter times).

Fig. 4. When cells are cooled slowly the extracellular medium freezes first causing the solutes to become more concentrated and the intracellular water to flow out of the cells by osmosis. The cells become shrunken as a result. Cells cooled rapidly do not have time to shrink and they freeze internally. Cell survival generally correlates with shrinkage. (Derived from Farrant, 1980.)

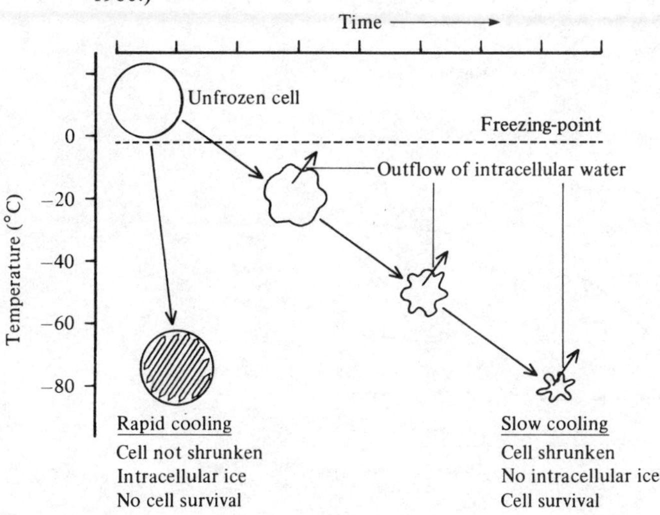

One of the most important effects of cryoprotectants is to reduce the amount of ice which is formed at any temperature during cooling thus reducing these increases in solute concentrations (Fig. 5).

Cell shrinkage

For most animal cells the loss of cell water produced by the rise in external solute concentration leads to cell shrinkage. This shrinkage becomes progressively more marked as the temperature is lowered (Fig. 6). Some plant cells retain their shape instead of becoming deformed through dehydration, but at the expense of shedding membrane to the surrounding medium (Grout & Morris, 1981). This renders them very susceptible to injury on thawing when a return to normal osmotic conditions accompanies the melting of the ice, causing water to flow into the cells. This leads to swelling and lysis of those cells which cannot accommodate the increase in volume by an expansion of their limiting membranes.

Because most plant cells have rigid cell walls the rise in external solute concentration leads to plasmolysis, i.e. shrinkage of the protoplast away from the cell wall (Morris, 1981). The cell wall may collapse at the same time. Ice may form in the space between the cell wall and protoplast causing the latter to become detached from the cell wall and thereby producing pseudoplasmo-

Fig. 5. The effect on the concentration of sodium chloride in Me_2SO–$NaCl$–H_2O mixtures of varying amounts of dimethylsulphoxide (Me_2SO) at different temperatures during freezing. The initial concentration of sodium chloride in each case is 0.154 M. An increase in the concentration of Me_2SO produces a reduction in the concentration of sodium chloride as ice separates during freezing. (From Farrant, 1965.)

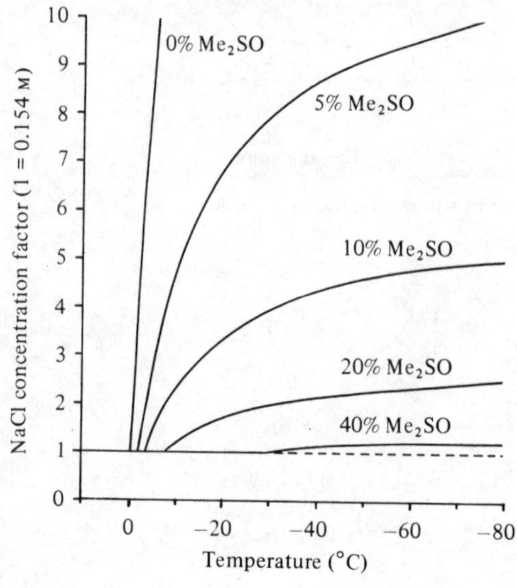

lysis. Some organisms have evolved mechanisms to encourage ice nucleation in their tissues in order to produce shrinkage by promoting water movement into the intercellular spaces and thus prevent the formation of damaging intracellular ice.

The rate at which a cell dehydrates depends on many interrelated factors. Water is driven across the membrane by the difference in vapour pressure between the intracellular and extracellular water. This will vary according to

Fig. 6. Electron micrographs of schistosomula of *Schistosoma mansoni*. (*a*) Control – unfrozen, longitudinal section. (*b*) Transverse section of a schistosomulum which was cooled slowly (1 °C min^{-1}) to −20 °C before rapid cooling to −196 °C and freeze-substituted at −80 °C. (*c*) Transverse section of a schistosomulum which was cooled slowly to −25 °C before rapid cooling to −196 °C. (*d*) Longitudinal section of a schistosomulum which was cooled slowly to −28 °C before rapid cooling to −196 °C. Tegument, T; longitudinal muscle blocks, L; circular muscle blocks, C; cell nuclei, N. The shrinkage produced by slow cooling has effectively been terminated in (*b*), (*c*) and (*d*) by rapid cooling to −196 °C. Note the presence of large ice cavities in (*b*), smaller ice cavities in (*c*) and the apparent absence of ice in (*d*). The extent of shrinkage is greatest in (*d*); note particularly the muscle blocks which have become shrunken with convoluted outlines and have pulled away from adjacent blocks. Survivors only occurred in treatment group (*d*). (From Walter & James, 1981.)

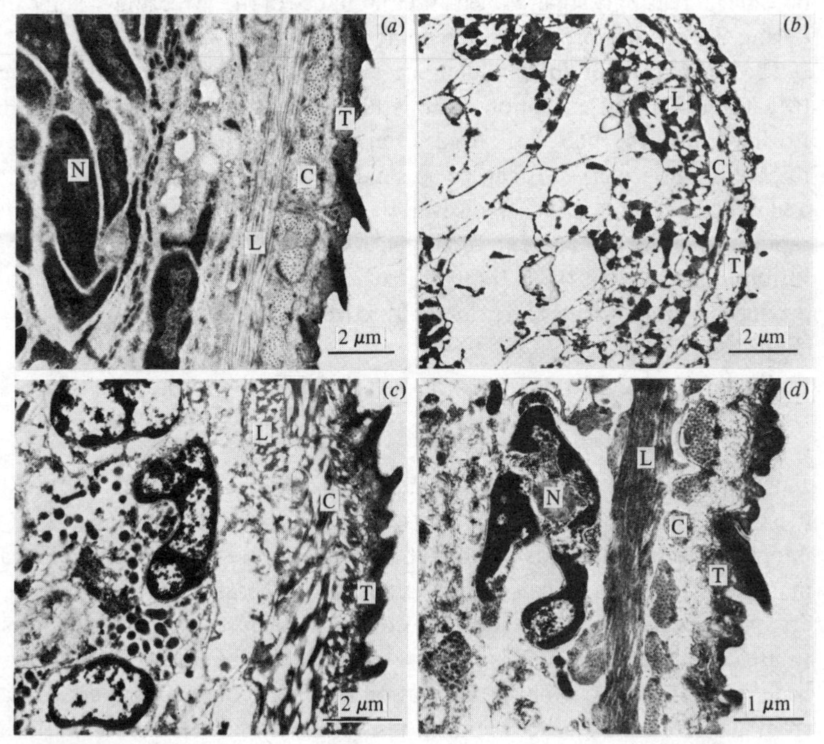

the amount of ice present and hence will depend on the temperature. This in turn will affect the solute concentration and thus the ionic strength and pH, which together with temperature will also affect the cell's membrane permeability. The traditional concepts of a pH range based on a water dissociation constant of 10^{-14} and of what constitutes 'normal pH' have to be abandoned when dealing with highly concentrated mixed aqueous solutions at low temperatures (Taylor, 1981).

Shrinkage proceeds at different rates at different temperatures and for different cooling rates, and also for different cell types having different membrane permeabilities, cell volumes and surface areas.

It has also been suggested that there is a minimum volume down to which a cell can shrink and beyond which it becomes physically damaged (Meryman, Williams & Douglas, 1977). Shrinkage to this critical level may cause the structural proteins to come into close proximity and to become denatured by the formation of disulphide bonds (Levitt, 1966).

Intracellular ice

If the movement of cell water across the membrane fails to keep pace with the change in extracellular vapour pressure the intracellular contents will become increasingly supercooled and the chance of their freezing at some time during the cooling process will become more likely.

The formation of intracellular ice is almost always damaging to cells (Mazur, 1970), but exceptions to this have been reported (Sherman, 1962; Lozina-Lozinsky, 1965; Rall, Reid & Farrant, 1980). At extremely rapid rates of cooling the ice crystals can be so small as to be microscopically invisible and structurally innocuous. However, these cells have to be warmed extremely rapidly in order to prevent further ice crystal growth (recrystallisation). Intracellular ice injury is thought to be due to the physical disruption produced by the crystals and principally to lesions produced in the membranes (Fujikawa, 1978).

Factors affecting survival of cells stored at low temperatures

Cooling rates

Both slow and rapid cooling can kill cells but many cells will survive if an intermediate cooling rate is used. What constitutes a slow, rapid or intermediate rate depends on the type of cell involved; these rates can vary markedly (Fig. 7) and can also be altered considerably by the addition of cryoprotectants. The theoretical aim of most cryopreservation schedules is to use a cooling rate which is slow enough to allow the cells to dehydrate in order to avoid intracellular ice formation but fast enough to prevent damage from the increased solute concentrations. The intermediate cooling rate is thus a compromise and rarely leads to 100% survival. It may still produce zero survival especially if the organism or tissue being frozen contains different types of cells each having widely varying optimum cooling rates. The optimum cooling rate may also depend on the warming rate being used.

With slow cooling, the phase change which occurs in membrane lipids as a result of lowered temperature proceeds slowly and the protein components become pushed into regions where the membrane is still fluid. Thus slow cooling produces lateral movement and aggregations of membraneous proteins while rapid cooling causes the proteins to become 'fixed' (Singer & Nicolson, 1972).

Two-step cooling. This technique involves 'prefreezing' the cells by cooling them rapidly to a sub-zero temperature and holding them at this temperature for a short period of time before further cooling them rapidly down to the storage temperature. The high concentrations of extracellular solutes produced during the initial cooling step appear to induce sufficient shrinkage during the period of prefreezing to protect the cells against damage from the second rapid cooling step (Luyet & Keane, 1955; Asahina, 1959; Walter, Knight & Farrant, 1975 and see Fig. 8).

Fig. 7. Variation of optimum cooling rate for survival with cell type. (From Mazur, 1980.) (Data from: Whittingham, Leibo & Mazur, 1972, for mouse embryos (filled triangles); Leibo *et al.*, 1970, for mouse marrow stem cells (open squares); Mazur & Schmidt, 1968, for yeast (filled circles); Mazur, Farrant, Leibo & Chu, 1969, for hamster tissue culture cells (open triangles); Rapatz, Sullivan & Luyet, 1968, for human erythrocytes (open circles).)

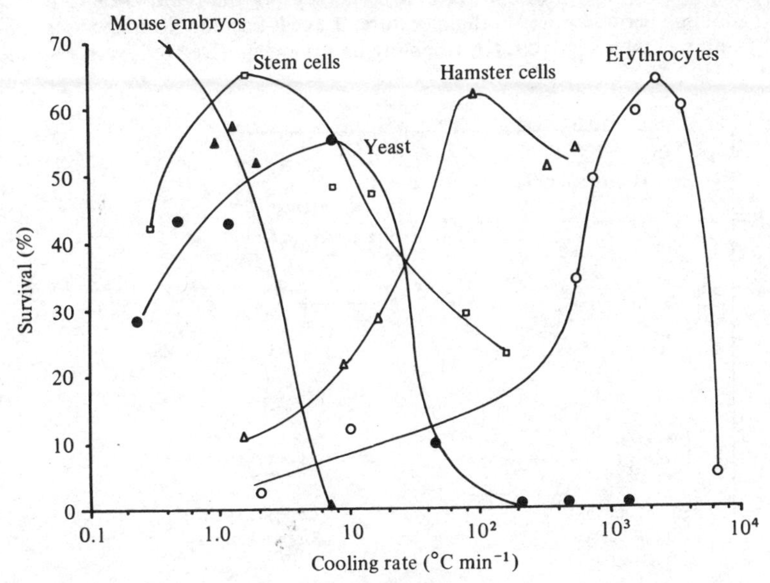

The optimum temperature and duration of the prefreezing period depend on the cell type and are related to the kinetics of water movement across the cell membrane. The addition of penetrating cryoprotectants causes the optimum prefreezing temperature to be lowered; this is because the amount of ice formed, and hence the concentration of the other solutes, is reduced.

Cryoprotectants

Cryoprotectants are a heterogeneous group of compounds which appear to act by a variety of different mechanisms to enhance the survival of cryopreserved cells. They are broadly grouped into penetrating and non-penetrating compounds. Glycerol, which was the first substance to be recognised as a cryoprotectant (Polge *et al.*, 1949), can act as a penetrating compound if added at room temperature or 37 °C and as a non-penetrating compound if added at 0 °C.

Cryoprotectants depress both the freezing-point (= melting-point) and the supercooling point of water (i.e. the temperature at which the homogeneous nucleation of ice occurs) and elevate the devitrification point (see Figs. 2 and 9). They thus reduce the amount of water removed from solution in the form of ice at any given temperature and reduce the concentrations of the other dissolved solutes. They also increase the viscosity of the solution, thus retarding the growth of ice crystals and making vitrification easier to achieve. Some glycoproteins present in the blood of polar fish appear to act by

Fig. 8. Two-step cooling. Cells cooled rapidly to below a critical temperature freeze internally and are killed. By interrupting the rapid cooling with a holding period above this temperature, the cells become shrunken with time and will then survive further cooling and thawing. (Derived from Farrant, 1980.)

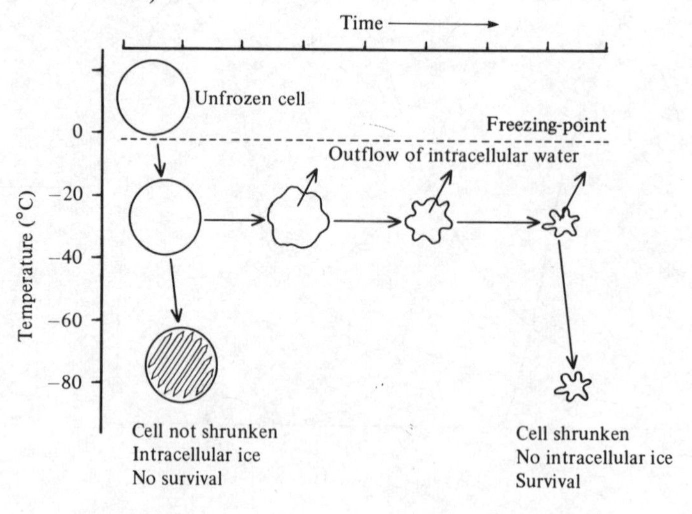

physically interfering with the addition of water molecules to an existing ice crystal lattice (De Vries, 1979). Dimethylsulphoxide (Me_2SO) appears to cause a redistribution of the intramembraneous particles (De Groot & Van Leene, 1979), which may enhance membrane permeability. Since the site of much of the damage produced by freezing is thought to be the cell membrane it is generally considered that non-penetrating cryoprotectants act specifically on the membrane.

The use of cryoprotectants introduces certain problems associated with their addition and removal. Except for the extremely rapidly penetrating compounds such as methanol and Me_2SO, the addition of a permeating cryoprotectant initially causes cell shrinkage; then, depending on the compound and the temperature of addition, the cell gradually regains its original shape over a period of time as the cryoprotectant penetrates. During removal of the cryoprotectant the process is reversed and the cell initially swells on being returned to isotonic conditions and then gradually returns to its normal size. If the concentration of additive required for cryoprotection is high these

Fig. 9. Glycerol 'phase diagram' giving the temperatures at which the glass (G), devitrification (D), recrystallisation (R) and melting-point (MP) transitions take place for glycerol solutions of various concentrations. Data derived from differential thermal analysis (see Fig. 10). (From Luyet & Rasmussen, 1968.)

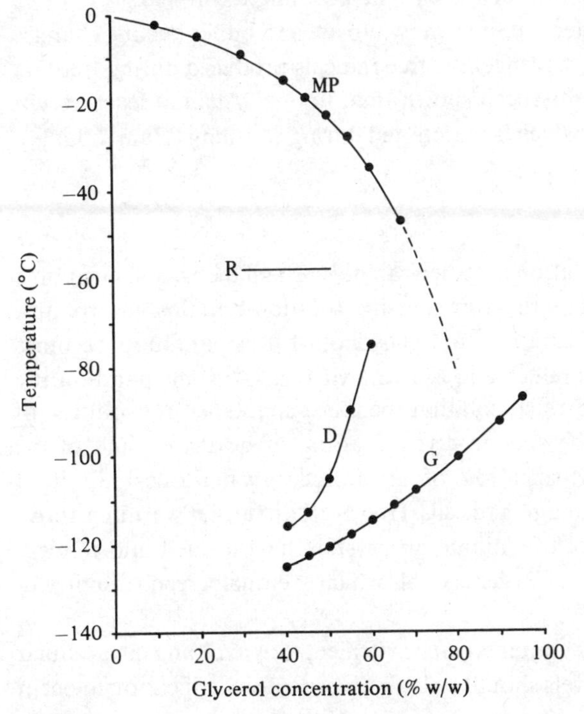

volume changes may be damaging and the cryoprotectant may have to be added in steps of increasing concentration and removed in steps of decreasing concentration or dialysed out. The stress of dilution is usually the more severe. It is possible to reduce dilution damage by incorporating into the diluting medium a non-penetrating compound which will prevent the cell from swelling while the cryoprotectant is diffusing out (Lovelock, 1952; Mutetwa, Liston & Last, 1980).

Some relatively large molecular weight compounds have been used to induce shrinkage and in this way provide protection (Gehenio & Luyet, 1953). Recently, a technique which uses an additive both as an internal cryoprotectant and as a shrinking agent, by its addition in two steps prior to cooling, has been used to significantly improve the survival of a parasitic worm (James, 1981). The organisms are exposed first to 10% v/v ethanediol at 37 °C, which penetrates. They are then exposed to 35% v/v ethanediol at 0 °C, which induces shrinkage since the ethanediol does not penetrate at this temperature. The pre-exposure to 10% ethanediol reduces the amount of shrinkage stress produced by the second addition of 35% ethanediol. The internal ethanediol concentration is effectively increased to 35% by shrinkage, and this is the concentration required for successful cryopreservation. This technique effectively parallels the cryodehydration which occurs during slow cooling or during the prefreezing period of the two-step cooling technique.

Serum can also be protective. It is known to act as a buffer against changes in pH and may also be a scavenger for free radicals produced during freezing and thawing. It has recently been shown that, in one system at least, serum protects against damage which is manifested during warming (Ham & James, 1982).

Vitrification

If an aqueous solution is cooled rapidly enough then insufficient time is available for ice crystals to grow and the solution remains amorphous. Below a certain temperature (i.e. the glass point), this condition becomes stable and the solution is referred to as being vitrified. At some point during the slow warming of a vitrified solution there is a spontaneous evolution of heat corresponding to the sudden crystallisation, or devitrification, of the solution (Fig. 10). Devitrification of intracellular water does not itself generally appear to be damaging to cells. However, on further warming above the recrystallisation point the minute ice crystals start to melt and refreeze resulting in the formation of larger crystals which eventually reach a sufficient size to be damaging.

In theory, if the cellular contents can be vitrified, thus avoiding intracellular ice formation, survival levels should be high. In order for vitrification to occur though, extremely rapid cooling rates are required. Vitrification of pure water is almost impossible, necessitating cooling at an estimated 10^{10} °C s^{-1} (Bruggeller & Mayer, 1980). There are two methods of overcoming this problem. The first is to use a very high concentration of a relatively high molecular weight cryoprotectant such as glycerol or ethanediol. The second method is to increase the barometric pressure. As the concentration of the cryoprotectant, or the pressure, is increased, so the cooling rate that is required to attain vitrification is reduced. Pressure can both prevent freezing damage and itself be injurious (Meryman, 1966). Some cryoprotectants used at high concentrations can protect against the deleterious effects of pressure and this approach forms the basis of a method which is being investigated for cryopreserving whole organs (Fahy & Hirsch, 1982).

Both the addition of cryoprotectants and an increase in pressure depress the freezing- and homogeneous nucleation points of a solution while at the same time elevating the glass point. At a particular concentration of cryoprotectant the freezing point and glass point will intersect, as occurs for 65% by weight PVP (MacKenzie, 1977 and also see under Ice nucleation), when the solution will be incapable of freezing. As the concentration of cryoprotectant is reduced from this point, ice nucleation and crystal growth require progressively shorter periods of time in which to occur, so that for

Fig. 10. Differential thermal analysis trace obtained during slow warming (5 °C min^{-1}) of a 50% w/w glycerol solution which had been precooled rapidly (> 4500 °C min^{-1}). The points G, D and MP mark the beginning of, respectively, glass transformation, devitrification and melting. (From Luyet & Rasmussen, 1968.)

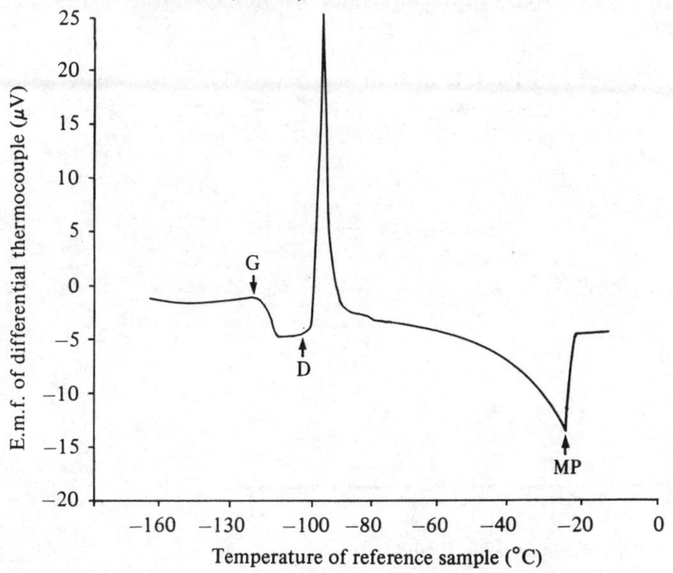

vitrification to be achieved the cooling rate has to be progressively increased. During continuous slow cooling the intracellular cryoprotectant may become sufficiently concentrated for vitrification to require a cooling rate of only a few degrees per minute and this may coincide with the cooling rate being used.

Warming

The thermal events during warming which are demonstrated by differential thermal analysis (Figs. 9 and 10) are paralleled by visual changes which can be observed under a cryomicroscope. As cells are warmed slowly they undergo a 'flashing' or darkening that corresponds to the sudden formation of intracellular ice, a further darkening associated with the migratory crystallisation of intracellular ice and the final disappearance of the dark material when the intracellular contents melt. However, actual damage to cells does not necessarily correlate with these events (Rall *et al.*, 1980).

Farrant has suggested that the importance of damage produced during warming has long been underrated. Few studies have paid much attention to the effects of warming although the warming rate that is used can be as important as the cooling rate (Fig. 11). It is generally considered that warming injury occurs primarily as a result of migratory recrystallisation, i.e. growth, of those small ice crystals which formed during the cooling phase (Mazur, 1966). Cells cooled rapidly are more likely to contain intracellular ice (Figs. 4 and 6) and thus rapid warming should give better survival than slow warming (Farrant, 1980). Direct evidence for the growth of intracellular ice during slow warming has been obtained (Asahina, 1965; Farrant, Walter, Lee & McGann, 1977).

Fig. 11. The effect of the rate of warming from −196 °C on the recovery of *Euglena gracilis* previously cooled at 0·3 °C min^{-1} using 10% v/v methanol as the cryoprotectant. (From Morris & Canning, 1978.)

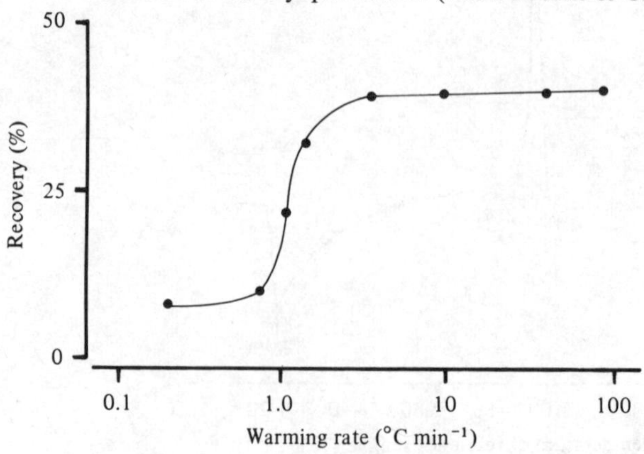

The most frequently used method of warming is that of placing the sample in a water-bath at 37 °C thereby producing a relatively rapid warming rate. This has the advantage of moving the sample quickly through the particularly damaging temperature range just below the melting-point. Occasionally, however, cells cooled slowly are damaged less by slow warming than by rapid warming. This has been reported for erythrocytes (Rapatz, Luyet & MacKenzie, 1975; Miller & Mazur, 1976) and embryos (Farrant *et al.*, 1977; Whittingham *et al.*, 1979). Embryos that were cooled slowly and continuously to −80 °C survived better if the warming rate was slow (20 °C min^{-1}) (Fig. 12). This work with embryos also demonstrated that injury from rapid (500 °C min^{-1}) warming is prevented if slow cooling (0·3–0·57 °C min^{-1}) is terminated at −40 °C and the embryos plunged from there into liquid nitrogen.

Another possible mechanism for warming injury has been outlined by Farrant (1977). He suggests that there is a difference in the chemical potential between the water inside and that outside the cell resulting from the difference in size of the ice crystals in the two compartments. This would constitute a driving force for moving water across the cell membrane and he further

Fig. 12. The effect of the warming rate on the survival (to the blastocyst stage in vitro) of mouse embryos cooled slowly (0.3–0.57 °C min^{-1}) to various intermediate temperatures before being plunged into liquid nitrogen at −196 °C, and recovered by rapid (500 °C min^{-1}) or slow (20 °C min^{-1}) warming. (From Whittingham *et al.*, 1979.)

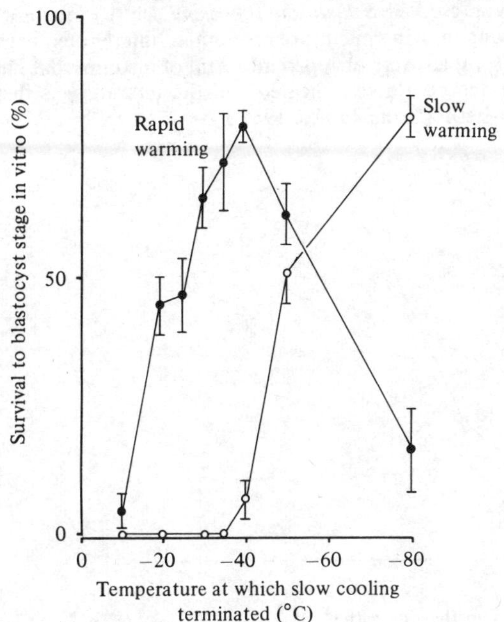

suggests that this water movement and/or the resultant increase in intracellular solute concentration is damaging.

Post-thaw treatment

Some cells, e.g. erythrocytes, are particularly sensitive to osmotic stresses and these can be exacerbated when combined with the stresses of freezing and thawing. Special care has to be taken over the post-thaw handling of these cells. The addition of certain substances, e.g. serum or a non-penetrating compound, to the medium can often significantly reduce the dilution stress by preventing a rapid influx of water by osmosis while the cryoprotectant diffuses out. However, in many cases the complete removal of the cryoprotectant may be unnecessary and continued washing only serves to further traumatise the cells.

The temperature and speed with which the diluting-out of the cryoprotectant is performed can be very important. When a slowly penetrating cryoprotectant is used, warming only to 0 °C may not allow sufficiently rapid diffusion of the additive out of the cell to prevent toxic injury. At room temperature or 37 °C, when cryoprotectants are more mobile, diffusion may be faster and survival levels consequently increased (Fig. 13). However, the toxicity of cryoprotectants increases with a rise in temperature and this may outweigh the advantage of an increase in cryoprotectant mobility.

Fig. 13. Survival of cryopreserved *Schistosoma mansoni* schistosomula (% normally motile) following thawing in culture medium at different temperatures. The 20-μl frozen samples were dropped into 2 ml of medium that had been prewarmed to the temperatures indicated and the mixture was then placed in a 37 °C water-bath. (From James, 1981.)

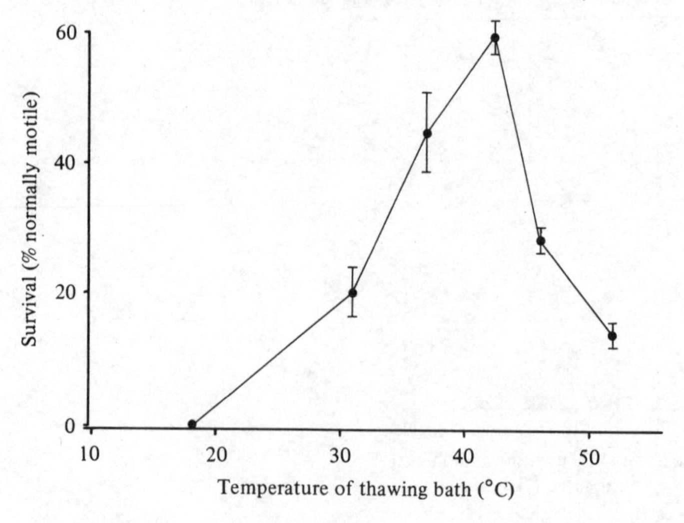

Viability assessment following cryopreservation

There are many specific techniques for measuring the survival of cells, tissues, organs or organisms following low-temperature preservation. Preserved material is usually assessed for viability immediately after thawing. This may, however, lead to a false calculation of survival. Many parasites, for example, appear to be morphologically and behaviourally normal after cryopreservation but following an infectivity assessment, which may take several days or weeks to complete, it can be shown that a significant proportion has in fact been damaged (James, 1981; Ham, James & Bianco, 1982). Other studies, which showed that increased survival occurred following modification of the post-thaw handling techniques, suggest that damage can sometimes be repaired after thawing (McGann, Kruuv & Frey, 1972).

Many techniques record the performance, following preservation, of populations rather than of individual cells. In those populations where a function of cell growth or cell interaction is measured, the response may not increase linearly with cell numbers. This has been clearly demonstrated by Knight & Farrant (1978) for human lymphocytes (Fig. 14). When measured at a cell concentration of 5×10^5 cells ml^{-1} the survival of cryopreserved cells was 16%, at $1{\cdot}3 \times 10^6$ cells ml^{-1} it was 100% and when assayed at higher cell concentrations the cryopreserved cells appeared to have survived better than

Fig. 14. The uptake of [^{3}H]thymidine by human lymphocytes in response to stimulation by phytohaemagglutinin on day 4 of in vitro culture. Unfrozen cells, open circles; dye-excluding frozen cells, filled circles; frozen cells, filled squares. The frozen cells were cooled at 0.4 °C min^{-1} with 5% v/v Me_2SO to −60 °C before being plunged into liquid nitrogen. (Data from Knight & Farrant, 1978.)

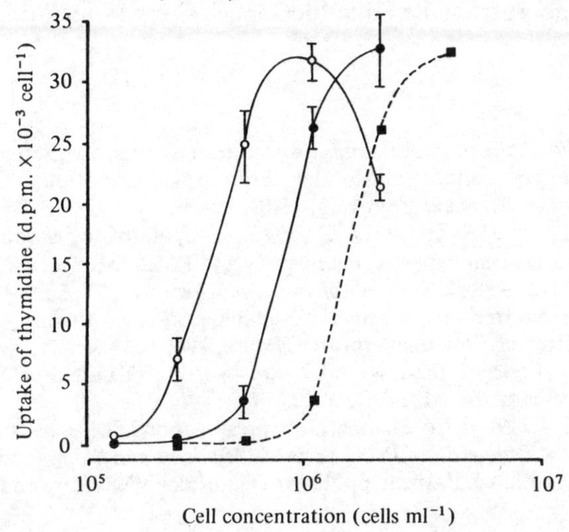

the controls. Clearly, the functioning of individual surviving cells had not been impaired but a reduction (by 80%) in the number of functional cells had occurred.

Conclusions

It cannot be emphasised strongly enough that the application of a published preservation schedule to a new system may well not be successful. Each different type of cell has a different tolerance to cryoprotectants and cooling and warming conditions. The route to successful preservation is likely to involve exhaustive testing of many different variables.

These variables, some of which are outlined in the sections above, are often interrelated and changing one parameter will almost certainly alter one or more of the others. For instance the incorporation of a cryoprotective additive may produce a shift towards a slower cooling rate for optimum survival; if the cooling rate is increased then the thawing rate may also have to be increased; and if cooling and/or warming conditions are suboptimal then the post-thaw handling techniques may have to be modified. Changes in some parameters may affect all the others; if the cells undergo different stages of development during their life or the life of the culture this may completely alter the cryopreservation schedule required.

The preceding sections have attempted to describe some of the important physical and chemical stresses that are imposed on cells during low-temperature preservation. A large number of different types of cells, tissues and organisms have been successfully cryopreserved, and in many fields cryopreservation has produced enormous benefits and has become an indispensable technology.

I wish to acknowledge support from the Edna McConnell Clark Foundation.

References

Aloia, R. C. (1978). Phospholipid composition of hibernating ground squirrel (*Citellus lateralis*) kidney and low temperature membrane function. *Comparative Biochemistry and Physiology*, **60B**, 19–26.

Angell, C. A., Shuppert, J. & Tucker, J. C. (1973). Cooperative behaviour in supercooled water: heat capacity, expansivity and PMR chemical shift anomalies from 0 to −38 °C. *Journal of Physical Chemistry*, **77**, 3092–9.

Asahina, E. (1959). Prefreezing as a method enabling animals to survive freezing at an extremely low temperature. *Nature*, **184**, 1003–4.

Asahina, E. (1965). Freezing process and injury in isolated animal cells. *Federation Proceedings*, **24**, S183–9.

Ashwood-Smith, M. J. (1980). Low temperature preservation of cells, tissues and organs. In *Low Temperature Preservation in Medicine and Biology*, ed. M. J. Ashwood-Smith & J. Farrant, pp. 19–44. Tunbridge Wells, England: Pitman Medical.

Blackshaw, A. W. (1954). The prevention of temperature shock of bull and ram semen. *Australian Journal of Biological Sciences*, **7**, 573–82.
Brandts, J. F. (1967). Heat effects on proteins and enzymes. In *Thermobiology*, ed. A. H. Rose, pp. 25–72. New York: Academic Press.
Bruggeller, P. & Mayer, E. (1980). Complete vitrification in pure liquid water and dilute aqueous solutions. *Nature*, **288**, 569–71.
Butler, W. J. & Roberts, T. K. (1975). Effects of some phosphatidyl compounds on boar spermatozoa following cold shock or slow cooling. *Journal of Reproduction and Fertility*, **43**, 183–7.
Cossins, A. R. (1983). Homeoviscous adaptation and its functional significance. *SEB Seminar Series: Cellular Acclimatization to Environmental Change*, ed. A. Cossins & P. Sheterline. Cambridge: Cambridge University Press.
De Groot, C. & Van Leene, W. (1979). The influence of cryoprotectants, temperature, divalent cations and serum proteins on the structure of the plasma membrane in rabbit peripheral blood lymphocytes. *European Journal of Cellular Biology*, **19**, 19–25.
De Vries, A. L. (1979). Structure and mechanism of action of fish glycopeptide and peptide antifreeze agents. *Cryobiology*, **16**, 585.
Fahy, G. M. & Hirsch, A. (1982). Prospects for organ preservation by vitrification. In *Organ Preservation, Present and Future*, ed. D. E. Pegg, I. A. Jacobsen & N. A. Halasz, pp. 399–404. Lancaster: MTP Press.
Farrant, J. (1965). Mechanism of cell damage during freezing and thawing and its prevention. *Nature*, **205**, 1284–7.
Farrant, J. (1977). Water transport and cell survival in cryobiological procedures. *Philosophical Transactions of the Royal Society of London, B*, **278**, 191–205.
Farrant, J. (1980). General Observations on Cell Preservation. In *Low Temperature Preservation in Medicine and Biology*, ed. M. J. Ashwood-Smith & J. Farrant, pp. 1–18. Tunbridge Wells, England: Pitman Medical.
Farrant, J. & Morris, G. J. (1973). Thermal shock and dilution shock as the cause of freezing injury. *Cryobiology*, **10**, 134–40.
Farrant, J., Walter, C. E., Lee, H. & McGann, L. E. (1977). Use of two-step cooling procedures to examine factors influencing cell survival following freezing and thawing. *Cryobiology*, **14**, 273–86.
Farrell, J. & Rose, A. H. (1968). Cold shock in a mesophilic and a psychrophilic pseudomonad. *Journal of General Microbiology*, **50**, 429–39.
Fujikawa, S. (1978). Morphology evidence of membrane damage caused by intracellular ice crystals. *Cryobiology*, **15**, 707.
Gehenio, P. M. & Luyet, B. J. (1953). The survival of myxamoebae after freezing in liquid nitrogen. *Biodynamica*, **7**, 175–80.
George, M. F., Burke, M. J., Pellett, H. M. & Johnson, A. G. (1974). Low temperature exotherms and woody plant distribution. *Horticultural Science*, **9**, 519–22.
Gerloff, E. D., Richardson, T. & Stahlmann, M. A. (1966). Changes in fatty acids of alfalfa roots during cold hardening. *Plant Physiology*, **41**, 1280–4.
Grout, B. & Morris, G. J. (1981). Membranes as a factor in cryoinjury. *Proceedings of the 4th Symposium on Problems of Low Temperature Preservation of Cells, Tissues and Organs*. May 20–23 1981, pp. 63–8. Berlin, DDR: Humboldt University.
Ham, P. J. & James, E. R. (1982). Protection of cryopreserved *Onchocerca* microfilariae (*Nematoda*) from dilution shock by the use of serum. *Cryobiology*, **19**, 448–57.

Ham, P. J., James, E. R. & Bianco, A. E. (1982). Separation of viable and non-viable *Onchocerca* microfilariae using an ion exchanger. *Transactions of the Royal Society of Tropical Medicine and Hygiene*, **76**, 758–67.

James, E. R. (1981). *Schistosoma mansoni*: Cryopreservation of schistosomula by two-step addition of ethanediol and rapid cooling. *Experimental Parasitology*, **52**, 105–16.

Jancz, E. R. & Maclean, F. I. (1973). The effect of cold shock on the blue green alga *Anacystis nidulans*. *Canadian Journal of Microbiology*, **19**, 381–7.

Knight, S. C. & Farrant, J. (1978). Comparing stimulation of lymphocytes in different samples: Separate effects of numbers of responding cells and their capacity to respond. *Journal of Immunological Methods*, **22**, 63–71.

Leibo, S. P., Farrant, J., Mazur, P., Hanna, M. G. Jr, & Smith, L. H. (1970). Effects of freezing on marrow stem cell suspensions; interactions of cooling and warming rates in the presence of PVP, sucrose or glycerol. *Cryobiology*, **6**, 315–32.

Levitt, J. (1966). Winter hardiness in plants. In *Cryobiology*, ed. H. T. Meryman. New York: Academic Press.

Levitt, J. (1972). *Responses of Plants to Environmental Stresses*. London and New York: Academic Press.

Lovelock, J. E. (1952). Resuspension in plasma of human red blood cells frozen in glycerol. *Lancet*, **1**, 1238.

Lovelock, J. E. (1953). The haemolysis of human red blood cells by freezing and thawing. *Biochemica et Biophysica Acta*, **10**, 414–26.

Lovelock, J. E. (1955). Haemolysis by thermal shock. *British Journal of Haematology*, **1**, 117–29.

Lozina-Lozinsky, L. K. (1965). Survival of some insects and cells following intracellular ice formation. *Federation Proceedings*, **24**, S206–11.

Lumsden, W. H. R. & Hardy, G. J. C. (1965). Nomenclature of living parasite material. *Nature*, **205**, 1032.

Luyet, B. J. (1967). On the possible biological significance of some physical changes encountered in the cooling and the rewarming of aqueous solutions. In *Cellular Injury and Resistance in Freezing Organisms, Proceedings of the International Conference on Low Temperature Science*, pp. 1–20, August 14–19 1966. Sapporo, Japan: Bunyeido Printing Corporation.

Luyet, B. J. & Gehenio, P. M. (1940). *Life and Death at Low Temperatures*. Normandy, Missouri, U.S.A.: Biodynamica.

Luyet, B. J. & Keane, J. (1955). A critical temperature range apparently characterised by sensitivity of bull semen to high freezing velocity. *Biodynamica*, **7**, 281–92.

Luyet, B. & Rasmussen, D. (1968). Study by differential thermal analysis of the temperatures of instability of rapidly cooled solutions of glycerol, ethylene glycol, sucrose and glucose. *Biodynamica*, **10**, 167–91.

MacKenzie, A. P. (1977). Non-equilibrium freezing behaviour of aqueous systems. *Philosophical Transactions of the Royal Society of London, B*, **278**, 167–89.

Macleod, R. A. & Calcott, P. H. (1976). Cold shock and freezing damage to microbes. In *Survival of Vegetative Microbes*, ed. T. R. G. Gray & J. P. Postgate, pp. 81–109. Symposium no. 26 of the Society for General Microbiology. Cambridge: Cambridge University Press.

Marr, A. G. & Ingram, J. L. (1962). Effect of temperature on the composition of fatty acids in *Escherichia coli*. *Journal of Bacteriology*, **84**, 1260–7.

Mazur, P. (1966). Physical and chemical basis of injury in single celled

microorganisms subjected to freezing and thawing. In *Cryobiology*, ed. H. T. Meryman, pp. 214–316. London and New York: Academic Press.
Mazur, P. (1970). Cryobiology: The freezing of biological systems. *Science*, **168**, 939–49.
Mazur, P. (1980). Limits to life at low temperatures and at reduced water contents and water activities. *Origins of Life*, **10**, 137–59.
Mazur, P., Farrant, J., Leibo, S. P. & Chu, E. H. Y. (1969). Survival of hamster tissue culture cells after freezing and thawing. Interactions between protective solutes and cooling and warming rates. *Cryobiology*, **6**, 1–9.
Mazur, P., Leibo, S. P. & Chu, E. H. Y. (1972). A two-factor hypothesis of freezing injury. *Experimental Cell Research*, **71**, 345–55.
Mazur, P. & Schmidt, J. (1968). Interactions of cooling velocity, temperature and warming velocity on the survival of frozen and thawed yeast. *Cryobiology*, **5**, 1–17.
McGann, L. E., Kruuv, J. & Frey, H. E. (1972). Repair of freezing damage in mammalian cells. *Cryobiology*, **9**, 496–501.
Meryman, H. T. (1966). Review of biological freezing. In *Cryobiology*, ed. H. T. Meryman, pp. 1–113. London and New York: Academic Press.
Meryman, H. T., Williams, R. J. & Douglas, M. St J. (1977). Freezing injury from solution effects and its prevention by natural or artificial cryoprotection. *Cryobiology*, **14**, 287–302.
Miller, R. H. & Mazur, P. (1976). Survival of frozen-thawed human red cells as a function of cooling and warming velocities. *Cryobiology*, **13**, 404–14.
Morris, G. J. (1975). Lipid loss and haemolysis by thermal shock: Lack of correlation. *Cryobiology*, **12**, 192–201.
Morris, G. J. (1976*a*). The cryopreservation of *Chlorella* II. Effect of growth temperature on freezing tolerance. *Archives of Microbiology*, **107**, 309–12.
Morris, G. J. (1976*b*). Effect of growth temperature on the cryopreservation of *Prototheca*. *Journal of General Microbiology*, **94**, 395–9.
Morris, G. J. (1980). Plant cells. In *Low Temperature Preservation in Medicine and Biology*, ed. M. J. Ashwood-Smith & J. Farrant, pp. 253–83. Tunbridge Wells, England: Pitman Medical.
Morris, G. J. (1981). *Cryopreservation, an Introduction to Cryopreservation in Culture Collections*. Cambridge: Institute of Terrestrial Ecology.
Morris, G. J. & Canning, C. E. (1978). Cryopreservation of *Euglena gracilis*. *Journal of General Microbiology*, **108**, 27–31.
Morris, G. J. & Clarke, A. (1978). The cryopreservation of *Chlorella* IV. Accumulation of lipid as a protective factor. *Archives of Microbiology*, **119**, 153–6.
Morris, G. J. & Farrant, J. (1973). Effect of cooling rate on thermal shock haemolysis. *Cryobiology*, **10**, 119–25.
Mutetwa, S., Liston, A. J. & Last, C. (1980). An improved method for the cryopreservation of *Plasmodium* populations. *Transactions of the Royal Society of Tropical Medicine and Hygiene*, **74**, 116.
Patterson, G. W. (1970). Effect of culture temperature on fatty acid composition of *Chlorella sorokiniana*. *Lipids*, **5**, 597–600.
Peters, J. & Syphert, P. S. J. (1978). Enrichment of mucin of *Mucor racemosus* by differential freeze killing. *Journal of General Microbiology*, **105**, 77–81.
Polge, C. (1980). Freezing of spermatozoa. In *Low Temperature Preservation in Medicine and Biology*, ed. M. J. Ashwood-Smith & J. Farrant, pp. 45–64. Tunbridge Wells, England: Pitman Medical.

Polge, C., Smith, A. U. & Parkes, A. S. (1949). Revival of spermatozoa after vitrification and dehydration at low temperatures. *Nature*, **164**, 666.

Polge, C., Wilmut, I. & Royson, L. E. A. (1974). The low temperature preservation of cow, sheep and pig embryos. *Cryobiology*, **11**, 560.

Polyansky, G. I. (1963). On the capacity of *Paramoecium caudatum* to stand the sub-zero temperatures. *Acta Protozoologica*, **1**, 165–75.

Rall, W. F., Reid, D. S. & Farrant, J. (1980). Innocuous biological freezing during warming. *Nature*, **286**, 511–14.

Rapatz, G., Luyet, B. & MacKenzie, A. (1975). Effect of cooling and rewarming rates on glycerolated human erythrocytes. *Cryobiology*, **12**, 293–308.

Rapatz, G., Sullivan, J. J. & Luyet, B. (1968). Preservation of erythrocytes in blood containing cryoprotective agents frozen at various rates and brought to a given final temperature. *Cryobiology*, **5**, 18–25.

Ring, R. A. (1980). Insects and their cells. In *Low Temperature Preservation in Medicine and Biology*, ed. M. J. Ashwood-Smith & J. Farrant, pp. 187–217. Tunbridge Wells, England: Pitman Medical.

Rotermund, A. J. Jr & Veltman, J. C. (1981). Modification of membrane-bound lipids in erythrocytes of cold-acclimated and hibernating 13-lined ground squirrels. *Comparative Biochemistry and Physiology*, **69B**, 523–8.

Salt, R. W. (1956). Cold hardiness of insects. *Proceedings of the 10th International Congress of Entomology*, Montreal, **2**, 73–7.

Sherman, J. K. (1962). Survival of higher animal cells after the formation and dissolution of intracellular ice. *Anatomical Record*, **144**, 171–89.

Siminovitch, D., Singh, J. & De La Roche, I. A. (1975). Studies on membranes in plant cells resistant to extreme freezing. 1. Augmentation of phospholipids and membrane substance without changes in unsaturation of fatty acids during hardening of black locust bark. *Cryobiology*, **12**, 144–53.

Singer, S. J. & Nicolson, G. L. (1972). The fluid mosaic model of the structure of cell membranes. *Science*, **175**, 720–31.

Smith, A. U. (1961). *Biological Effects of Freezing and Supercooling*. London: Edward Arnold.

Taylor, M. J. (1981). The meaning of pH at low temperatures. *Cryo Letters*, **2**, 231–9.

Walter, C. A. & James, E. R. (1981). Ultrastructural appearance of freeze-substituted schistosomula of *Schistosoma mansoni* frozen by a two-step cooling schedule. *Cryobiology*, **18**, 125–32.

Walter, C. A., Knight, S. C. & Farrant, J. (1975). Ultrastructural appearance of freeze-substituted lymphocytes frozen by interrupted rapid cooling with a period at −26 °C. *Cryobiology*, **12**, 103–9.

Whittingham, D. G., Leibo, S. P. & Mazur, P. (1972). Survival of mouse embryos frozen to −196 and −269 °C. *Science*, **178**, 411–14.

Whittingham, D. G., Wood, M., Farrant, J., Lee, H. & Halsey, J. A. (1979). Survival of frozen mouse embryos after rapid thawing from −196 °C. *Journal of Reproduction and Fertility*, **56**, 11–21.

Willemot, C., Hope, H. J., Williams, R. J. & Michand, R. (1977). Changes in fatty acid composition of winter wheat during frost hardening. *Cryobiology*, **14**, 87–93.

Withers, L. A. & Street, H. E. (1977). The freeze-preservation of cultured plant cells: III. The pregrowth phase. *Physiologia Plantarum*, **39**, 171–8.

L. A. WITHERS

Germplasm storage in plant biotechnology

Introduction

The uniqueness of germplasm has long been recognised in animal and plant biology. With recent progress in certain techniques, new areas of expertise are opening up and bringing with them new types of valuable germplasm. Thus, in animal husbandry where the storage of sperm in liquid nitrogen is a routine technique, attempts are now turning to similar storage of embryos for genetic conservation and research applications as well as for breeding (Polge, 1978; Whittingham, 1980). Improvements in transplantation techniques have led to a demand for long-term storage of human tissues and organs (Ashwood-Smith, 1980). In no other field is the identity of the specimen more critical. However, I do not think that it would be too partisan a view to suggest that in the development of plant biotechnology we are seeing a most dramatic evolution of new techniques with valuable applications in agriculture and industry. Good germplasm storage techniques are essential to underpin these applications.

If we survey the topics covered in this symposium, and indeed in any other treatment of plant biotechnology, it is evident that a very wide range of in vitro systems is involved. The range extends from the non-cellular (e.g. plastids and DNA) through protoplasts, cells and tissues to highly organised cultures such as shoot-tips and plantlets. Therefore, although dealing with a speciality within tissue culture, we in no way concentrate upon a limited number of specimen types.

It might be assumed that in wishing to apply germplasm storage methodology to the new subject of biotechnology rather than to the more fundamental one of the conservation of germplasm worldwide (see Henshaw & O'Hara, this volume), the constraints and practical approaches might be different. However, this will be seen not to be the case. Germplasm, in whatever its form, is the irreplaceable genetic building material from which essential plant products are derived. Special efforts must be made to store this germplasm when a number of circumstances obtain. These can be summarised as (a) lack of a propagule that is amenable to long-term storage by conventional means, (b) instability and (c) multiplicity.

As serious as the above problems are in a global context, in biotechnology they may be more acute. The lack of a natural propagule that is amenable to long-term storage (e.g. by low-moisture and low-temperature storage of orthodox seeds) is commonly encountered. In some cases, this is a carry-over from the in vivo situation, as with *Dioscorea* (the producer of the important pharmaceutical compound diosgenin) where the normal storage propagule is a tuber with limited viability and a tendency to deteriorate during storage. *Theobroma cacao*, which is of interest from the point of view that it is a source of flavour compounds, has recalcitrant seeds (Roberts & King, 1982). They can be stored for only very brief periods. Many other examples could be cited from within the categories of vegetatively propagated and recalcitrant seed-producing species.

To compound existing problems, some cultures, used for example for the synthesis of secondary products, will not necessarily be morphogenic: they may be cell or callus cultures which have lost their capacity to regenerate plants, thereby precluding the possibility of seed production. In any case, when the culture induction and selection processes have been laborious, it would be undesirable to have to reverse the process for storage and then retrace one's steps to produce the culture again when required. Even then, there would be no certainty of the regenerated culture carrying all of the desirable characters of the original culture. As a general rule, storage should be carried out on material which is as similar as possible in all respects to the experimental or production cultures.

The instability of plant cell cultures has many facets. Some culture systems are ephemeral, i.e. they are not stable and self-reproducing in the way that, say, an ideal cell or callus culture is. They follow a progression which could usefully be suspended at certain stages. Some typical examples are protoplasts which develop into cells, tissues and organised cultures, pollen which will undergo androgenesis or callus formation, material undergoing mutagenesis and selection, and embryogenic cultures exposed to inductive conditions. In all of these cases, intermediate stages in the procedures can only be retained unchanged for indefinite periods by deliberate intervention in the normal course of events. A further example which comes within this category involves germplasm which has been freed of pathogens by tissue culture techniques. It is metastable since it is prone to reinfection once returned to the field. It would therefore be desirable to store this germplasm in a 'clean' state for future utilisation.

Stability in terms of viability is a problem in any culture system which requires a constant input of nutrients. In the present context, we may be dealing with genotypes which have very special requirements, for example the supply of certain amino-acids. Unless the supply is maintained, viability will be lost and, perhaps more seriously, selection pressures may operate. This takes us to the critical point of genetic stability. As already stated, biotechnology involves all culture systems, including those most prone to instability, i.e. cells and calluses (D'Amato, 1978). Some variant lines which have been selected using time-consuming and exacting procedures may well be particularly susceptible to loss by further mutation or selection within heterogeneous populations. They may be outgrown by more vigorous variants and the 'wildtype'.

Although a detailed treatment is outside the scope of this chapter, we should not overlook the special situation of non-cellular germplasm, i.e. DNA sequences, plasmids, organelles etc. The former two may not present serious problems since storage in a laboratory refrigerator may be adequate, but for organelles, a means of preventing deterioration and loss of function will be required.

Finally, there is the problem of the multiplicity of samples. For germplasm storage on a global scale, the prime aim in conservation is the maintenance of pre-existing genotypes. Variability deriving from stages in tissue culture procedures is positively undesirable. However, in biotechnology both pre-existing and newly generated genotypes are involved and their potential number is infinite. The task of storage could, in quantitative terms, be formidable. In a programme involving several species, or varieties within a species, the labour and materials inputs required to maintain cultures can be a sizeable drain on resources. When further variables such as mutagenesis or somaclonal variation become involved (Larkin & Scowcroft, 1981), the task can become practically impossible. Much valuable genetic variability will be lost if we are unable to keep all cultures until they can be screened for desirable characters.

In the future, in vitro germplasm will become an increasingly important commodity. It will be necessary to lodge it in safe storage for patent reasons as well as for the protection of the investment of effort expended in its production. Further, it will be desirable to have secure methods for exchanging cultures in a state of suspended animation, which is far superior an approach to that of attempting to maintain environmental stability and the input of essential nutrients for growth at a normally rapid rate. These requirements have implications for the actual choice of storage method. Let us hope that for all of the reasons outlined above, attempts are made now to develop good storage technology before the task of maintaining and protecting precious germplasm becomes impossible.

Technical approaches to storage

The technical approaches to storage which are available can be divided into those involving storage of germplasm in the growing state and those involving the suspension of growth. The general principles involved will be given in this section and specific examples with references in the following sections.

Many tissue cultures grow at a relatively rapid rate and require transfer to fresh medium at intervals. Typically, a cell suspension culture would need to be transferred every 7–10 days and a callus culture every 14–30 days. Immediately after transfer, the culture may enter a lag phase before growth is resumed, and, after a period of growth (often exponential), it may enter a stationary phase induced by limitation of one or more nutrients. Both lag and stationary phases may involve the imposition of stresses on the culture and this may, in turn, result in the loss of viability and/or selection. Therefore, if storage of a culture in the growing state is to be carried out, it is desirable to maintain steady growth. In consequence, if the frequency of subculturing is not to be too burdensome, the growth rate during storage must be reduced significantly.

There are several ways of limiting growth. The most obvious and widely used is a reduction in the culture temperature. Alternatively, retardant chemicals may be added to the culture medium. These may act osmotically, as in the case of mannitol, or they may be of a hormonal nature, e.g. abscisic acid. Mineral oil overlay has been used for the storage of microbial cultures but has been explored only to a limited extent for higher plant cultures. Another possible approach involves modification of the atmosphere to which the culture is exposed, e.g. by a reduction in the level of oxygen.

Typical storage periods for cultures under growth limitation are in the region of one year, which is satisfactory for short-term storage but may present problems in the longer term. The potential for growth limitation is finite. If attempts are made to reduce growth beyond a certain, as yet undefined, rate, then the consequence is death of the culture. Therefore, if the demands of subculturing are to be avoided completely, it is necessary to turn to an approach to storage which involves the suspension of all metabolism. Freeze-drying is widely used in microbiology but it is not, at present, an available technique for higher plant tissue cultures. The only serious option is cryopreservation, i.e. storage at the temperature of liquid nitrogen (or similar). The basic principles of cryopreservation have been detailed by James (this volume) and only a brief outline of the stages involved in cryopreservation procedures will be given here.

During the period of time preceding cryopreservation, the freeze-tolerance of a culture can be improved either by timing the harvesting of the culture to be at the most amenable stage of growth, or by modifying the culture conditions. Very few botanical specimens other than orthodox seeds have any intrinsic freeze-tolerance to the temperatures which are involved in cryopreservation. Therefore, both chemical cryoprotection and optimisation of the freezing and thawing rates are essential.

Shortly before preservation, various cryoprotectants can be applied. Both single cryoprotectants and mixtures are used. They are normally prepared in culture medium (but see under Cell suspension cultures in the following section), adjusted to a standard pH and sterilised before being applied to the culture. Although autoclaving can be used safely for some preparations, it is inadvisable for mixtures of high molarity (which can be up to 4 M). Filter sterilisation should be employed for these instead. For specimens suspended in liquid medium, double-strength cryoprotectants are prepared and added slowly to an equal volume of culture medium over a period of *c*. 1 h. For specimens which are sufficiently large and sturdy to be transferred, for example by forceps, a series of concentrations of cryoprotectants, which increases to full strength, can be prepared.

As primary enclosures for specimens in storage, shatter-proof polypropylene ampoules are generally preferred to glass. They can be obtained pre-sterilised, have close-fitting screw tops and a labelling area. For some specimens, foil envelopes have been used. These can be sterilised by autoclaving. In a few cases, naked specimens are involved since they are exposed directly to liquid nitrogen in order to maximise freezing rates (see under Shoot-tip cultures in the following section). As yet, there is no completely convenient way of storing these.

Cooling rates range widely from very slow (1 or 2 °C min^{-1}) for cell suspensions, calluses and various organised cultures to very rapid (from 50 to > 1000 °C min^{-1}) for the entire cooling of some specimens (e.g. certain shoot-tips) and for the final quenching of initially slowly frozen specimens. Practical details of how to carry out freezing will be dealt with, using illustrative examples, in the following section.

Under the conditions which obtain in the cryopreservation of most tissue culture specimens, it is necessary to maintain a very low temperature during storage to ensure absolute stability (see Meryman & Williams, 1982). A liquid nitrogen refrigerator running at − 150 or − 196 °C (in the vapour or liquid phases, respectively) is technically ideal and convenient in use.

Thawing is usually carried out rapidly, using a warm water-bath (35–40 °C). As a precaution against microbial contamination, it is advisable to place an inner container (e.g. a beaker) of sterile water within the bath. In a few exceptional cases, specimens are thawed slowly by exposure to air at room temperature.

Careful handling during the post-thaw phase is essential for sustaining viability in the vulnerable, freshly isolated specimen and for encouraging a rapid resumption of growth. This phase, like that of pregrowth, has been neglected in studies until relatively recently but is of enormous importance in determining the final outcome of cryopreservation attempts. Few generalisations can be made beyond stating that it is not always essential, or even advisable, to carry out post-thaw washing, and that the conditions used for culture prior to freezing are not necessarily suitable after thawing.

Thus, we have two main options when choosing storage methods: growth limitation, which requires periodic renewal but which can be very useful in the short and medium term, and cryopreservation, which can in theory store material for indefinite periods. It would be desirable to be able to select the most appropriate approach to storage according to practical requirements and circumstances, and this may well be the case in the not-too-distant future. However, for the present we do not have good, reproducible methods within each category for all types of culture. In the following section, each culture system will be examined in turn, and both the storage methods which are available to date and the areas in which further experimentation is required before there can be a completely unhindered choice of method will be indicated. Results are summarised in Table 1.

Reported successes in the storage of plant tissue cultures

Protoplasts

Protoplast technology plays a unique and critical role in a number of aspects of biotechnology. The naked protoplast, isolated either mechanically or enzymatically from cultures or explants, is the nearest equivalent that we have to the cultured animal cell. It is bounded by a plasma membrane (the plasmalemma) which forms a direct interface between the cytoplasm and the environment. The uptake of exogenous material into cells, including nucleic acids, microbial cells or organelles, and the capacity of these to fuse with other protoplasts to form somatic hybrids depend upon the properties of the plasmalemma. Thus, if protoplasts are to be stored for such applications, it is essential that wall regeneration has not occurred by the time that storage is carried out. For this reason, it is unlikely that any form of growth limitation could be successful or satisfactory. Indeed, no references can be found in the literature to such attempts. Where storage is carried out *after* various manipulations this point will not be as critical. However, once the process of wall regeneration is well under way, we are dealing with a cell rather than a protoplast and techniques should be chosen accordingly (see Cell suspension cultures).

Protoplasts have been the subjects of a number of low-temperature biochemical and biophysical studies since they provide an opportunity to observe the behaviour of the osmotically active cellular unit free of the influences of the cell wall (e.g. Wiest & Steponkus, 1978; Levin, Ferguson, Dowgert & Steponkus, 1979; Singh, 1979). However, much as this has enabled us to understand the nature of freezing damage, particularly in terms of membrane lesions, none of the studies have involved the continued culture of the protoplasts in order to determine whether physical integrity could be matched by recovery and growth. Nonetheless, there are two studies which do give some indication of the potential for cryopreserving protoplasts, and one of these clearly demonstrates recovery.

Mazur & Hartmann (1979) have compared the behaviour of cells and enzymatically isolated protoplasts of *Bromus inermis* and *Daucus carota*. All specimens could survive freezing carried out in four sequential steps of +4, −20, −60 and −196 °C after cryoprotection with dimethylsulphoxide (DMSO) or glycerol (each at 0·7 M), although there were differences in the relative survival levels with the two cryoprotectants. Viability was assessed by trypan blue exclusion. The interesting points which emerge from this study are that protoplasts may require a period of equilibration between isolation and freezing, and that protoplasts may sometimes survive better than cells under identical freezing conditions. The former point implies that there may be a necessity to compromise between amenability to freezing and suitability for uptake, fusion, etc. as discussed above.

Takeuchi, Matsushima & Sugawara (1980) have succeeded in regenerating cryopreserved protoplasts of *D. carota* and *Marchantia* sp. (a liverwort). The protoplasts were isolated enzymatically from suspension-cultured cells and callus respectively. There is no suggestion of a period of equilibration being required before freezing. Various cryoprotectants were surveyed, including some high molecular weight polymers. The most effective were found to be 10% DMSO or 5% DMSO plus 10% glucose. Freezing was carried out slowly at *c.* 1 °C min^{-1} either to −196 °C or just to −30 or −40 °C followed by quenching in liquid nitrogen, storage and then rapid thawing. The thawed protoplasts were washed in fresh medium and returned to culture. Wall regeneration commenced in surviving protoplasts within two days of thawing. This was monitored by staining with Calcofluor White and observation under ultraviolet light. Cell division occurred within five days of thawing and eventually, actively growing cultures formed. In the case of *Marchantia*, cultures placed under inductive conditions formed protonemata, rhizoids and thalli. Morphogenesis was not demonstrated in the culture of *D. carota* but in a preliminary study Withers (1982*a*) has shown that protoplasts isolated from an embryogenic culture of this species are capable of embryogenesis after cryopreservation (See also p. 217 for *Note added in proof*.)

Table 1. *A summary of reports of the storage of plant tissue cultures*[a] (See also p. 217 for *Note added in proof.*)

Species	Culture system	Conditions for storage by growth limitation	Conditions for storage by cryopreservation	References
Acer pseudoplatanus	Cell suspension	—	LN[b]	Withers & King (1980); Withers & Street (1977*a*);[c]
A. saccharum	Cell suspension	—	−40 °C[d]	[c]
Arachis hypogaea	Shoot-tip	—	LN	[c]
Asparagus officinalis	Shoot-tip	—	LN	[c]
Atropa belladonna	Anther	—	LN	Bajaj (1978*a*)
A. belladonna	Pollen embryo	—	LN	Bajaj (1977*a*, 1978*a*, *b*)
A. belladonna	Cell suspension	—	LN	[c]
Beta vulgaris	Shoot-tip	LT[e]	—	[c]
Brassica napus	Shoot-tip	—	LN	Withers (1982*c*)
Bromus inermis	Protoplast	—	LN	Mazur & Hartmann (1979)
Capsicum annuum	Cell suspension	—	−30 °C	Withers & Street (1977*a*)
C. annuum	Cell suspension	—	LN	Withers & Street (1977*a*)
C. annuum	Callus	LT	—	Withers (1978*c*)
Chrysanthemum morifolium	Callus	—	−3.5 °C	Bannier & Steponkus (1972)
C. morifolium	Shoot-tip	LO_2[f]	—	Bridgen & Staby (1981)
Cicer arietinum	Shoot-tip	—	LN	[c]
Dactylis sp.	Shoot-tip	LT	—	[c]
Datura stramonium	Cell suspension	—	LN	[c]

Daucus carota	Protoplast	—	LN	Mazur & Hartmann (1979); Takeuchi *et al.* (1980);[c]
D. carota	Cell suspension	—	LN	Nag & Street (1973); Dougall & Wetherell (1974); Withers & Street (1977*a*);[c]
D. carota	Callus	MO[g]	—	Caplin (1959)
D. carota	Callus	Des.[h] + Ret.[i]	—	Nitzsche (1978)
D. carota	Callus	Des. + Ret.	−80 °C	Nitzsche (1978)
D. carota	Somatic embryo	Des. + LS[j]	—	Jones (1974)
D. carota	Somatic embryo/ clonal plantlet	—	LN	Withers & Street (1977*b*); Withers (1979)
Dianthus caryophyllus	Cell suspension	—	LN	Anderson (1979)
D. caryophyllus	Shoot-tip	—	LN	Seibert (1976); Seibert & Wetherbee (1977); Uemura & Sakai (1980)
Festuca sp.	Shoot-tip	LT	—	[c]
Fragaria × *ananassa*	Shoot-tip	—	LN	Sakai *et al.* (1978); Kartha *et al.* (1980)
Fragaria spp.	Shoot-tip	LT	—	[c]
Glycine max	Cell suspension	—	LN	Bajaj & Reinert (1977);[c]
Haplopappus ravenii	Cell suspension	—	−20 °C	[c]
Hordeum vulgare	Zygotic embryo	—	LN	Withers (1982*c*)
Hyoscyamus muticus	Cell suspension	—	LN	Withers & King (1980)
Ipomoea batatas	Shoot-tip	LT	—	[c]
Ipomoea sp.	Cell suspension	—	−40 °C	[c]
Lactuca sativa	Shoot-tip	—	LN	[c]
Linum usitatissimum	Cell suspension	—	−50 °C	Quatrano (1968)
Lolium multiflorum	Shoot-tip	LT	—	[c]
Lolium sp.	Seedling (shoot-tip)	—	LN	[c]
Lycopersicon esculentum	Zygotic embryo (shoot-tip)	—	LN	Grout (1979)

Table 1. (*cont.*)

Species	Culture system	Conditions for storage by growth limitation	Conditions for storage by cryopreservation	References
L. esculentum	Seedling (shoot-tip)	—	LN	Grout *et al.* (1978)
Malus domestica	Shoot-tip	LT	—	[c]
Manihot esculentum	Shoot-tip	LT	—	[c]
M. esculentum	Shoot-tip	—	LN	Henshaw *et al.* (1980);[c]
M. usitatissimum	Shoot-tip	—	LN	[c]
Medicago sativa	Callus	—	LN	Finkle *et al.* (1979)
M. sativa	Shoot-tip	LT	—	[c]
Nicotiana sylvestris	Cell suspension	—	LN	[c]
N. tabacum	Anther	—	LN	Bajaj (1978*a*)
N. tabacum	Pollen embryo	—	LN	Bajaj (1977*a*, 1978*a*, *b*)
N. tabacum	Cell suspension	—	LN	Bajaj & Reinert (1977);[c]
N. tabacum	Callus	LO_2	—	Bridgen & Staby (1981)
N. tabacum	Shoot-tip	LO_2	—	Bridgen & Staby (1981)
Oryza sativa	Anther	—	LN	Bajaj (1980)
O. sativa	Cell suspension	—	LN	Cella *et al.* (1978); Sala *et al.* (1979)
O. sativa	Callus	—	LN	Finkle *et al.* (1979, 1983)
Petunia hybrida	Anther	—	LN	Bajaj (1978*a*)
Phleum sp.	Shoot-tip	LT	—	[c]
Phoenix dactylifera	Callus	—	LN	Finkle *et al.* (1979, 1982); Tisserat *et al.* (1981)
Pisum sativum	Shoot-tip	—	LN	Kartha *et al.* (1979)
Populus euramericana	Callus	—	LN	Sakai & Sugawara (1973)

Primula obconica	Anther	—	LN	Bajaj (1981)
Rosa (Paul's Scarlet Rose)	Cell suspension	—	LN	Withers & King (1980)
Saccharum sp.	Cell suspension	—	LN	[c]
Saccharum sp.	Callus	—	−23°C	Ulrich *et al.* (1979)
Saccharum sp.	Callus	—	LN	Ulrich *et al.* (1979)
Sambucus racemosa	Callus	—	LN	[c]
Solanum spp.	Shoot-tip	LT+/or Ret.	—	Henshaw *et al.* (1980);[c]
Solanum spp.	Shoot-tip	—	LN	Grout & Henshaw (1978); Henshaw *et al.* (1980); Towill (1981);[c]
Sorghum bicolor	Cell suspension	—	LN	Withers & King (1980)
Trifolium repens	Shoot-tip	LT	—	[c]
Triticum aestivum	Zygotic embryo	—	LN	Withers (unpublished observations)
Vitis rupestris	Shoot-tip	LT	—	[c]
Zea mays	Cell suspension	—	LN	Withers & King (1979, 1980); Withers (1980*d*)
Z. mays	Callus	—	LN	Withers (1978*c*)
Z. mays	Zygotic embryo	—	LN	Withers (1980*c*)

[a] Other reports, particularly of storage by growth limitation, conveyed to the author in response to a survey, are detailed in Withers (1981) and Withers (1982*b*).
[b] Storage in liquid nitrogen (−196 °C) or liquid nitrogen vapour (*c.* −150 °C).
[c] For references see Withers (1980*c*, 1982*a*) or Withers & Street (1977*b*).
[d] Storage in the frozen state at a temperature other than −150 °C or −196 °C.
[e] Storage at a low growth temperature.
[f] Storage under conditions of low oxygen supply.
[g] Storage under mineral oil.
[h] Tissue desiccated prior to storage.
[i] Storage in the presence of/after application of retardants.
[j] Storage under conditions of sucrose limitation.

A useful incidental observation made by Takeuchi *et al.* (1980) is that protoplasts frozen in heat-sealed foil envelopes survive at a slightly higher level than those frozen in glass tubes. This may be due to better heat conductivity, a more uniform freezing rate or to the fact that less severe compressive forces may be generated during extracellular freezing.

Drawing together all of the above observations, it seems that the cryopreservation of protoplasts is a realistic possibility and that high survival levels in the region of 70–90% can be anticipated. A combination of cryoprotection (with e.g. 10% DMSO), slow or step-wise freezing and rapid thawing is recommended. No special pregrowth or post-thaw treatments need be applied, although care should obviously be taken to preserve the integrity of the protoplasts by the use of appropriate osmotic stabilisers.

Before leaving this area, mention should be made of an interesting observation reported by Cella *et al.* (1978). They found that freshly frozen and thawed cell suspension cultures of *Oryza sativa* were incapable of yielding stable protoplasts. After a recovery period of *c.* two days in liquid culture, the yield of protoplasts from the cells returned to approximately that of the unfrozen controls. This observation should be borne in mind when there is a need to provide consistent material for protoplast studies if the introduction of a storage stage *before* isolation seems most convenient.

Pollen, pollen embryos and anthers

The often brief period of viability of pollen presents difficulties in carrying out hybridisations between parents with different maturity times and between parents in different geographical locations. Therefore, even before tissue culture considerations became important, there was interest in storing pollen. The subject has been reviewed elsewhere (e.g. Roberts, 1975) and will be dealt with only briefly here. It appears that for a number of species, storage may be possible at low humidity and low temperature by freezing in liquid nitrogen or by lyophilisation (freeze-drying). However, there are definite interspecific differences and the Graminae in general are difficult subjects. In a recent study Weatherhead, Grout & Henshaw (1979) have demonstrated the superiority of storage in liquid nitrogen over storage at −20 °C for pollen of *Solanum* spp. Other studies with pollen have thrown light upon some of the biochemical changes which can occur as a result of freezing at very low temperatures (e.g. Anderson, Nath & Harner, 1978).

In a plant tissue culture context, the motives for storage are rather different. Pollen can be considered to constitute a single cell system which is useful for genetic manipulations. Haploid pollen plants capable of being ‘doubled-up’ into homozygous diploid plants are of immense value in speeding up breeding programmes. Haploid cell and callus cultures have applications in mutagenesis studies. Therefore it is not difficult to justify the development of storage methods for these specimens.

Rather than being concerned with loss of viability over a relatively short period of time, as occurs in the case of pollen required for fertilisation, one is more concerned here with the loss of microspores at a suitable stage of development for androgenesis and the loss of haploid cells and calluses through polyploidisation. Again, growth limitation is intrinsically unsuitable for these specimens and all efforts at storage to date have concentrated upon cryopreservation. The only contributor in this area has been Bajaj. His work is summarised below.

It would appear to be technically feasible to cryopreserve material from *Atropa belladonna*, *Nicotiana tabacum*, *O. sativa*, *Petunia hybrida* and *Primula obconica*, but survival rates are variable and often very low (Bajaj, 1977*a*, *b*; 1978*a*, *b*; 1980; 1981). Survival of freshly isolated anthers is virtually nil. Only very rapid freezing is effective and lag periods of 2–3 months intervene before recovery growth, which results in the production of callus and the development of multicellular pollen grains, commences at a much reduced level. However, anthers which have been precultured for 3–5 weeks to initiate androgenesis are able to survive rapid freezing (direct immersion in liquid nitrogen) in some cases and slow freezing (1–3 °C min^{-1}) in others. After washing in fresh liquid medium anthers are returned to culture on semi-solid medium or occasionally in liquid medium. Chemical cryoprotection is necessary and a number of applications are effective (e.g. 5–7% DMSO; 15% glycerol; 7% DMSO plus 7% sucrose; 5% DMSO plus 5% sucrose plus 5% glycerol). There is an apparent toxic effect of some cryoprotectants as revealed by a drop from 45% of the *Primula* anthers yielding plants in the controls to 25% in DMSO-treated (unfrozen) specimens. Recovery, after lag periods of up to three months, takes the form of continued pollen embryo development or callus formation. Recovery rates range from 0.5% to 5% in terms of anthers yielding regenerating material. Occasionally, as in the case of *O. sativa*, callus recovered after cryopreservation may only be able to regenerate roots, not entire plants. There is a little evidence that the rates of recovery may be increased by cutting the anthers transversely before freezing.

Pollen embryos isolated from cultured anthers of *N. tabacum* and *A. belladonna* can survive slow freezing after treatment with DMSO or glycerol. They show the greatest freeze-tolerance during the earlier stages of development as revealed by staining with fluorescein diacetate immediately after thawing. Survival levels of the various stages are as follows: globular embryos 31%, early heart-shaped embryos 9% and late heart-shaped embryos 2%. There is no survival in embryos which show cotyledonary development. Globular embryos and early heart-shaped embryos are capable of continuing normal development but the late heart-shaped embryos produce only calluses or malformed plants. These observations are in accord with others made during the development of cryopreservation methods for somatic embryos produced from cell suspension cultures (see below).

For the future application of cryopreservation techniques, it is possible to make some very general recommendations. Preculture to induce androgenesis seems advisable. Cryoprotection with DMSO alone (at *c.* 7%) or with other compounds as listed earlier, should be followed by slow freezing (*c.* 2 °C min^{-1}), rapid thawing, washing and reintroduction into culture on semi-solid medium. However, it is clear that with present techniques, viabilities after cryopreservation, and therefore yields of pollen plants and haploid cultures, are extremely low. In heterogeneous specimens there will be the additional risk of selection for freeze-tolerant genotypes. Clearly there is considerable scope for further research into methodology here.

Cell suspension cultures

Cell suspension cultures are of great importance in secondary product work, in studies of physiology and morphogenesis, and in genetic engineering. Although not single-celled in the true sense, many cell suspension cultures are very finely dispersed and apparently uniform and most grow rapidly. They offer convenience in handling and an opportunity to carry out quantitative work based on the model of microbial systems. However, balanced against these advantages is the serious problem of genetic instability. Examples of changes in ploidy and the development of aneuploidy are well documented (see D'Amato, 1978). Although we know a little about the effects of media composition and other environmental factors on the genotypes present in a culture (Bayliss, 1977*a*, *b*), means of control are not readily available. Therefore, good, reliable, storage methods are required.

For this system also, growth limitation is not really suitable as an approach to storage since the problems of continued growth are simply spread over a longer period of time and the stresses accompanying growth limitation may exacerbate the selection of abnormal genotypes. It is inconvenient and relatively difficult technically to maintain large stocks of suspension cultures at a reduced temperature. The possibility of adding growth inhibitors to the liquid culture medium has not been explored and the results may reward investigation. Suspension cultures can be transferred to semi-solid medium for storage and then the methods used for the limited growth of callus cultures can be employed (see below). However, for the storage of suspension cultures without adopting an alternative growth habit, cryopreservation is the most appropriate method. In fact, progress over recent years has been such that the prospects for developing a generally applicable method are very good and there may soon be no need to seek methods for growth limitation.

Cell suspension cultures have been the subject of cryopreservation studies for more than a decade. Quatrano first reported the storage of a culture of *Linum usitatissimum* at a temperature of −50 °C in 1968. The first report of storage in liquid nitrogen appeared in 1973 when Nag & Street successfully preserved an embryogenic culture of *D. carota* and demonstrated the regeneration of embryos from the recovered culture. Since that time more than 15 other species have yielded a culture that is capable of a positive viability test response after cryopreservation, and many of these are further capable of regenerating a healthy culture. Details of the development of the methodology used for cryopreserving cell suspension cultures have been well documented in recent years (see Bajaj & Reinert, 1977; Withers, 1980*a*, *b*, *c*; 1982*a*) and will not be repeated here. However, the work in this area is usefully considered in two contexts: the information gathered concerning the improvement of freeze-tolerance by manipulation of pregrowth and post-thaw conditions, and the possibility of developing general cryopreservation methodology.

It is clear from a number of studies that a prime requirement for successful cryopreservation is the selection of material at an appropriate stage of growth. Cells in early exponential phase seem to possess the highest freeze-tolerance; cells in early lag phase and stationary phase are particularly susceptible to cryoinjury (see e.g. Withers & Street, 1977*a*). It is likely that a large individual cell size and increased water content contribute to this susceptibility. Withers (1978*a*) has shown that there is also a relation between the cell cycle stage and survival. Cells of *Acer pseudoplatanus* in early G_1 (or G_0) phase survive freezing far better than those in other phases. This can be explained only partly in terms of relative cell sizes. Despite this observation, it should be said that synchronisation of cultures is a laborious procedure, particularly if tight synchrony is sought and other means of improving freeze-tolerance are preferable.

It is possible to eliminate both lag and stationary phases from the growth cycle, thereby eliminating the need to pin-point the most suitable stage for preservation, by reducing the passage duration. Over a number of passages of, for example, seven days' duration as in the case of *A. pseudoplatanus*, this will lead to a net reduction in the average cell size (Withers & Street, 1977*a*).

For some species, the above procedure may be adequate for the optimisation of freeze-tolerance, but for others the cell size may still be far too large and further measures must be taken. Pregrowth in medium supplemented with osmotically active additives such as mannitol (at *c.* 6%) or proline (at *c.* 10%) is widely effective in increasing freeze-tolerance (Withers & Street, 1977*a*; Withers & King, 1979; 1980 and unpublished observations; Pritchard, Grout, Reid & Short, 1982). Light and electron microscopical observations indicate that a reduction in cell size often occurs and that the single large vacuole present in control cells may be replaced by several smaller vesicles (Withers, 1978*b*; Pritchard *et al.*, 1982). H. W. Pritchard & B. W. W. Grout (personal communication, 1982) have further shown that cells pregrown in the presence of mannitol show differences in oxygen uptake, wall flexibility and thickness, and in their behaviour during the early stages of freezing as revealed by observation in the cryomicroscope.

The minimum effective period for pregrowth in supplemented medium appears to be 3–4 days (Withers & King, 1979, 1980). No maximum period is evident for pregrowth in the presence of mannitol but prolonged growth in the presence of proline leads to a progressive reduction in viability and growth rate and darkening of the cells (Withers, unpublished observations). DMSO, which is effective as a pregrowth additive for shoot-tips (see Shoot-tip cultures, below), appears to be of little value for cell suspension cultures.

As well as the size of a cell being important, the size of an aggregate can affect survival for two possible reasons. In large aggregates, the proportion of highly meristematic, rapidly dividing cells is reduced, and there may be significant differences in freezing rates and the degree of protective dehydration in different regions of the aggregate. Despite the popular image of a fine cell suspension, highly aggregated cultures, often containing malformed roots or shoots, are formed by a number of species including many cereals (King, Potrykus & Thomas, 1978). Such cultures may still respond to a good cryopreservation protocol but it may be necessary to break up large aggregates mechanically and/or filter out the largest cell masses (Dougall & Wetherell, 1974). Separation by enzymatic means is not recommended.

A wide range of cryoprotectants from single applications of DMSO or proline to complex mixtures of these and other compounds, has been employed for cell suspension cultures. Although 5–10% DMSO or 10% proline can be effective in a limited number of cases, the most generally effective cryoprotectants appear to be three-component mixtures based on DMSO, glycerol and a third component such as proline or sucrose, the total molarity of the mixture being in the region of 2 M.

Routinely, cryoprotectants are prepared in culture medium and until the role of culture medium components was specifically examined recently, their importance was not realised. Withers (1980*d*) compared the performance of cells of *Zea mays* that were cryoprotected with either DMSO, glycerol and proline or DMSO, glycerol and sucrose, each mixture having been prepared in water or standard culture medium. Cells cryoprotected with DMSO, glycerol and proline in water demonstrated post-thaw viability as indicated by the fluorescein diacetate test (see Withers, 1980*a*) but were incapable of recovery growth. Inclusion of medium components restored the capacity to recover. However, this difference was not evident for the cells cryoprotected with the sucrose-containing mixture, suggesting that the presence of the sucrose substituted for the medium components. Comparison with other single cryoprotectants and two-component mixtures prepared in water or medium confirms the superiority of the sucrose-containing mixture.

As well as improving the survival potential under optimum freezing conditions, the use of DMSO, glycerol and sucrose increases the differentials between under- and over-dehydration, and damage caused by the formation of ice or the 'solution effects' (described by James, previous chapter), respectively. This can be demonstrated by comparing the performance of cells cryoprotected with this mixture with those cryoprotected with DMSO and glycerol (as widely used in early studies, e.g. Withers & Street, 1977*a*), under different freezing conditions. The cells are frozen to different sub-zero temperatures before quenching in liquid nitrogen and then warmed rapidly or slowly, or are warmed rapidly or slowly direct from the intermediate transfer temperature. For the cells treated with DMSO and glycerol, quenching in liquid nitrogen is lethal except when they are warmed rapidly from intermediate transfer temperatures (Fig. 1*a*). However, the cells treated with DMSO, glycerol and sucrose show no significant difference between rapid and slow thawing and no decline in survival at lower transfer temperatures (Fig. 1*b*). Throughout, recovery potentials are very good and sometimes higher in

Fig. 1. The response of cells of *Zea mays* to freezing and thawing under a range of conditions after cryoprotection with (*a*) 1 M DMSO + 1 M glycerol or (*b*) 0.5 M DMSO + 0.5 M glycerol + 1 M sucrose. Both cryoprotectant mixtures were prepared in water. The cells were frozen at a rate of 1 °C min^{-1} to the temperatures indicated and then either thawed directly (light lines indicated by arrows; open circles) or quenched in liquid nitrogen and then thawed (heavy lines; solid circles). Thawing was carried out either rapidly (continuous lines) or slowly (broken lines). Lines indicate initial post-thaw viability levels as determined by fluorescein diacetate staining. The size of the circles indicates the relative efficiency of recovery growth in terms of the number and size of colonies produced after 60 days of regrowth. Where no circle is shown, this indicates that a post-thaw viability may have been recorded but recovery growth did not take place. (Modified from Withers, 1980*d*.)

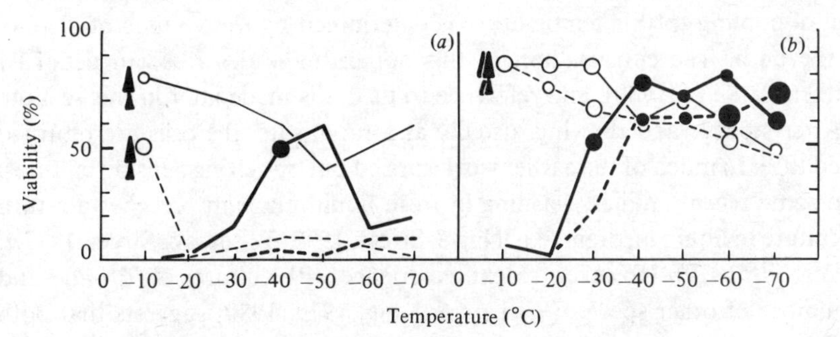

the slowly thawed cells, unlike in former preparations where good recovery occurred only exceptionally. Clearly, the cells cryoprotected with the three-component mixture are 'buffered' against the deleterious effects of freezing in liquid nitrogen once the early stages of cooling have been passed. The cryopreservation procedure is thus greatly simplified. The advantages of this are obvious.

Cells about to undergo cryopreservation are dispensed into ampoules, or exceptionally into foil envelopes (Takeuchi *et al.*, 1980), and then cooled by a programme which will lead to adequate extracellular freezing. Rapid freezing is always lethal. The actual cooling conditions used reflect upon the apparatus available. For example, Sala, Cella & Rollo (1979) have used a non-linear rate of *c.* 0·1 °C min^{-1} overall which was achieved by placing the specimen inside a vacuum flask that was itself within a deep-freeze running at −70 °C. Other workers have used apparatus in which the specimen is suspended in the vapour above liquid nitrogen and exposed to various non-linear cooling rates the values of which are dependent upon the distance between the body of liquid nitrogen and the specimen (e.g. Nag & Street, 1973). Yet others have used automatic pre-programmed freezing apparatus which can give exact linear rates of cooling, the most commonly applied being 1 or 2 °C min^{-1} (e.g. Withers & Street, 1977*a*).

There can be practical disadvantages in using slow freezing to develop cryopreservation protocols that are dependent upon extracellular freezing since it is necessary to conduct one experiment per rate of freezing. Accordingly, it can be very useful to adopt the alternative of step-wise freezing (sometimes termed 'prefreezing') wherein the specimen is taken to a sub-zero temperature and held at this temperature for various periods of time to bring about protective dehydration. Withers & King (1979, 1980) have combined slow freezing with step-wise freezing to investigate the development of cryopreservation methods for cell suspension cultures. Such a programme is easily achieved using a simple, improvised apparatus consisting of an alcohol bath cooled by a dip-cooling coil. At the commencement of cooling the refrigeration is switched on and a thermostat is set to the desired holding temperature. The rate of cooling to this temperature is determined by the volume of alcohol in the bath. The construction of this apparatus is described in detail by Withers & King (1980), and reference to its use is made later in this section.

After storage and thawing, usually at a rapid rate, the cells are returned to culture. In much of the earlier work carried out on cell suspension cultures, and some recent studies, washing in fresh liquid medium precedes a return to culture in liquid medium (e.g. Nag & Street, 1973; Withers & Street, 1977*a*; Sala *et al.*, 1979). However, recent experience with cultures of *Z. mays* and a number of other species (Withers & King, 1979, 1980) suggests that both washing and liquid culture may be deleterious, even to the point of preventing recovery in otherwise viable cultures. The harmful effects may relate to deplasmolysis injury and to disturbances in the solute balance of the freshly thawed, and probably 'leaky', cells. It appears that the continued presence of the cryoprotectant-containing medium in which the cells were frozen aids recovery. This medium may contain solutes lost from the cells during freezing and thawing as well as the cellular water removed during extracellular freezing. Thus, the simplest procedure, that of layering the cells over semi-solid medium, is best for recovery. After a few days under these conditions, the cells can be transferred to liquid medium, cell division and growth having been resumed.

It is interesting to note, in the light of this apparent requirement for a period of equilibration before normal culture conditions are resumed, that Cella *et al.* (1978) found that protoplasts could not be isolated from freshly thawed cells (as discussed earlier). The same workers have also found that a period of 2–4 days ensues before normal respiration resumes in frozen and thawed rice cells (F. Sala, R. Cella and colleagues, personal communication, 1982).

Taking the most suitable treatments at all stages of the cryopreservation procedure, Withers & King have compiled a routine method which can be applied to a wide range of species including *A. pseudoplatanus*, *D. carota*, *Hyoscyamus muticus*, *N. tabacum*, *O. sativa*, *Rosa* cv. Paul's Scarlet Rose, *Sorghum bicolor* and *Z. mays* (Withers & King, 1979, 1980 and unpublished observations). Those familiar with such cultures will realise that they cover a wide range of morphologies as well as genotypes. The composite method involves pregrowth for three or four days in standard medium or medium containing 6% mannitol or 10% proline, cryoprotection with 0·5 M DMSO, 0·5 M glycerol plus 1 M proline or sucrose, freezing at *c.* 1 °C min^{-1} to −35 °C, holding for 30–40 min at −35 °C, quenching in liquid nitrogen, rapid thawing and then layering over semi-solid medium. After about two days, cell division and growth are resumed and within one or two weeks a healthily growing culture is available for use. This procedure, using the simple apparatus described above, is in routine use in King's laboratory at the Friedrich Miescher-Institut, Basel, Switzerland, for the maintenance of stock cultures and variant lines produced by mutagenesis (Table 2). The prospects for wider use by other workers are very good.

The general procedure was arrived at empirically but it is interesting to note that electron microscopical examination supports its superiority. Cells cryopreserved by a suboptimal procedure, e.g. cryoprotection with DMSO or DMSO and glycerol, and returned to culture in liquid medium exhibit various degrees of ultrastructural damage, the most common form of damage being dilation of cytoplasmic organelles (Withers, 1978*b*). During the days

Table 2. *Cell suspension culture lines stored in liquid nitrogen at the Friedrich Miescher-Institut, Basel, Switzerland (March 1982)*[a]

Species	Line designation	Special features[b]
Acer pseudoplatanus	AM	—
Berberis dictyophylla	—	—
Catharanthus roseus	—	—
Corydalis sempervirens	—	—
Daucus carota	Wild	—
D. carota	CΔ6A3	$5MT^R$
D. carota	AlC5/I	—
D. carota	CV8/1	AEC^R
Glaucium flavum	—	—
Glycine max	—	—
Hyoscyamus muticus	wt	wt
H. muticus	VA5	his^-
H. muticus	VIIIB9	trp^-
H. muticus	IVH2	nic^-
H. muticus	MA2	NR^-
Nicotiana tabacum	S3	AEC^R
N. tabacum	TX^4	wt
N. tabacum	MFPr8	MFP^R
Onobrydus viciifolia	—	—
Oryza sativa	Sask.	—
Pennisetum americanum	—	—
Rhazia americanum	—	—
R. stricta	—	—
Rosa (Paul's Scarlet Rose)	wt	wt
Rosa (Paul's Scarlet Rose)	OMT^R	OMT^R
Rosa (Paul's Scarlet Rose)	aca^R	aca^R
Solanum melongena	—	—
Sorghum bicolor	—	—
Triticum monococcum	—	—
Zea mays	B73	—
Z. mays	BMS	wt
Z. mays	BMS-C	ADH^-

[a] List kindly supplied by Dr P. J. King. Please note that these cell lines are not available for distribution.

[b] Key to special features. wt, wildtype cells. ^{R} Cell lines resistant to: 5MT, 5-methyl tryptophan; AEC, aminoethylcysteine; MFP, metafluorophenylalanine; OMT, *ortho*-methylthreonine; aca, azetidine-2-carboxylic acid. $^-$Cell lines deficient in: his, histidine; trp, tryptophan; nic, nicotinamide; NR, nitrate reductase; ADH, alcohol dehydrogenase.

following thawing many of the cells continue to degenerate and recovery ensues from only a minority of the cell population. Within the degenerating cells elaborate processes of membrane breakdown and reutilisation may be observed (Withers, 1980*a*, *b*). Cells frozen by the improved procedure show very little damage immediately after thawing. Within about two days there is a clear demarcation between those which are lethally damaged and those which are recovering and undergoing mitosis. See also *Note added in proof*, p. 217.

Callus cultures

Callus cultures are of rather less obvious importance in biotechnology than are cell suspensions, although as other contributions to this volume will indicate, they do have a role in morphogenesis studies and in the establishment of cultures for secondary product synthesis. A callus phase is commonly involved in the initiation of various types of culture and, as stated above, suspension cultures may be transferred to semi-solid medium for maintenance as callus stocks.

Callus culture is the first culture system so far considered which can realistically be stored by both the growth limitation method and cryopreservation. However, the problem of genetic instability, which is particularly marked in some callus cultures (see Novak, 1974, 1980; D'Amato, 1978), would lead one to favour cryopreservation.

Studies on storage at reduced growth temperatures and growth in the presence of inhibitors have not been reported in detail in the literature but there are some anecdotal reports of workers simply placing cultures in the laboratory refrigerator at or near 0 °C (e.g. Potrykus, 1978). Bannier & Steponkus (1972) have exposed callus cultures of *Chrysanthemum morifolium* to a temperature of −3·5 °C for 28 days. Survival declined throughout the 'storage' period and could only be maintained at any acceptable level by supplementation of the culture medium with 10% sucrose.

Bridgen & Staby (1981) have investigated the use of atmospheric modification for the storage of callus cultures of *N. tabacum*. Both reduced atmospheric pressure and a reduced partial pressure of oxygen were effective in reducing growth, and to a similar extent. Reducing the available oxygen by up to *c.* 60% reduced growth by 60–80% over a period of six weeks. Greater reductions in the oxygen supply reduced growth so severely that deleterious effects would be likely in storage for prolonged periods. Available oxygen is the limiting factor in the mineral oil overlay method for callus storage investigated by Caplin (1959). Cultures cannot survive if immersed in liquid medium but can continue to grow at approximately a quarter of the rate of controls when immersed in mineral oil due to the greater solubility of oxygen in the latter. Unfortunately this promising technique does not appear to have been adopted or further developed by other workers.

In another isolated study, Nitzsche (1978) has stored callus cultures of *D. carota* in the desiccated state after pregrowth in medium containing additional sucrose and abscisic acid. The dried callus could survive for seven or 14 days and could tolerate freezing to −80 °C.

Cryopreservation has been applied successfully to callus cultures of several species. As with cell suspension cultures, slow or step-wise freezing is required and, with one exception, so is cryoprotection. The most widely successful cryoprotection has been given by two- or three-component mixtures of cryoprotectants such as that applied to several species by Finkle and colleagues, i.e. 10% polyethylene glycol (carbowax 6000), 8% glucose and 10% DMSO (Finkle *et al.*, 1979; Ulrich, Finkle, Moore & Ginoza, 1979; Tisserat, Ulrich & Finkle, 1981).

When morphogenic callus has been preserved, plant regeneration has usually been possible after thawing but observations with *Saccharum* sp. (Ulrich *et al.*, 1979) suggest that this cannot be assumed. It is possible that freezing to the temperature of liquid nitrogen did not bring about a genotypic change but destroyed primordial structures which would have given rise to shoots. However, no such problem was encountered when these workers attempted to regenerate plants from a cryopreserved callus culture of *Phoenix dactylifera*, which may in fact have consisted of a mass of pro-embryos rather than callus proper (Tisserat *et al.*, 1981). In the latter study, the material recovered from cryopreservation was compared with unfrozen controls in terms of the patterns of extracted isozyme proteins. No differences were found. Such examination will be invaluable in the future in cases where it is essential to verify the nature of the recovered cultures.

Although cold-hardening preceded cryopreservation in one study (Sakai & Sugawara, 1973), no general recommendations can be made about either pregrowth or post-thaw treatments. One interesting observation reported by Finkle, Ulrich, Schaeffer & Sharpe (1983) may, however, be worth exploring with a range of species. They found that callus cultures of *O. sativa* survived best when cryoprotectants were applied to 0 °C and removed by washing with medium at 22 °C.

It is clear that there is still some way to go before a general method for cryopreserving callus cultures emerges but as a starting-point one can suggest cryoprotection with a three-component mixture such as is used for cell suspension cultures or as detailed in this section, slow freezing at *c.* 2 °C min^{-1}, rapid thawing and returning to culture with or without washing in warm medium.

Embryos

Zygotic embryo culture is a very useful technique for rescuing otherwise inviable embryos from very wide crosses or from mutagenised seeds. Rescue may be necessary owing, for example, to failure of the endosperm. More authentic tissue culture applications are involved when zygotic embryos are used for protoplast isolation or for the initiation of morphogenic cultures, particularly in cereals.

Secondary embryogenesis, the formation of somatic embryos or embryoids, is an important stage in the regeneration of some species from culture. Examples can be found of embryos developing from protoplasts, pollen, cells, calluses and zygotic embryos.

Although mature zygotic embryos present little challenge in their storage since most are highly dehydrated and can be stored like seeds, immature zygotic embryos are very similar to somatic embryos in their water content and consequent storage requirements. We can consider the two types of specimen together.

All but one of the reports of storage involve cryopreservation. The exception relates to the storage of globular and heart-shaped somatic embryos of *Daucus carota* on medium lacking sucrose but otherwise under normal culture conditions for two years (Jones, 1974). Upon the addition of a solution of 2% sucrose (w/v), the embryos 'germinated' and proliferated, forming new secondary embryos. The desiccation and sucrose limitation during storage can be likened to processes involved in normal seed maturation.

Turning to cryopreservation: zygotic embryos still contained within the seed have been frozen successfully in a number of cases (see Sakai & Noshiro, 1975; King & Roberts, 1979). Detail would be unjustified here but it is interesting to note that in one study on the cryopreservation of imbibed seeds of *Lycopersicon esculentum*, tissue culture techniques were used to recover meristems which were incapable of survival if left in the entire seed (Grout, 1979).

Immature embryos and somatic embryos have received little attention but there are indications that cryopreservation is feasible in both cases (Withers, unpublished observations). Early attempts to cryopreserve somatic embryos of *D. carota* using a method which was successful with cell suspension cultures of the species failed. Only when washing was omitted from the post-thaw treatment did any recovery take place. Even then it occurred infrequently and involved secondary embryogenesis from the superficial cells of some of the original embryos (Withers & Street, 1977*b*). However, further modification of the procedure led to dramatic improvements in recovery rates. The technique which developed has been termed 'dry freezing' (Withers, 1979) and involves treating the specimen with 5–10% DMSO, then removing it from the cryoprotectant solution, blotting it on sterile filter-paper to remove superficial moisture and enclosing it in a foil envelope. Freezing at a slow rate (e.g. 1 °C min^{-1} to −100 °C), quenching in liquid nitrogen and then thawing relatively slowly are followed by recovery on semi-solid medium. Globular and heart-shaped embryos survive this procedure and resume growth rapidly. Late torpedo-stage embryos and plantlets recover by secondary embryogenesis from the meristem regions only but recovery rates approach 100% in all cases. The technique can be further refined by desiccating the embryos briefly before freezing, prolonging the thawing stage to minimise physical damage to the brittle frozen embryos and incorporating activated charcoal in the recovery medium to minimise the formation of callus. Attempts to adopt this method for immature zygotic embryos of *Z. mays* were successful (Withers, 1978*c*). More conventional slow freezing and rapid freezing failed with this species but recent studies with the smaller embryos of *Hordeum vulgare* and *Triticum aestivum* reveal that several different methods may be used for any one specimen. However, the pattern of recovery may also vary from method to method. The objective of this study was to incorporate a cryopreservation stage into the procedure of culture initiation from immature cereal embryos, thereby relieving the cost and inconvenience of maintaining growth facilities all the year round for embryo production.

Preliminary findings suggest that all stages from the entire seed head through to the embryo dissected from the surface-sterilised grain and cultured for two days can survive freezing (Withers, 1982*c* and unpublished observations). However, the former specimens must be frozen rapidly and survival levels are poor (10–15%). For all stages after dehusking of the individual grain, rapid, slow and dry freezing are effective and survival levels can approach 100%. It is interesting to note that rapid freezing tends to promote germination under appropriate post-thaw conditions whereas slow and dry freezing promote callus formation. This suggests that the former favours survival of the embryo axis and the latter that of the scutellum. This work is proceeding. In future experiments the morphogenic potential of callus developed from frozen embryos, and the phenotype of plants regenerated from this callus, will be compared with unfrozen controls.

Despite the limited amount of work which has been carried out to date on embryos, it does appear that once a suitable method has been determined, very high survival levels can be anticipated. The dry-freezing method in particular may provide the basis for a generally applicable cryopreservation technique.

Shoot-tip cultures

Within the category shoot-tip cultures are included specimens which are variously termed shoot-apices, shoot apical meristems and shoot-tips. Their importance in studies of morphogenesis and in any area of biotechnology where regeneration of the whole plant is involved need not be emphasised. However, storage of shoot-tips probably has the most significant role to play in the conservation of pre-existing germplasm worldwide. Since this is treated in detail by Henshaw & O'Hara (this volume), only a brief consideration covering points which reflect upon the general subject of germplasm storage will be given here.

The growth limitation method of storage has been most highly developed for this type of system. Reports of the maintenance of cultures at low temperature, in medium containing osmotic or hormonal inhibitors and under oxygen limitation cover a wide range of species (see Withers, 1981, 1982*a*, *b*). It would appear to be feasible in most cases to maintain the cultures without transfer for periods of one year. However, an observation of G. Staritsky (personal communication, 1980) suggests that by alternating limitation with normal growth (e.g. by alternating exposures to low and normal temperatures), incipient lethal stress can be avoided and overall storage periods can be prolonged.

It appears that a careful choice of growth temperature for low-temperature storage and careful timing of transfers may be important for ensuring high survival levels. The normal habitat of the plant in question (i.e. whether tropical or temperate) will give some guidance on temperature but these factors will generally have to be determined empirically.

The first report of successful cryopreservation of shoot-tips appeared in 1976 (Seibert). From this and a number of subsequent studies it did seem that rapid freezing would be the most suitable technique for cryopreserving shoot-tips. However, we can now find examples of the same species responding to both rapid and slow (or step-wise) freezing (*Dianthus caryophyllus*: Seibert, 1976; Seibert & Wetherbee, 1977; cf. Anderson, 1979; Uemura & Sakai, 1980; *Fragaria* × *ananassa*: Sakai *et al.*, 1978; cf. Kartha, Leung & Pahl, 1980; *Solanum* spp. Grout & Henshaw, 1978; cf. Towill, 1981). In at least two of the examples cited, the rapid freezing method involved direct exposure of the shoot-tip to liquid nitrogen. Whilst this presents no practical difficulties at the time of freezing other than having to handle the specimens individually, there are problems in storing unenclosed specimens securely and efficiently.

It is undeniable that rapid freezing is the least demanding approach technically but it should be remembered that very rapid thawing rates will be required and this may present the risk of physical damage to brittle specimens. Therefore the effort expended in developing adequate slow or step-wise methods may be rewarded by a greater reproducibility of freezing and thawing conditions and a reduced level of damage during handling.

It is not anticipated that many other culture systems, except perhaps embryos and anthers, could benefit directly from the development of both rapid and slow methods for the same specimen. However, information on pregrowth and post-thaw conditions gained from studies on shoot-tips may have wider implications. In several cases, a pregrowth treatment involving culture on basal medium or medium supplemented with 5% DMSO has been essential to survival (Kartha, Leung & Gamborg, 1979; Henshaw, Stamp & Westcott, 1980; Kartha *et al.*, 1980; Withers, 1982*c* and unpublished observations). The requirement for a period of time in culture between excision and cryopreservation may relate to healing of the excision wound. The effect of pregrowth in DMSO-supplemented medium is more difficult to explain. Preliminary studies by Withers (unpublished observations) on the pregrowth requirements of shoot-tips of *Brassica napus* suggest that a sequence of one day on callus-inducing medium lacking DMSO and two days on basal medium containing DMSO may be optimal, and further that other solutes, including proline, may be as effective as DMSO in promoting freeze-tolerance.

Cryopreserved shoot-tips have a tendency to form callus after thawing rather than to continue organised growth (see e.g. Grout, Westcott & Henshaw, 1978) even under apparently suitable culture conditions. Modification of the post-thaw culture medium by the addition of gibberellic acid may control this formation of callus and promote shoot outgrowth (Grout *et al.*, 1978). More dramatically, manipulation of the hormonal content of the post-thaw medium can lead to enormous improvements in overall recovery rates, as seen in *Solanum* spp. (Henshaw *et al.*, 1980; and Henshaw & O'Hara, this volume).

A final point to be made in connection with post-thaw recovery is that even when recovery rates are high and plant regeneration presents no difficulty, it cannot be assumed that recovery is proceeding by resumption of the normal growth of the apex. Developing shoots may in fact be adventitious as shown by examination of recovering shoot-tips of *Pisum sativum* (Haskins & Kartha, 1980).

Conclusions

Examination of the list of species for which storage methods have been reported to date (Table 1) gives some cause for optimism. With each new addition to the list, the time draws closer when general methods can be applied without recourse to a lengthy, empirical determination of optimum conditions. Added to this is the apparent facility of exchanging cultures between research laboratories when these are maintained under growth limitation as described by Henshaw & O'Hara (this volume). Although largely untried, there should be no serious technical problems in exchanging frozen cultures.

For cell cultures, it may well be possible now to adopt cryopreservation as both a long- and short-term storage method. Storage of shoot-tips in the short and medium term by growth limitation is at a similar state of development. For most other systems and storage methods it is clear that a substantial input of work will be required before such a point can be reached.

It is sometimes surprising that more interest is not shown in cryopreservation. Perhaps as long as even the most inconvenient, inefficient, expensive and risky culture maintenance methods are available, workers will continue to use them through familiarity thus shunning the possible security and saving in time and costs which can be offered, especially by cryopreservation. It should be stressed that even the best conventional culture maintenance procedure cannot prevent genotypic change or drift in the composition of heterogeneous populations; nor can it suspend time such that the worker rather than the culture dictates the scheduling of experimental work. Plant biotechnology is moving forward at a very rapid rate and the current level of activity is intense, as demonstrated by the contributions to this symposium. Failure to safeguard the raw material and products of this new field would be scientifically and economically disastrous. So let us hope that sufficient effort will be devoted to storage methodology in the near future before valuable genotypes are irretrievably lost.

The support of the Agricultural Research Council and the Science and Engineering Research Council is gratefully acknowledged.

References

Anderson, J. O. (1979). Cryopreservation of apical meristems and cells of carnation (*Dianthus caryophyllus*) (Abstract). *Cryobiology*, **16**, 583.

Anderson, J. O., Nath, J. & Harner, E. J. (1978). The effects of freeze-preservation on some pollen enzymes. II. Freezing and freeze-drying stresses. *Cryobiology*, **15**, 469–77.

Ashwood-Smith, M. J. (1980). Low temperature preservation of cells, tissues and organs. In *Low Temperature Preservation in Medicine* and *Biology*, ed. M. J. Ashwood-Smith & J. Farrant, pp. 19–44. Tunbridge Wells: Pitman Medical.

Bajaj, Y. P. S. (1977*a*). Survival of *Nicotiana* and *Atropa* pollen embryos frozen at −196 °C. *Current Science* (India), **46**, 305–7.

Bajaj, Y. P. S. (1977*b*). Clonal multiplication and cryopreservation of cassava through tissue culture. *Crop Improvement*, **4**, 198–204.

Bajaj, Y. P. S. (1978*a*). Effect of superlow temperature on excised anthers and pollen embryos of *Atropa*, *Nicotiana* and *Petunia*. *Phytomorphology*, **28**, 171–6.

Bajaj, Y. P. S. (1978*b*). Regeneration of plants from pollen embryos frozen at ultralow temperature: A method for the preservation of haploids. In *Proceedings of the Fourth International Palynological Conference, Lucknow*. **1**, 343–6.

Bajaj, Y. P. S. (1980). Induction of androgenesis in rice anthers frozen at −196 °C. *Cereal Research Communications*, **8**, 365–9.

Bajaj, Y. P. S. (1981). Regeneration of shoots from ultra-low frozen anthers of *Primula obconica*. *Scientia Horticulturae*, **14**, 93–5.

Bajaj, Y. P. S. & Reinert, J. (1977). Cryobiology of plant cell cultures and establishment of gene banks. In *Applied and Fundamental Aspects of Plant Cell, Tissue and Organ Culture*, ed. J. Reinert and Y. P. S. Bajaj, pp. 757–77. Berlin: Springer.

Bannier, L. J. & Steponkus, P. L. (1972). Freeze-preservation of callus cultures of *Chrysanthemum morifolium* Ramat. *HortScience*, **7**, 194.

Bayliss, M. W. (1977*a*). Factors affecting the frequency of tetraploid cells in a predominantly diploid cell suspension culture of *Daucus carota*. *Protoplasma*, **92**, 109–15.

Bayliss, M. W. (1977*b*). The causes of competition between two cell lines of *Daucus carota* in mixed culture. *Protoplasma*, **92**, 117–27.

Bridgen, M. P. & Staby, G. L. (1981). Low pressure and low oxygen storage of plant tissue cultures. *Plant Science Letters*, **22**, 177–86.

Caplin, S. M. (1959). Mineral-oil overlay for conservation of plant tissue cultures. *American Journal of Botany*, **46**, 324–9.

Cella, R., Sala, F., Nielsen, E., Rollo, F. & Parisi, B. (1978). Cellular events during the regrowth phase after thawing of freeze-preserved rice cells. Abstracts: Federation of European Societies of Plant Physiology, pp. 127–8. Edinburgh, July 1978.

D'Amato, F. (1978). Chromosome number variations in cultured cells and regenerated plants. In *Frontiers of Plant Tissue Culture*, ed. T. A. Thorpe, pp. 287–96. Calgary: Calgary University.

Dougall, D. K. & Wetherell, D. F. (1974). Storage of wild carrot cultures in the frozen state. *Cryobiology*, **11**, 410–15.

Finkle, B. J., Ulrich, J. M., Rains, D. W., Tisserat, B. B. & Schaeffer, G. W. (1979). Survival of alfalfa, rice and date-palm callus after liquid nitrogen freezing. (Abstract.) *Cryobiology*, **16**, 583.

Finkle, B. J., Ulrich, J. M., Schaeffer, G. W. & Sharpe, F. (1983). Cryopreservation of cells of rice and other plant species. In *Potentials of Cell and Tissue Culture Techniques in the Improvement of Cereal Plants*. Science Press, Beijing: Academia Sinica & International Rice Research Institute, in press.

Grout, B. W. W. (1979). Low temperature storage of imbibed tomato seeds: a model for recalcitrant seed storage. *Cryo Letters*, **1**, 71–6.

Grout, B. W. W. & Henshaw, G. G. (1978). Freeze-preservation of potato shoot-tip cultures. *Annals of Botany*, **42**, 1227–9.

Grout, B. W. W., Westcott, R. J. & Henshaw, G. G. (1978). Survival of shoot-meristems of tomato seedlings frozen in liquid nitrogen. *Cryobiology*, **15**, 478–83.

Haskins, R. H. & Kartha, K. K. (1980). Freeze-preservation of pea meristems: Cell survival. *Canadian Journal of Botany*, **58**, 833–40.

Henshaw, G. G., Stamp, J. A. & Westcott, R. J. (1980). Tissue culture and germplasm storage. In *Developments in Plant Biology, Vol. 5. Plant Cell Cultures: Results and Perspectives*, ed. F. Sala, B. Parisi, R. Cella & O. Cifferi, pp. 277–82. Amsterdam, New York and Oxford: Elsevier North Holland.

Jones, L. H. (1974). Long term survival of embryos of carrot (*Daucus carota* L.) *Plant Science Letters*, **2**, 221–4.

Kartha, K. K., Leung, N. L. & Gamborg, O. L. (1979). Freeze-preservation of pea meristems in liquid nitrogen and subsequent plant regeneration. *Plant Science Letters*, **15**, 7–15.

Kartha, K. K., Leung, N. L. & Pahl, K. (1980). Cryopreservation of strawberry meristems and mass propagation of plantlets, *Journal of the American Society for Horticultural Science*, **105**, 481–4.

King, M. W. & Roberts, E. H. (1979). *The Storage of Recalcitrant Seeds – Achievements and Possible Approaches.* International Board of Plant Genetic Resources (IBPGR) Report, AGP: IBPGR/79/44. Rome: IBPGR Secretariat.

King, P. J., Potrykus, I. & Thomas, E. (1978). *In vitro* genetics of cereals: problems and perspectives. *Physiologie Végétale*, **16**, 381–99.

Larkin, P. J. & Scowcroft, W. R. (1981). Somaclonal variation – a novel source of variability from cell cultures. *Theoretical and Applied Genetics*, **60**, 197–214.

Levin, R. L., Ferguson, J. R., Dowgert, M. F. & Steponkus, P. L. (1979). Cryobiology of isolated plant protoplasts: I. Thermodynamic considerations (Abstract). *Cryobiology*, **16**, 592.

Mazur, R. A. & Hartmann, J. X. (1979). Freezing of plant protoplasts. In *Plant Cell and Tissue Culture: Principles and Applications*, ed. W. R. Sharp, P. O. Larsen, E. F. Paddock & V. Raghavan, p. 876. Columbus: Ohio State University Press.

Meryman, H. T. & Williams, R. J. (1982). The mechanisms of freezing injury and natural tolerance, and the principles of artificial cryopreservation. In *Crop Genetic Resources – The Conservation of Difficult Material*, ed. L. A. Withers & J. T. Williams, pp. 5–37. Paris: International Union of Biological Sciences, IBPGR.

Nag, K. K. & Street, H. E. (1973). Carrot embryogenesis from frozen cultured cells. *Nature*, **245**, 270–2.

Nitzsche, W. (1978). Erhaltung der Lebensfahigkeit in getrocknetern Kallus. *Zeitschrift für Pflanzenphysiologie*, **87**, 469–72.

Novak, F. J. (1974). The changes of karyotype in callus cultures of *Allium sativum* L. *Caryologia*, **27**, 45–54.

Novak,F. J. (1980). Phenotype and cytological status of plants regenerated from callus cultures of *Allium sativum* L. *Zeitschrift für Pflanzenzüchtung*, **84**, 250–60.

Polge, C. (1978). Embryo transfer and embryo preservation. *Symposia of the Zoological Society of London*, **43**, 303–16.

Potrykus, I. (1978). Discussion comment. In *Production of Natural Compounds by Cell Culture Methods*, ed. A. W. Alfermann & E. Reinhard. Munich: Gesellschaft für Strahlen- und Umweltforschung mbH.

Pritchard, H. W., Grout, B. W. W., Reid, D. S. & Short, K. C. (1982). The effect of growth under water stress on the structure, metabolism and cryopreservation of cultured sycamore cells. In *Biophysics of Water*, ed. F. Franks & S. F. Mathias, pp. 315–18. Chichester: John Wiley.

Quatrano, R. S. (1968). Freeze-preservation of cultured flax cells using DMSO. *Plant Physiology*, **43**, 2057–61.

Roberts, E. H. (1975). Problems of long-term storage of seed and pollen for genetic resources conservation. In *Crop Genetic Resources for Today and Tomorrow*, ed. O. H. Frankel & J. G. Hawkes, pp. 269–96. Cambridge: Cambridge University Press.

Roberts, E. H. & King, M. W. (1982). Storage of recalcitrant seeds. In *Crop Genetic Resources – the Conservation of Difficult Materials*, ed. L. A. Withers & J. T. Williams, pp. 39–48. Paris: International Union of Biolgical Sciences, IBPGR.

Sakai, A. & Noshiro, M. (1975). Some factors contributing to the survival of crop seeds cooled to the temperature of liquid nitrogen. In *Crop Genetic Resources for Today and Tomorrow*, ed. O. H. Frankel & J. G. Hawkes, pp. 317–26. Cambridge: Cambridge University Press.

Sakai, A. & Sugawara, Y. (1973). Survival of Poplar callus at super-low temperatures after cold acclimation. *Plant and Cell Physiology*, **14**, 1201–4.

Sakai, A., Yamakawa, M., Sakato, D., Harada, T. & Yakuwa, T. (1978). Development of a whole plant from an excised strawberry runner apex frozen to −196 °C. *Low Temperature Science, Series B*, **36**, 31–8.

Sala, F., Cella, R. & Rollo, F. (1979). Freeze-preservation of rice cells. *Physiologia Plantarum*, **45**, 170–6.

Seibert, M. (1976). Shoot initiation from carnation shoot apices frozen to −196 °C. *Science*, **191**, 1178–9.

Seibert, M. & Wetherbee, P. J. (1977). Increased survival and differentiation of frozen herbaceous plant organ cultures through cold treatment. *Plant Physiology*, **59**, 1043–6.

Singh, J. (1979). Freezing of protoplasts isolated from cold-hardened and non-hardened winter rye. *Plant Science Letters*, **16**, 195–201.

Takeuchi, M., Matsushima, H. & Sugawara, Y. (1980). Long term freeze-preservation of protoplasts of carrot and *Marchantia*. *Cryo Letters*, **1**, 519–24.

Tisserat, B., Ulrich, J. M. & Finkle, B. J. (1981). Cryogenic preservation and regeneration of date-palm tissue. *HortScience*, **16**, 47–8.

Towill, L. E. (1981). *Solanum etuberosum*: a model for studying the cryobiology of shoot-tips of the tuber bearing *Solanum* species. *Plant Science Letters*, **20**, 315–24.

Uemura, M. & Sakai, A. (1980). Survival of carnation (*Dianthus caryophyllus* L.) shoot apices frozen to the temperature of liquid nitrogen. *Plant and Cell Physiology*, **21**, 85–94.

Ulrich, J. M., Finkle, B. J., Moore, P. H. & Ginoza, H. (1979). Effect of a mixture of cryoprotectants in attaining liquid nitrogen survival of callus cultures of a tropical plant. *Cryobiology*, **16**, 550–6.

Weatherhead, M. A., Grout, B. W. W. & Henshaw, G. G. (1979). Advantages of storage of potato pollen in liquid nitrogen. *Potato Research*, **21**, 97–100.

Whittingham, D. G. (1980). Principles of embryo preservation. In *Low Temperature Preservation in Medicine and Biology*, ed. M. J. Ashwood-Smith & J. Farrant, pp. 65–83. Tunbridge Wells: Pitman Medical.

Wiest, S. C. & Steponkus, P. L. (1978). Freeze-thaw injury to isolated spinach protoplasts and its simulation at above freezing temperature. *Plant Physiology*, **62**, 699–705.

Withers, L. A. (1978*a*). Freeze-preservation of synchronously dividing cultured cells of *Acer pseudoplatanus* L. *Cryobiology*, **15**, 87–92.

Withers, L. A. (1978*b*). A fine-structural study of the freeze-preservation of plant tissue cultures. II. The thawed state. *Protoplasma*, **94**, 235–47.

Withers, L. A. (1978*c*). Freeze-preservation of cultured cells and tissues. In *Frontiers of Plant Tissue Culture*, ed. T. A. Thorpe, pp. 297–306. Calgary: Calgary University.

Withers, L. A. (1979). Freeze-preservation of somatic embryos and clonal plantlets of carrot (*Daucus carota* L.). *Plant Physiology*, **63**, 460–7.

Withers, L. A. (1980*a*). Low temperature storage of plant tissue cultures. In *Advances in Biochemical Engineering, Plant Cell Cultures II*, ed. A. Fiechter, vol. 18, pp. 102–50. Berlin: Springer.

Withers, L. A. (1980*b*). Preservation of germplasm. In *International Review of Cytology, Supplement 11B, Perspectives in Plant Cell and Tissue Culture*, ed. I. K. Vasil, pp. 101–36. New York: Academic Press.

Withers, L. A. (1980*c*). *Tissue Culture Storage for Genetic Conservation*, IBPGR Technical Report, AGP: IBPGR/80/8. Rome: IBPGR Secretariat.

Withers, L. A. (1980*d*). The cryopreservation of higher plant tissue and cell cultures – an overview with some current observations and future thoughts. *Cryo Letters*, **1**, 239–50.

Withers, L. A. (1981). *Institutes Working on Tissue Culture for Genetic Conservation.* IBPGR Consultant Report, AGP: IBPGR/81/30. Rome: IBPGR Secretariat.

Withers, L. A. (1982*a*). Storage of plant tissue cultures. In *Crop Genetic Resources – The Conservation of Difficult Material*, ed. L. A. Withers & J. T. Williams, pp. 49–82. Paris: International Union of Biological Sciences, IBPGR.

Withers, L. A. (1982*b*). *Institutes Working on Tissue Culture for Genetic Conservation* (revised edition). IBPGR Consultant Report, AGP: IBPGR/82/30. Rome: IBPGR Secretariat.

Withers, L. A. (1982*c*). The development of cryopreservation techniques for plant cell, tissue and organ cultures. In *Plant Tissue Culture 1982*, ed. A. Fujiwara, pp. 793–4. Tokyo: Japanese Association for Plant Tissue Culture.

Withers, L. A. & King, P. J. (1979). Proline – a novel cryoprotectant for the freeze-preservation of cultured cells of *Zea mays* L. *Plant Physiology*, **64**, 657–78.

Withers, L. A. & King, P. J. (1980). A simple freezing-unit and routine cryopreservation method for plant cell cultures. *Cryo Letters*, **1**, 213–20.

Withers, L. A. & Street, H. W. (1977*a*). The freeze-preservation of cultured plant cells: III. The pregrowth phase. *Physiologia Plantarum*, **39**, 171–8.

Withers, L. A. & Street, H. E. (1977*b*). Freeze-preservation of plant cell cultures. In *Plant Tissue Culture and its Bio-technological Application*, ed. W. Barz, E. Reinhard & M.-H. Zenk, pp. 226–44. Berlin: Springer.

Note added in proof

A number of recent publications are noted below since they extend the range of species to which cryopreservation has been applied successfully (Binder & Zaerr, 1980*a*, *b*; Kartha, 1982; Maddox, Gonsalves & Shields, 1983), support the feasibility of cryopreserving protoplasts (Hauptmann & Widholm, 1982; Takeuchi, Matsushima & Sugawara, 1982) and examine the stability of variant cell lines through a cryopreservation cycle (Hauptmann & Widholm, 1982).

Binder, W. & Zaerr, J. B. (1980*a*). Freeze preservation of suspension cultured cells of a gymnosperm, Douglas-fir. *Cryobiology*, **17**, 624.

Binder, W. & Zaerr, J. B. (1980*b*). Freeze preservation of suspension cultured cells of a hardwood poplar. *Cryobiology*, **17**, 624–5.

Hauptmann, R. M. & Widholm, J. M. (1982). Cryostorage of cloned amino acid analog-resistant carrot and tobacco suspension cultures. *Plant Physiology*, **70**, 30–4.

Kartha, K. K. (1982). Cryopreservation of periwinkle, *Catharanthus roseus* cells cultured *in vitro*. *Plant Cell Reports*, **1**, 135–8.

Maddox, A. D., Gonsalves, F. & Shields, R. (1983). Successful preservation of suspension cultures of three *Nicotiana* species at the temperature of liquid nitrogen. *Plant Science Letters*, **28**, 157–62.

Takeuchi, M., Matsushima, H. & Sugawara, Y. (1982). Totipotency and viability of protoplasts after long-term freeze preservation. In *Plant Tissue Culture 1982*, ed. A. Fujiwara, pp. 797–8. Tokyo: Japanese Association for Plant Tissue Culture.

G. G. HENSHAW and J. F. O'HARA

In vitro approaches to the conservation and utilisation of global plant genetic resources

Introduction

It has been stated that many highly successful plant breeding programmes are liable to enter a period of decline after 50 to 70 years unless deliberate steps are taken to maintain a flow of genes into the breeding population from other sources (Simmonds, 1979). Locally-adapted ancient varieties and their wild relatives have been the traditional sources of breeding material and, until fairly recently, they have been preserved in primitive agricultural systems around the world as well as in their natural habitats. Within the last 20 years it has been recognised that this supply is becoming supplanted by highly-bred modern varieties that have been produced by plant breeders working in association with internationally coordinated programmes of agricultural improvement deemed necessary to provide the increasing world population with an adequate food supply (Frankel & Bennett, 1970; Frankel & Hawkes, 1975). In addition, changes in natural habitats have led to losses of locally-adapted populations of the wild ancestral species.

It is now recognised that there is some urgency for preserving ancient strains for future use. There are two possible approaches to the problem of how these genetic resources might be conserved on a long-term basis: either there should be a deliberate attempt to conserve, *in situ*, examples of relevant primitive agricultural systems or natural habitats, or there should be some means of conserving the material *ex situ* in gene banks. Either way, there can be formidable political, economic and technical problems, but the seriousness of the situation has led to the development over the last 20 years of an internationally coordinated programme for the conservation and utilisation of plant genetic resources. The initiative for this programme came from the Food and Agriculture Organisation of the United Nations (FAO) which in the early 1960s set up panels of experts to advise on the problem, and in 1974 the International Board for Plant Genetic Resources (IBPGR) was established with the mandate '...to create and coordinate a worldwide collaborative network for genetic resource conservation and to mobilise financial support for such a programme' (Anon., 1975). IBPGR is one of the family of international research boards supported by the Consultative Group on International Agricultural Research (CGIAR) which is sponsored by the FAO, the World Bank and the United Nations Development Programme (UNDP). IBPGR contains members from more than a dozen countries and has a defined function to further the collection, conservation, documentation, evaluation and use of germplasm.

Germplasm conservation and utilisation

Since for any one economically important species many hundreds of genotypes may exist, the burden of conservation is considerable. While *in situ* conservation policies clearly have an important role to play in the overall strategy (Jain, 1975), especially with regard to forest species, it is quite obvious that a complex network of *ex situ* collections will always make a major contribution. Coordination of such a network on an international scale must balance a number of conflicting interests including considerations of both the type and location of the collections and the nature of the germplasm to be conserved.

Existing collections of germplasm are diverse. They range from those relatively small working collections assembled by plant breeders for their own immediate needs, which are not necessarily intended for long-term conservation, to the large-scale conservation collections comprising whole ranges of genotypes, including those with no immediately discernible value. The present need is to ensure that the critical material is at least represented in conservation collections and that it is freely available to satisfy the more specific requirements of the working collections. The plant germplasm collections must therefore be dynamic; regular international exchange of material is both desirable and inevitable.

A strong case can be made for collections to be located in the vicinity of the regions in which the material is collected, so that at times of seed stock regeneration or growth of vegetative stocks in the field, selection pressures are least likely to shift the balance of genes within the population. By implication, the collections would be located in the regions of greatest genetic diversity for any particular crop and, moreover, since these regions tend to be centres for several crop species, they would generally have multiple crop responsibilities. On the other hand, there is the important argument that the conservation of genetic resources must not be an end in itself and that, ultimately, proper exploitation depends upon the germplasm being maintained and evaluated by centres specialising in the breeding of particular species. This means that collections are likely to be duplicated in regional and crop-specific centres; again, international exchange is implicit. Duplication is also desirable as an insurance against loss of unique germplasm.

The long-term storage of germplasm as seed is now technologically feasible for the majority of crop species (Roberts, 1972; 1975) to the extent that the IBPGR has been able to specify recommended storage conditions and seed-bank management protocols (Ellis, Roberts & Whitehead, 1980). However, the methodology cannot be applied to those species whose seeds are 'recalcitrant' in the sense that they cannot withstand long storage periods (King & Roberts, 1979), or to others which do not even produce viable seed. In these cases, it is essential that satisfactory methods are available for the long-term preservation of germplasm in a vegetative form. Clonal maintenance can also be of importance with species producing 'orthodox' seeds, especially those which are out-breeders, where there is the danger of recombination disrupting valuable gene complexes that have been assembled by coadaptive processes possibly operating over thousands of years. Thus, clonal maintenance has an important subsidiary role in overall genetic conservation strategy, but it undoubtedly presents some of the greatest technological problems. The long-term maintenance of a collection of vegetative material involves more constant attention than does low-temperature storage of a seed collection. In addition, there are plant health considerations; losses of material through disease must be avoided, and, further, the material must be acceptable for international distribution so that it can be incorporated into breeding programmes. This last aspect is important in view of the fact that international regulations can be particularly stringent concerning the transfer of vegetative plant material (Kahn, 1977; Roca, Bryan & Roca, 1979).

Different types of vegetative material present different problems. Those species producing recalcitrant seeds are predominantly long-lived woody perennials (see King and Roberts, 1979, for a list of these) which are relatively well suited to long-term conservation in field plantations. However, they do require large areas of land and regular tending. Herbaceous perennial species generally demand a fair amount of attention, especially when perennation involves the annual storage of perishable vegetative propagules. Crop species vary in their vulnerability to plant health problems, but none can be assumed to be immune and some are severely affected. It can be seen, therefore, that conventional methods for the conservation of vegetative germplasm may be expensive and of limited value if plant health considerations restrict the utilisation of the stored material. It is particularly in this area that an alternative approach, using in vitro techniques, is proving invaluable.

Factors influencing the in vitro approach to genetic conservation

Before an in vitro system can be adopted as an aid to the conservation and utilisation of plant genetic resources, it must satisfy the criteria which are discussed below. Neglect of any of these aspects can only undermine any advantages gained over conventional methods, and while certain in vitro systems apparently do meet all of these requirements, it is important to define the conditions under which they might be employed.

Genetic stability

The maintenance of an acceptable level of genetic stability is the *sine qua non* for any germplasm storage system. Unfortunately, however, the standards for this are not easily defined since absolute stability is unattainable as long as cell division is continuing and there is as yet no practical means of describing the total genotype. The working standards defined by the IBPGR for seed storage are pragmatically based on the observation that there is a correlation between loss of viability and the incidence of both chromosome aberrations in the seed and mutant phenotypes in succeeding generations (Abdalla & Roberts, 1968, 1969). The standards have been defined in terms of seed viability and it is recommended that seed stock be regenerated as soon as a significant reduction in percentage viability is evident (Anon., 1976).

Imperfect though these standards might be, they are superior to any employed for vegetatively-propagated germplasm, where there is usually complete reliance on the subjective roguing of off-types. It is important to note that some genetic variation must inevitably arise within clones propagated by conventional means and that this is tolerated and coped with within certain limits. This has implications for the sampling procedures both prior to and during the conservation process. When seed stocks are replenished, a population size of 50 plants is recommended to produce the next supply of seed in order to minimise genetic drift yet allow full expression of variation. Although clonal propagation by definition implies genetic homogeneity among the propagules, it would be dangerous to replenish stocks from too small a group of plants since any bud-mutations or chimeras would have a strong chance of being perpetuated.

It is against this rather unsatisfactory background concerning the question of genetic stability that the suitability of in vitro methods for genetic conservation must be assessed. It is known from chromosome studies (D'Amato, 1975, 1978) that certain in vitro systems can be particularly unstable, especially the less organised tissue and cell cultures. Whether the mutation rate in these cultures is necessarily greater than that in the somatic tissues of normal plants is a debatable point, but the consequences are likely to be more serious since new meristems can develop adventitiously from the mutant cells whereas diplontic selection (i.e. selection among diploid cells) in vivo will tend to exclude such mutant cells from the non-adventitious shoot meristems which are generally used for vegetative propagation.

Genetic instability in disorganised cell and tissue cultures may be a product of three factors: genetic variability among cells of the explant, the mutagenic effect of the in vitro culture conditions and in vitro selection pressures. The relative contributions of these factors will vary with the plant species and the actual culture conditions. At present, the prospects are poor for making substantial improvements where the stability of a growing culture is in question. Cryopreservation techniques, which involve storage at ultra-low temperatures where all intracellular biological and chemical processes cease, offer the possibility of long-term stabilisation of this type of culture (Withers, 1980 and this volume), but even so the system would be vulnerable to change during the periods of active growth before and after storage.

Thus, organised cultures in which multiplication is based on the production of axillary shoot meristems are generally more appropriate for the long-term genetic conservation of plant germplasm. With the present state of knowledge, it seems wise to rely on natural mechanisms which maintain stability both within shoot meristems and from one non-adventitious meristem to another. Where variants do arise among populations of vegetatively-propagated plants, understanding of the phenomenon is so poor that it is not always clear whether the cause is genetic or epigenetic, i.e. under nuclear or extranuclear control, respectively. Whatever the explanation might be, it must be emphasised that sampling procedures, both in vivo and in vitro, should be carefully controlled when vegetative propagation is employed for genetic conservation over very long periods.

Culture initiation

To facilitate inclusion of a species in a storage programme, initiation of cultures must be possible from a wide range of genotypes by means of a reasonably uniform procedure. The initiation of shoot-tip cultures seems to present few problems with the herbaceous perennial dicotyledons; many species respond well to one or other of the standard basal media, and frequently without any hormone supplement. The response of herbaceous monocotyledonous species is more variable and, of necessity, rather more attention has been paid to methods involving the production of adventitious buds. It might be assumed from the fairly widespread use of the latter techniques for the commercial multiplication of ornamental species that good levels of genetic stability are maintained, as has been claimed for *Lilium* (Sheridan, 1968) and *Freesia* (Davies, 1972). However, as it is sometimes the practice to replace stock cultures at regular intervals it can also be assumed that problems concerning trueness-to-type can be encountered, making it very unlikely that such cultures would be suitable for genetic conservation purposes. Again, with woody species more interest has centred around the production of adventitious buds, leaving the organ culture technique relatively unexploited, especially with the forest trees (Bonga, 1974). In principle, the response of woody species to in vitro techniques seems to be little different to that from other species, except that the juvenile/adult phase change is an added complication and difficulties can be encountered in establishing cultures from mature buds (Jones, this volume). In many woody species the evaluation of characters that are useful for breeding programmes can only be completed at the adult phase, entailing a lengthy waiting period. This is obviously a great handicap, particularly if the species concerned also produce recalcitrant seed. A shoot-tip culture technique for these species could provide a valuable alternative to the otherwise inevitable field plantation for the purpose of genetic conservation.

Plant regeneration

In vitro systems from which plants cannot be regenerated are of little interest for genetic conservation, at least when conventional methods of plant breeding are contemplated. Disorganised cultures present the greatest plant regeneration problems: in some species regeneration has yet to be achieved using cultures derived from mature tissues, in others (including several crops, e.g. many cereals, grain legumes, cassava and potato) it has only been achieved with specific genotypes (Thomas & Wernicke, 1978). Even in those species from which plantlets can be regenerated, the capacity for prolonged regeneration may be difficult to guarantee for reasons which could be either genetic or physiological. Assuming a non-genetic explanation, the loss of regenerative capacity could be due either to a general loss of morphogenetic competence in the cells or to the in vitro conditions giving a selective advantage to non-competent cells. Either way, it will probably be difficult to prevent this process in growing cultures showing this tendency, except by very careful monitoring of the conditions, to an extent that would probably be too expensive for genetic conservation purposes. Alternatively, the regenerative capacity of a species may be more stable in a non-growing culture where the chances of selection are reduced. This approach would be feasible if suitable cryogenic techniques (see below) were to prove successful with a wide range of species.

In contrast to these difficulties which can be encountered with disorganised cultures, plant regeneration problems rarely occur with shoot-tip cultures once they are established, even when they are maintained over long periods of time.

Culture storage

For economic reasons alone, it is desirable that cultures should require minimal attention whilst in store. Storage in a non-growing state of 'suspended animation' would offer considerable advantages in terms of maintenance costs and the possible reduction of mutation rates.

Cryopreservation in liquid nitrogen provides a means of attaining this state, but the techniques are not yet sufficiently developed for universal application with organised cultures, although more success has been achieved with disorganised cell and tissue cultures (see Withers, this volume). Cryopreservation techniques generally involve several basic stages. First the organ to be frozen is excised, then the cells are treated with a cryoprotectant to aid preservation of the components during exposure to low temperatures. After this, the organ is taken through the freezing stage, which may be carried out at different controlled rates, down to the temperature of liquid nitrogen (−196 °C). Finally, the organ is stored in a liquid nitrogen refrigerator. One of the most critical factors for cell survival during freezing to −196 °C is water content, since ice formation is one of the major causes of irreparable cellular damage. In a multicellular plant organ such as a shoot-tip, the water content will obviously vary according to the cell type and the state of differentiation. The rationale behind most freezing protocols is the avoidance of ice damage. This can be achieved either by reducing the amount of available water, by partially dehydrating the cells before immersion in liquid nitrogen (e.g. by using an initial slow cooling rate), or by freezing the cells ultra-rapidly (e.g. by direct immersion in liquid nitrogen) so that the water vitrifies because it has insufficient time to form nucleation points for ice crystals. It is becoming apparent from the diversity of freezing protocols used for successful cryopreservation (see Kartha, 1981, for a summary of these) that each species will demand particular treatments. It is to be hoped that it will be possible to formulate a generally applicable method for the freeze-preservation of shoot-tips which will give adequate levels of survival and permit the conservation of a wide range of crop species. At the temperature of liquid nitrogen, all metabolic physical and chemical processes cease, and it is theoretically possible to store deep-frozen material indefinitely provided that the storage conditions are strictly maintained.

A practical alternative to a state of non-growth is the manipulation of culture conditions so that only an extremely slow rate of growth is permitted. Inevitably, such 'minimal-growth' procedures involve regular attention to the cultures for the purpose of replenishing the medium supply, albeit at considerably extended intervals, usually in excess of one year. The application of minimal-growth techniques can involve shifting the culture conditions away from optimal, so that the growth rate is limited by factors such as lowered temperature, reduced nutrient supply or increased osmolarity. Alternatively, growth may be inhibited by agents such as growth regulators. Once minimal growth has been established and the subculture interval subsequently lengthened, the limiting factors tend to be physical, e.g. the rate of the evaporation of the medium. Careful attention must be paid to the type of storage vessel and the mode of closure used. One advantage of minimal-growth storage over that of zero-growth storage is that the material continues to grow, and thus portions taken for evaluation or breeding purposes will be replaced by the part left in storage. However, there is a greater risk of genetic drift, although this should be no greater than that associated with conventional methods of seasonal planting.

Multiplication rates

After the storage period, adequate plant multiplication rates should be attainable both to replenish stocks and to provide breeding material when required. Generally, much higher multiplication rates can be obtained by in vitro methods than by conventional propagation. Often, the theoretical rates calculated by extrapolation from relatively short-term in vitro experiments are so high that it is logistic factors which are most likely to become rate limiting. The multiplication rates that can be achieved by 'non-adventitious' methods are generally lower than those that can be achieved by 'adventitious' methods (see Hussey, this volume), and although they might then be too low to be considered for commercial multiplication, they are likely to be more than adequate for any multiplication that is associated with replenishment and utilisation of the germplasm stocks. A particular advantage of in vitro propagation is the independence from seasonal restrictions: stocks can be built up at any time of the year, and dormancy problems do not usually arise.

Elimination and exclusion of pests and pathogens

Ideally, germplasm should be free of pests and pathogens at least to the extent that stocks are not lost and international distribution is not restricted. With many, but not all, species these standards are readily achieved by the use of suitably treated and stored seeds. Vegetatively-propagated germplasm presents special problems because of exposure to pests and pathogens and the risk of systemic contaminants being transmitted from generation to generation. Some seed-propagated species can also present problems when they are known to harbour seed-transmitted pathogens (Bos, 1977; Neergard, 1977; Hewett, 1979). With such germplasm, satisfactory storage standards are only really achieved if 'clean' plants are grown under protected conditions which exclude important pests and pathogens. However, it is not always easy to achieve these standards by conventional methods and the procedure can be very expensive without being entirely reliable. On the other hand, the elimination and exclusion of contaminants can be regarded as an integral part of the standard in vitro process. In practice, however, the normal precautions involving the surface decontamination of explants and the rejection of obviously contaminated cultures are only likely to eliminate organisms which induce effects that are very apparent; those organisms which have a more balanced relation with the plant cells can escape detection. These include important systemic pathogens such as viruses, some bacteria and mycoplasmas.

Standard in vitro procedures do not, therefore, guarantee the production of plants which are free from pests and pathogens, but, used with other precautions, they can often provide the most effective means of producing such material. This is especially so for those species in which the seed stage cannot be used as a filter to prevent the transmission of contaminants between generations. The extra precautions usually involve the isolation of explants from those parts of the plant most likely to be contaminant-free (in practice the shoot meristematic dome plus one or two leaf-primordia). Pretreatment of the donor plant, for example, by heat therapy increases the chances of obtaining 'clean' material, and further treatments such as the use of high and low temperatures, or of antibiotics and anti-viral agents may be applied in vitro. Even so, the use of a satisfactory indexing procedure for the detection of contaminants is crucial and it may need to be more thorough than is often appreciated: some contaminants can only be detected in the mature plant, in which case it is necessary to test regenerated plants and then to reisolate cultures from those indexing negatively.

Once a culture has been declared free of known pests and pathogens, the standard aseptic procedures used with in vitro systems should prevent recontamination.

International distribution

Vegetative plant material and some seeds present particular phytosanitary problems with regard to international distribution. Apart from other considerations, the sheer bulk of vegetative material could render a quarantine procedure either ineffective or uneconomic. On these grounds alone plant tissue cultures are an attractive alternative means of distributing vegetative material (Kahn, 1977). As the guaranteed quality of the in vitro material is only as good as the indexing procedure employed, the use of such material certainly does not eliminate the need for adequate quarantine arrangements. With experience, however, confidence can be expected to increase and the quarantine procedure might well be simplified. It is in this area that use of in vitro techniques has the major advantage over more conventional methods.

Economic acceptability of in vitro storage

A detailed study has not yet been made of how in vitro methods and conventional methods compare from an economic point of view. The novelty of in vitro techniques for storage inevitably implies a need for personnel training. However, the aseptic manipulations involved are very similar to the shoot-tip dissection and culture which have long been used in virus-elimination procedures. The associated equipment required, including constant-temperature rooms and laboratories equipped for aseptic working conditions, may necessitate an initially high capital outlay. This is essential for the efficient operation of in vitro systems and the cost may be compared favourably with that of the sophisticated greenhouse and storage facilities employed in seed banks, where plants are grown under conditions in which they are protected from pests and pathogens.

However, if there were no plant health considerations affecting the vegetative storage of germplasm, the in vitro techniques could well prove to be uneconomic in comparison with conventional methods. Yet the serious implications of phytosanitary problems must favour these techniques, especially since the standard procedure of shoot-tip culture can be advantageously employed for rapid multiplication, disease elimination and exclusion, and international distribution, as well as for the actual storage. Support for the economic acceptability of in vitro techniques comes from the fact that at least two major international centres, i.e. Centro Internacional de la Papa (CIP) and Centro Internacional de Agricultura Tropical (CIAT), now routinely use these methods for the storage of germplasm. Cryopreservation techniques for the time being remain an exception, since they are proving expensive in terms of the initial research and development that is needed before the methods can begin to be implemented.

Some applications of in vitro storage techniques

Several facets of plant germplasm conservation and utilisation have been considered above in general terms and can be further illustrated by reference to four major crop species: potato, cassava, sweet potato and yam. These are all tuber crops which are normally propagated vegetatively and hence have some requirement for germplasm conservation in clonal form. It is useful to compare these species because they are taxonomically diverse, representing as they do one monocotyledonous and three dicotyledonous families, namely Dioscoreaceae, Solanaceae, Euphorbiaceae and Convolvulaceae. These families differ with respect to the methods conventionally used by man to propagate them; they exhibit varying degrees of seed sterility and they include both temperate and tropical species. The success achieved in using in vitro storage techniques for these diverse crops should indicate the feasibility of extending the procedures employed to other crop species.

Potato (Solanum tuberosum)

As the most important tuber crop on a global scale, 'potato' includes a range of related species and varieties among which several ploidy levels are represented (Hawkes, 1978). The potatoes of northern temperate countries have been developed from a narrow genetic base and are mostly tetraploid. Although largely self-compatible, they tend to be intolerant of inbreeding and rather infertile. It is important to conserve the wild, predominantly diploid South American progenitorial species as these are a source of new genes for breeding programmes (Glendinning, 1979). Potato is conventionally propagated by tubers, which are used for planting on a commercial scale, and also by stem cuttings. In vitro micropropagation has been suggested as a rapid and convenient alternative for multiplying stock plants under disease-free conditions (Hussey & Stacey, 1981).

While nodal explants with intact axillary buds can be grown up to plantlets on basal MS medium (Murashige & Skoog, 1962) without growth hormones, isolated shoot-tips (the dome plus up to four leaf primordia) can frequently be established more readily with a hormone supplement (Westcott, Henshaw & Roca, 1977), although the requirements may differ according to the genotype (Westcott, Henshaw, Grout & Roca, 1977). In these studies by Westcott *et al.*, cytokinin elicited the most uniform response but often gave rise to multiple shoots which were produced from callus at the base of the explant. This introduced the risk of genetic change or the disruption of otherwise stable chimeras which regularly occur among potato varieties (Howard, 1970). However, a comparison (by electrophoresis) of tuber proteins from second-generation tubers of plants that had been regenerated either via shoot-tip cultures or from basal callus, with tuber proteins from virus-free plants showed neither detectable changes in band pattern nor any significant differences in phenotype (Roca, Espinoza, Roca & Bryan, 1978). A study comparing plants regenerated directly from shoot-tip cultures with plants grown from tubers did disclose significant morphological differences, as might be expected, but again there were no differences in tuber proteins. This suggested that the plants were genetically, if not phenotypically, true-to-type (Denton, Westcott & Ford-Lloyd, 1977).

The response of shoot-tip cultures from different potato genotypes to minimal growth conditions has been studied in some detail (Westcott, 1981*a*). Although there were differences between the cultivars tested, it was generally possible to increase the passage length to at least one year by adopting a regime of alternating low temperatures (12 °C during the day and 6 °C during the night). A steady temperature of 10 °C together with an increased sucrose supply (8%) had a similar effect. Other successful approaches included the use of growth inhibitors such as abscisic acid (5.0–10.0 ml l^{-1}) or a medium with increased osmolarity produced by the addition of 3%

mannitol (Westcott, 1981*b*). It seems, therefore that there are several ways of achieving the same objectives of reduced growth rates and increased subculture intervals (Henshaw, O'Hara & Westcott, 1980) which should permit evaluation of the most economical conditions for routine long-term storage under a particular set of circumstances.

Cryopreservation, which theoretically should permit storage for possibly decades with only minimal maintenance, seems feasible for potato germplasm. One successful method involves the use of dimethylsulphoxide (DMSO) as cryoprotectant followed by ultra-rapid freezing achieved by plunging the sample into liquid nitrogen (Grout & Henshaw, 1978; O'Hara & Henshaw, 1980). In another study (Bajaj, 1981), a similar method which used as cryoprotectant a mixture of DMSO, sucrose and glycerol gave comparable levels of survival (a maximum of 27% survival). These ultra-fast freezing methods are suitable for some potato genotypes, but perhaps a more generally applicable procedure may come from the two-step method. Towill (1981) reported that shoot-tips which were cooled slowly to −40 °C before immersion in liquid nitrogen had survival levels of around 70%.

The general points to emerge from these studies are that survival rates are variable according to the cultivar and that many other factors can affect the response, especially those which influence the physiological state of the shoot-tip both before freezing and after thawing. Thus the present protocols for the cryopreservation of potato shoot-tips are not thought to be optimal and many refinements need to be carried out (Henshaw *et al.*, 1980) before it can be considered for routine storage. One possible problem is the fact that not all the cells of each shoot-tip survive and regeneration of thawed plants often occurs via adventitious buds originating from clumps of viable cells (Grout & Henshaw, 1980).

No complete investigation has yet been made into the genetic stability of stored material. In one study of three diploid varieties, *Solanum goniocalyx* plus two cultivars of *S. stenotomum*, which had been stored under minimal growth conditions for 34 months without disturbance, root-tip squashes revealed over 90% of the cells to be normal diploids (the remainder being plus or minus one chromosome). Tuber proteins were also investigated using eight genotypes from seven species derived from cultivars that had been stored for up to nine months, and these were found to be more or less indistinguishable from proteins of the parental plants (R. J. Westcott, personal communication). Tests are continuing with more general agronomic characters. In looking for variation, it is important to compare like with like, i.e. plants from pathogen-free stored material must have undergone a complete growth cycle after retrieval from the in vitro system before comparison is made with pathogen-free non-stored plants of the same clone grown under identical conditions.

Use is now being made of in vitro culture to distribute potato germplasm internationally. This has been particularly valuable where the aseptic techniques have first been used for virus elimination and hence have enabled 'clean' propagules to be disseminated without the danger of reinfestation. CIP alone has exchanged such material with more than 12 quarantine authorities, and found that it expedited the quarantine period (Roca *et al.*, 1978).

At CIP, in vitro technology is beginning to be used routinely for virus elimination, maintenance of pathogen-tested accessions in long- and short-term storage, maintenance of the breeders' collection in long-term storage (at 22 °C) and shipment of material for international distribution (L. Schilde, personal communication).

Cassava (Manihot esculenta)

Cassava is a perennial shrub with tuberous roots. It no longer exists in the wild state but is grown as a crop in the lowland tropics. It is fertile but the seeds are highly heterozygous and propagation is usually carried out by stem cuttings. Much effort has been put into initiating cultures from cassava and in vitro methods are now being exploited. Several Latin-American and Asian countries have developed commercial varieties from material selected in the breeding programmes of the main centre in Columbia (CIAT) and the exchange of this material has been greatly facilitated by the use of in vitro meristem culture (Anon., 1981).

Similarly to sweet potato and yam, cassava shoot-tips need the addition of auxin plus cytokinin and gibberellic acid in order to develop into plantlets and there are genotypic differences in the rooting response (Kartha, Gamborg, Constabel & Shyluk, 1974; Anon., 1978; Kartha & Gamborg, 1979; Nair, Kartha & Gamborg, 1979). Nodal cuttings also require the addition of these growth hormones: 90% of the varieties tested grew to plantlets (Anon., 1978) and produced on average 3–6 plants per nodal cutting per month. Increasing the cytokinin supply and altering the physical conditions from static culture on solid medium to rotatory culture in liquid medium produced *c.* 20 shoots per node per month owing to stimulation of the production of multiple shoots. Theoretically, therefore, very high annual rates of multiplication can be achieved, and logistic considerations, rather than the in vitro procedures, are likely to become limiting. Bearing this in mind, it is perhaps important to take into account the fact that a non-tissue culture method of propagation using single leaf-bud cuttings can now yield over 300000 cuttings per mother plant per year (Roca *et al.*, 1980).

Long-term minimal growth storage involving subculture intervals of up to 15 months has been implemented using reduced temperature (20 °C) with either higher sucrose levels (4%) plus 0.01 mg l^{-1} 6-benzylaminopurine (BAP) or lower sucrose levels (2% w/v) plus 0.05 mg l^{-1} BAP (Anon., 1979) and it was found that 95% of the cultures remained viable. Some attempts have been made to preserve shoot-tips in liquid nitrogen, but so far success has been limited (J. Stamp, personal communication) with a 5% survival level after ultra-rapid freezing of shoot-tips treated with 10% DMSO, and the survival that did occur was only dedifferentiated callus tissue from which shoots could not be regenerated.

In vitro techniques are being fully exploited for the storage and utilisation of cassava germplasm. CIAT now has all of its accessions stored in vitro and an active programme of global dissemination is under way: in 1981, 61 lines were sent as meristem cultures to eight countries, including several in SE Asia. Also, several publications report on the use of tissue culture and heat therapy to eradicate disease (primarily cassava mosaic virus) in cassava (Berbee, Berbee & Hildebrandt, 1973; Kartha & Gamborg, 1979; Adejare & Coutts, 1981).

Sweet potato (Ipomoea batatas)

Sweet potato is a perennial herb of the tropics. It is conventionally propagated either by stem-cuttings, or in cooler regions, via the edible tubers. The crop is prone to attack from numerous pests and diseases.

Attempts have been made to eliminate virus from *Ipomoea* plants by meristem culture techniques (see Nielsen & Terry, 1977 for summary) and have met with moderate success (up to 45% of the plants from the meristem tips of infected cultivars indexed virus-free on *I. setosa*).

A detailed study has been made into the storage of sweet potato germplasm (Alán, 1979). Nodal cuttings taken directly from the plant needed no additives to the basic MS medium beyond a sucrose supply in order to develop into plantlets. The much smaller shoot-tip explants required an auxin and cytokinin supply for their development. Physical conditions were also important. For example, the height of the filter-paper wick above the level of liquid medium was critical, with normal shoot development being favoured when the wick was close to the surface of the liquid. Axillary buds from nodal cuttings developed into plantlets with four nodes within six weeks and were serially cultured for 15 passages.

It was possible to use the reduced growth rate found at 22 °C, compared with that at 28 °C, to increase the subculture interval from six weeks to 55 weeks. The main problems associated with the increased passage duration were a result of medium evaporation. This was overcome and the passage length increase to 89 weeks either by the use of agar medium or by employing suitable culture vessel closures (silicone rubber caps or polythene adhesive tape).

Tissue culture techniques are not yet used on a wide scale with sweet potato, but the results of studies so far indicate that exploitation should be feasible.

White yam (Dioscorea rotundata)

The white yam is grown over a greater acreage than all other yam species. It is native to West Africa and is no longer found in the wild. Each locality grows its own cultivar, which has led to problems of nomenclature and classification. Yam requires a long growing season, yields are low and the tubers store badly. It is usually propagated via the edible tubers; although seeds are produced, they are very heterozygous and a large proportion are non-viable. The white yam is thus a prime target for germplasm conservation.

Microcuttings that include the whole or half of the node will develop to plants in vitro on basal MS medium plus sucrose only. Shoot-tips dissected from the axillary buds require a low level (0.1 mg l^{-1}) of both naphthaleneacetic acid (NAA) and BAP for shoot development (they give a 50–70% response). A new plantlet was formed within two weeks if nodal explants were cultured on medium at 25 °C (all data from Mathias, 1981). By using the optimal passage length of four weeks and lowering the sucrose level to 1%, it was estimated that nearly 200 plants per year could be produced per cutting. Yam was propagated in vitro for seven passages and no deterioration in multiplication rate was noted. It seemed that the growth rate under these conditions was naturally slow, perhaps owing in part to strong apical dominance: even after 43 weeks on basal MS medium, not all the nutrients in the medium had been exhausted. Thus growth inhibitors for long-term storage were unnecessary. The best approach was thought to be to increase the number of nodes per explant without altering the growth rate, i.e. to increase the multiplication potential by axillary buds while in storage.

Previous reports have been concerned with the clonal propagation of several *Dioscorea* species using nodal cuttings (Mantell, Haque & Whitehall, 1978), and it seems that the use of tissue culture techniques for germplasm conservation and utilisation should be feasible.

Conclusion

The examples which we have considered in some detail above represent crops at different stages of exploitation with potato being the most advanced. It was fully appreciated several years ago that access to primitive species related to potato was becoming difficult so the programme of collection, evaluation and conservation of potato was one of the first to be set in motion. The collection and evaluation of potato germplasm are now well under way. At CIP, a major centre for potato germplasm, in vitro techniques are routinely used to free material from disease and to clone accessions when increased numbers are needed for breeding or evaluation. With cassava also, in vitro techniques are beginning to play a significant role: at CIAT representatives of all the cassava accessions are duplicated as shoot-tip cultures and the cultures are now being used routinely for international exchange of disease-tested material. In the case of sweet potato and white yam, the techniques are only beginning to be used.

From the response of the four major crops considered, it seems that in vitro culture of microcuttings such as nodes and shoot-tips can be effected with simple medium and conditions. In fact, the conditions are not as exacting as those already employed for elimination of viruses by meristem excision and culture. Obviously in vitro methods do have great potential as a useful alternative when conventional storage methods are unsuitable. However, the local resources and financial support of the individual gene banks have to be taken into account. It can be foreseen that certain techniques, particularly cryopreservation in liquid nitrogen, might only be justifiable for unique or extremely valuable specimens because of the relatively expensive and labour-intensive methodology that is at present needed to put material into store. These costs would be offset to some extent by the subsequent low cost of maintenance over many years.

In vitro techniques could offer a viable alternative for the conservation of species which produce recalcitrant seed and which at present can only be maintained as vegetative plants requiring constant attention and replenishment. While such seeds are unlikely to survive well in their intact state in liquid nitrogen (being generally large seeds with a high water content), tissue culture methods could help through procedures such as minimal growth storage (using the natural tendency of the seed to germinate and then slowing down subsequent growth of the embryo axis) or through the initiation of suitable cultures which can be stored in liquid nitrogen and from which plants can be regenerated when required. This has recently been achieved using embryogenic cultures of date palm (Tisserat, Ulrich & Finkle, 1981) and the method holds great promise, even though a caution must be inserted about the possibility of genetic variation among the somatic embryos generated.

In conclusion, the role of in vitro techniques in the global scheme for germplasm storage is now well established for some tuber crops and the principles have been laid down for other vegetatively propagated species including herbaceous and woody ornamentals, soft fruits and pomaceous fruits. Future work is still needed for forest species and other species producing recalcitrant seed.

It can be envisaged that tissue culture techniques will be particularly valuable where there is difficulty in using conventional methods of propagation or where it is necessary to solve phytosanitary problems. However, conventional methods that are dependent upon cheap and available land, labour and technology will not be usurped unless there is a very positive advantage in doing so, since in economic terms the high technology necessary for tissue culture may not always be justifiable.

References

Abdalla, F. H. & Roberts, E. H. (1968). Effects of temperature, moisture and oxygen on the induction of chromosome damage in seeds of barley, broad beans and peas during storage. *Annals of Botany*, **32**, 119–36.

Abdalla, F. H. & Roberts, E. H. (1969). The effects of temperature and moisture on the induction of genetic changes in seeds of barley, broad beans and peas during storage. *Annals of Botany*, **33**, 153–67.

Adejare, G. O. & Coutts, R. H. A. (1981). Eradication of cassava mosaic disease from Nigerian cassava clones by meristem-tip culture. *Plant Cell, Tissue and Organ Culture*, **1**, 25–32.

Alán, J. J. (1979). Tissue culture storage of sweet potato germplasm. PhD Thesis, University of Birmingham.

Anon. (1975). *The Conservation of Crop Genetic Resources*. Rome: International Board for Plant Genetic Resources.

Anon. (1976). *Report of the International Board of Plant Genetic Resources Working Group on Engineering, Design and Cost Aspects of Long-Term Seed Storage Facilities*. Rome: International Board for Plant Genetic Resources Secretariat.

Anon. (1978). *Cassava Programme. Centro Internacional de Agricultura Tropical Report* (*1978*) (hereafter CIAT Report). Cali, Colombia: CIAT.

Anon. (1979). *Cassava Programme. CIAT Report* (1979). Cali, Colombia: CIAT.

Anon. (1981). *Cassava Programme. CIAT Report* (1981). Cali, Colombia: CIAT.

Bajaj, Y. P. S. (1981). Regeneration of plants from potato meristems freeze-preserved for 24 months. *Euphytica*, **30**, 141–5.

Berbee, F. M., Berbee, J. G. & Hildebrandt, A. C. (1973). Induction of callus and virus-symptomless plants from stem-tip cultures of cassava. *In Vitro* (Annual Meeting Abstracts), **8**, 421.

Bonga, J. M. (1974). Vegetative propagation: tissue and organ culture as an alternative to rooting cuttings. *New Zealand Journal of Forestry Science*, **4**, 253–60.

Bos, L. (1977). Seed-borne viruses. In *Plant Health and Quarantine in International Transfer of Genetic Resources*, ed. W. B. Hewitt & L. Chiarappa, pp. 39–69. Ohio: CRC Press Inc.

D'Amato, F. (1975). The problem of genetic stability in plant tissue and cell cultures. In *Crop Genetic Resources for Today and Tomorrow*. Ed. O. H. Frankel & J. G. Hawkes, pp. 333–48. Cambridge: Cambridge University Press.

D'Amato, F. (1978). Chromosome number variation in cultured cells and regenerated plants. *Frontiers of Plant Tissue Culture*, ed. T. A. Thorpe, pp. 287–95. Calgary: Calgary University.

Davies, D. R. (1972). Speeding up the commercial propagation of freesias. *The Grower*, **77**, 711.

Denton, I. R., Westcott, R. J. & Ford-Lloyd, B. V. (1977). Phenotypic variation of *Solanum tuberosum* L. cv. Dr. McIntosh regenerated directly from shoot-tip culture. *Potato Research*, **20**, 131–6.

Ellis, R. H., Roberts, E. H. & Whitehead, J. (1980). A new, more economic and accurate approach to monitoring the viability of accessions during storage in seed banks. *Plant Genetic Resources Newsletter*, **41**, 3–18.

Frankel, O. H. & Bennett, E. (Eds.) (1970). *Genetic Resources in Plants – their Exploration and Conservation*. International Biological Programme Handbook, No. 11. Oxford and Edinburgh: Blackwell.

Frankel, O. H. & Hawkes, J. G. (Eds.) (1975). *Crop Genetic Resources for Today and Tomorrow*. Cambridge: Cambridge University Press.

Glendinning, D. R. (1979). The potato gene-pool, and benefits deriving from its supplementation. In *Proceedings of the Conference Broadening the Genetic Base of Crops*, ed. A. C. Zeven & A. M. Van Harten, pp. 187–94. Wageningen: Pudoc.

Grout, B. W. W. & Henshaw, G. G. (1978). Freeze preservation of potato shoot-tip cultures. *Annals of Botany*, **42**, 1227–9.

Grout, B. W. W. & Henshaw, G. G. (1980). Structural observations on the growth of potato shoot-tip cultures after thawing from liquid nitrogen. *Annals of Botany*, **46**, 243–8.

Hawkes, J. G. (1978). Biosystematics of the potato. In *The Potato Crop*, ed. P. M. Harris, pp. 15–69. London: Chapman & Hall.

Henshaw, G. G., O'Hara, J. F. & Westcott, R. J. (1980). Tissue culture methods for the storage and utilization of potato germplasm. In *Tissue Culture Methods for Plant Pathologists*, ed. D. S. Ingram & J. P. Helgeson, pp. 71–6. Oxford, London, Edinburgh, Boston and Melbourne: Blackwell Scientific Publications.

Hewett, P. D. (1979). Regulating seed-borne disease by certification. In *Plant Health*, ed. D. L. Ebbels & J. E. King, pp. 163–73. Oxford, London, Edinburgh and Melbourne: Blackwell Scientific Publications.

Howard, H. W. (1970). *The Genetics of the Potato, Solanum tuberosum*. London: Logos Press.

Hussey, G. & Stacey, N. J. (1981). *In vitro* propagation of potato (*Solanum tuberosum* L.). *Annals of Botany*, **48**, 787–96.

Jain, S. K. (1975). Genetic Reserves. *Crop Genetic Resources for Today and Tommorow*, ed. O. H. Frankel & J. G. Hawkes, pp. 379–98. Cambridge: Cambridge University Press.

Kahn, R. P. (1977). Plant quarantine: principles, methodology and suggested approaches. *Plant Health and Quarantine in International Transfer of Genetic Resources*, ed. W. B. Hewitt & L. Chiarappa, pp. 289–307. Cleveland, Ohio: Chemical Rubber Company Press Incorporated.

Kartha, K. K. (1981). Meristem culture and cryopreservation – methods and applications. In *Plant Tissue Culture*, ed. T. A. Thorpe, pp. 181–211. New York: Academic Press Incorporated.

Kartha, K. K. & Gamborg, O. L. (1979). Cassava tissue culture – principles and applications. In *Plant Cell and Tissue Culture: Principles and Applications*, ed. W. R. Sharp, P. O. Larsen, E. F. Paddock & V. Raghavani, pp. 712–25. Columbus, Ohio: Ohio State University Press.

Kartha, K. K., Gamborg, O. L., Constabel, F. & Shyluk, J. P. (1974). Regeneration of cassava plants from apical meristems. *Plant Science Letters*, **2**, 107–13.

King, M. W. & Roberts, E. H. (1979). *The storage of recalcitrant seeds –*

achievements and possible approaches. Rome: International Board for Plant Genetic Resources Secretariat.

Mantell, S. H., Haque, S. Q. & Whitehall, A. P. (1978). Clonal multiplication of *Dioscorea alata*, L. and *D. rotundata* Poir. Yams by tissue culture. *Journal of Horticultural Science*, **53**, 95–8.

Mathias, S. F. (1981). *Dioscorea rotundata* Poir: storage and propagation through tissue culture. MSc Thesis, University of Birmingham.

Murashige, T. & Skoog, F. (1962). A revised medium for rapid growth and bioassays with tobacco tissue cultures. *Physiologia Plantarum*, **15**, 473–97.

Nair, N. G., Kartha, K. K. & Gamborg, O. L. (1979). Effect of growth regulators on plant regeneration from shoot apical meristems of cassava (*Manihot esculenta* Crantz) and on the culture of internodes *in vitro*. *Zeitschrift für Pflanzenphysiologie*, **95**, S51–6.

Neergard, P. (1977). Methods for detection and control of seed-borne fungi and bacteria. In *Plant Health and Quarantine in International Transfer of Genetic Resources*, ed. W. B. Hewitt & L. Chiarappa, pp. 33–8. Ohio: CRC Press Inc.

Nielsen, L. W. & Terry, E. R. (1977). Sweet potato (*Ipomoea batatas* L.). *Plant Health and Quarantine in International Transfer of Genetic Resources*, ed. W. B. Hewitt & L. Chiarappa, pp. 271–6. Ohio: CRC Press Inc.

O'Hara, J. F. & Henshaw, G. G. (1980). Cryopreservation and strategies for plant germplasm conservation. *CryoLetters*, **1**, 261–6.

Roberts, E. H. (ed.) (1972). *Viability of Seeds.* London: Chapman & Hall Ltd.

Roberts, E. H. (1975). Problems of long-term storage of seed and pollen for genetic resources conservation. *Crop Genetic Resources for Today and Tomorrow*, ed. O. H. Frankel & J. G. Hawkes, pp. 269–96. Cambridge: Cambridge University Press.

Roca, W. M., Bryan, J. E. & Roca, M. R. (1979). Tissue culture for international transfer of potato genetic resources. *American Potato Journal*, **56**, 1–11.

Roca, W. M., Espinoza, N. O., Roca, M. R. & Bryan, J. E. (1978). A tissue culture method for the rapid propagation of potatoes. *American Potato Journal*, **55**, 691–701.

Roca, W. M., Rodriguez, A., Pateña, L. F., Barba, R. C. & Toro, J. C. (1980). Improvement of a propagation technique for cassava using single leaf-bud cuttings: a preliminary report. *Cassava Newsletter*, **8**, 4–5. Cali, Colombia: CIAT.

Sheridan, W. F. (1968). Tissue Culture of the monocot *Lilium*. *Planta*, **82**, 189–92.

Simmonds, N. W. (1979). *Principles of Crop Improvement.* London and New York: Longman.

Thomas, E. & Wernicke, W. (1978). Morphogenesis in herbaceous crop plants. *Frontiers of Plant Tissue Culture*, ed. T. A. Thorpe, pp. 403–10, Calgary: Calgary University.

Tisserat, B., Ulrich, J. M. & Finkle, B. J. (1981). Cryogenic preservation and regeneration of date palm tissue. *Hort Science*, **16**, 47–8.

Towill, L. E. (1981). *Solanum etuberosum*: A model for studying the cryobiology of shoot-tips in the tuber-bearing Solanum species. *Plant Science Letters*, **20**, 315–24.

Westcott, R. J. (1981*a*). Tissue culture storage of potato germplasm. 1. Minimal growth storage. *Potato Research*, **24**, 331–42.

Westcott, R. J. (1981*b*). Tissue culture storage of potato germplasm. 2. Use of growth retardants. *Potato Research*, **24**, 343–52.

Westcott, R. J., Henshaw, G. G., Grout, B. W. W. & Roca, W. M. (1977). Tissue culture methods and germplasm storage in potato. *Acta Horticulturae*, **78**, 45–9.

Westcott, R. J., Henshaw, G. G. & Roca, W. M. (1977). Tissue culture storage of potato germplasm: culture initiation and plant regeneration. *Plant Science Letters*, **9**, 309–15.

Withers, L. A. (1980). Low temperature storage of plant tissue cultures. *Advances in Biochemical Engineering, Plant Cell Cultures II*, ed. A. Fiechter, **18**, pp. 102–50. Berlin, Heidelberg and New York: Springer-Verlag.

PART IV

In vitro approaches to the genetic manipulation of plants

E. C. COCKING

Genetic transformation through somatic hybridisation

Introduction

Because biotechnology is a difficult subject to define, it is important to emphasise that it is not synonymous with so-called 'genetic engineering'. There are many facets of biotechnology that are equally exciting from an industrial standpoint. However, it is the new concept of the cell as a 'factory' that pervades so much of the current thinking and practice, and plant cells in particular are especially attractive in this respect.

In this survey of genetic transformation through somatic hybridisation the term 'transformation' will be used in the classical bacterial sense to refer to the uptake and expression of foreign DNA by a cell as has been used in discussions on gene transfer into mammalian cells (Wright, Lewis & Parfett, 1980). Genes have been introduced into mammalian cells using donor material in a variety of forms and these have been summarised by Wright *et al.* (1980). Purified DNA is also capable of transforming recipient cells but the frequency of transformation is almost three orders of magnitude lower than that obtained using cell fusion (Table 1).

As discussed by Flavell (1981) the breakthrough in yeast studies came when

Table 1. *Transfer of the dehydrofolate reductase gene into Chinese hamster ovary cells*[a]

Genetic donor	Colonies/10^7	Genetic material retained
Whole cells	1000	Complete genome
Microcells	100	One or a few chromosomes
Chromosomes	50	Subchromosomal fragments
DNA	2	? (less than with chromosomes)

[a] A methotrexate-resistant Chinese hamster ovary cell line was used as the genetic donor, and the recipient hamster clones were selected for their ability to grow in the presence of high concentrations of methotrexate after transfer. Untreated recipient cells produce no colonies when more than 10^8 cells are plated in the presence of methotrexate. (Adapted from Wright *et al.*, 1980.)

cells lacking a known specific enzyme as a result of a simple genetic defect were presented with a high concentration of DNA fragments carrying the gene which specified the missing enzyme, isopropylmalate dehydrogenase. Protoplasts were used because the absence of the cell wall facilitated the uptake of DNA enriched with the yeast gene for isopropylmalate dehydrogenase; there was no difficulty in obtaining suitable yeast auxotrophic recipients for transformation with these wildtype genes.

For higher plants several methods, which are somewhat comparable to those for mammalian cells shown in Table 1, exist for the introduction of foreign genes. Theoretically, auxotrophic mutants are an additional alternative, as will be discussed later, but very few such mutants are available.

There is an extensive literature on both plant cell culture (Evans & Cocking, 1975) and the regeneration of plants that can be used in breeding programmes (Vasil, Ahuja & Vasil, 1979). This in vitro culture of plants is fundamental to the implementation of the new plant biotechnology.

Gene transfer by whole plant cell fusion

Nuclear genomes

Normally the somatic cells of higher plants do not fuse together and whole cell fusion is dependent on the use of cells from which the cell wall has been removed using suitable enzymatic procedures. Such protoplasts can now be readily produced from a wide range of species and different tissue types (including roots and cotyledons as well as in vitro plant cell cultures). Protoplasts can be induced to fuse by a variety of different fusogens, and practical details of the various procedures involved in protoplast isolation and fusion to produce heterokaryons have been fully described previously (Power & Davey, 1979). More recently the importance for some species of a simple purification of the enzymes employed for protoplast isolation has been highlighted (Patnaik, Wilson & Cocking, 1981).

Ever since the first demonstration that such enzymatically isolated protoplasts could be induced to fuse to form readily identifiable heterokaryons (Power, Cummins & Cocking, 1970), it has been progressively established that such protoplasts can be hybridised to form cells, and in some instances whole plants, containing approximately two complete nuclear genomes. Initially interest centred on the development of the techniques required and the use of species that could be hybridised sexually. In the studies within the *Petunia* genus carried out by the Nottingham group (Cocking, 1979), we produced large numbers of amphidiploid somatic hybrid plants by fusing wildtype leaf protoplasts of one species with albino protoplasts from cell suspension cultures of the other species. This resulted in the production of flowering plants with 28 chromosomes of the somatic hybrid *Petunia parodii*

(2n = 14)⊗*P. hybrida* (2n = 14) and of the hybrid *P. parodii* (2n = 14)⊗ *P. inflata* (2n = 14).

In recent years, the fact that somatic plant cell fusion permits genomes of sexually incompatible species and genera to be brought together has greatly stimulated plant breeders and plant biotechnologists to use such procedures to try to overcome the restriction on gene flow that is inherent in conventional breeding and crossing procedures. Somatic hybrid plants have been produced between species that are difficult or impossible to hybridise conventionally, e.g. *Lycopersicon esculentum* and *Solanum tuberosum* (Melchers, Sacristan & Holder, 1978), *Datura innoxia* and *Atropa belladonna* (Krumbiegel & Schieder, 1979), *Arabidopsis thaliana* and *Brassica campestris* (Gleba & Hoffmann, 1979) and *P. parodii* and *P. parviflora* (Power, Berry, Chapman & Cocking, 1980). Taken overall, if fertile seed-producing somatic hybrids are desired, it would seem likely that the main opportunities for the plant breeder will centre on the use of protoplast fusions for hybridisations within the same genus or between closely related genera.

An ongoing problem in the detailed analysis of the consequences of fusion of protoplasts has been the lack of a suitable selection system which facilitates the ready identification and isolation of heterokaryons and hybrid cells. Until recently most of the biochemical complementation/selection methods available tended to lead to the preferential recovery of amphiploid somatic hybrids to the exclusion of other potentially very useful plant types: as pointed out by Cocking, Davey, Pental & Power (1981), those potential plant hybrids possessing one complete genome with only a few chromosomes of the other parent are probably lost during the process of selection, since they may not survive the strong selection conditions. Currently, such problems are being circumvented by the development of procedures for the isolation of heterokaryons following fusion of wild type protoplasts which are suitably labelled with either bright field or fluorescence markers (Patnaik, Cocking, Hamill & Pental, 1982). The heterokaryons are isolated either manually (Fig. 1) or by using automated fluorescence selection systems. Sometimes the percentage heterokaryon formation may not be sufficiently high for the ready application of these procedures. Further basic studies on protoplast fusion including the electrical depolarisation of membranes (Vienken, Ganser, Hampp & Zimmermann, 1981) could greatly help in this respect.

One of the objectives of the fusion of protoplasts of widely divergent plant species is to attempt to incorporate into a recipient crop species some limited genetic attribute of a donor species. In the case of mammalian cells, it is possible by X-ray irradiation of the parental cells and other chromosome-destabilising procedures to bring about directional chromosome elimination after fusion, and there are indications from the work of Dudits *et al.* (1980)

that comparable elimination with the retention of reconstructed chromosomes, or only a few genes (Power, Frearson, Hayward & Cocking, 1975), may also be possible when using protoplasts. When very high doses of irradiation are employed (20000 to 100000 rad) and the irradiated protoplasts are fused with normal protoplasts, the transfer of a few genes into a suitable recipient species may be possible (Cocking, 1981). For adequate assessment suitable auxotrophic mutants that are capable of regenerating into plants are required:

Fig. 1. Apparatus for the manual isolation of heterokaryons. Heterokaryons are identified with bright field illumination using an inverted microscope and isolated by means of a simple micro-manipulator and capillary pipette coupled to a specially constructed syringe. Aseptic conditions are maintained by carrying out all the procedures inside a laminar air flow cabinet.

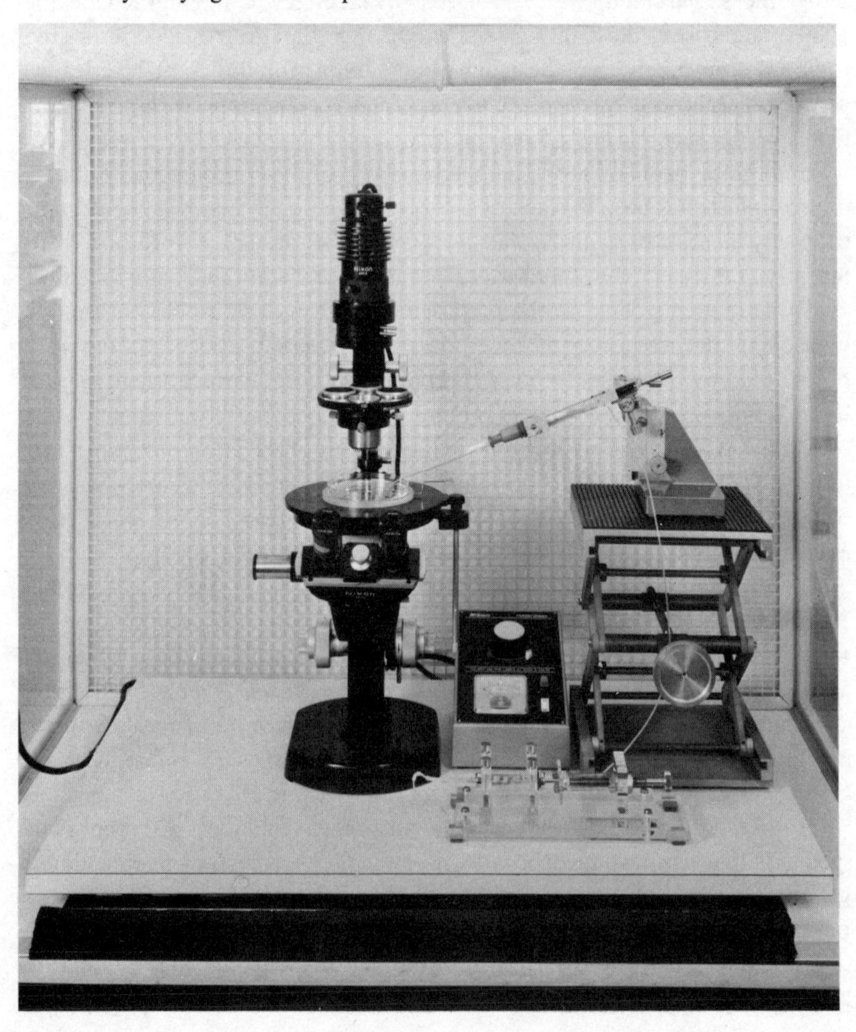

a nitrate-reductase-deficient mutant of tobacco has recently been isolated and its protoplasts were shown to be capable of regenerating into plants. This system should be able to detect any gene transfer from, say, soybean into tobacco even if it occurs at a very low frequency (Pental *et al.*, 1982).

Cytoplasmic genomes

Of importance for plant biotechnology is the fact that the cytoplasmic mix obtained from protoplast fusions is novel and provides a means of obtaining heterozygosity of extrachromosomal genes. As far as chloroplasts are concerned the characterisation of chloroplast organelles indicates that the two types of parental chloroplasts in a somatic hybrid eventually sort out into one or the other, or in some cases only one, of the parental types (Fig. 2). As described by Kumar, Cocking, Bovenberg & Kool (1982) the sorting out process of the parental chloroplasts in the somatic hybrid population is likely to be greatly influenced by many factors encountered during protoplast fusion and the subsequent selection and regeneration of somatic hybrids. For instance, analysis of the chloroplast DNA (cp DNA) using various restriction endonucleases showed that only the cp DNA of *P. parodii* was present in three somatic hybrids which had been produced by the fusion of protoplasts from wildtype mesophyll cells of *P. parodii* with protoplasts from cell suspension cultures of albino mutants of *P. hybrida*, *P. inflata* and *P. parviflora*. It was suggested that this unidirectional sorting out in favour of *P. parodii*

Fig. 2. Protoplast fusion and some possible chloroplast segregations.

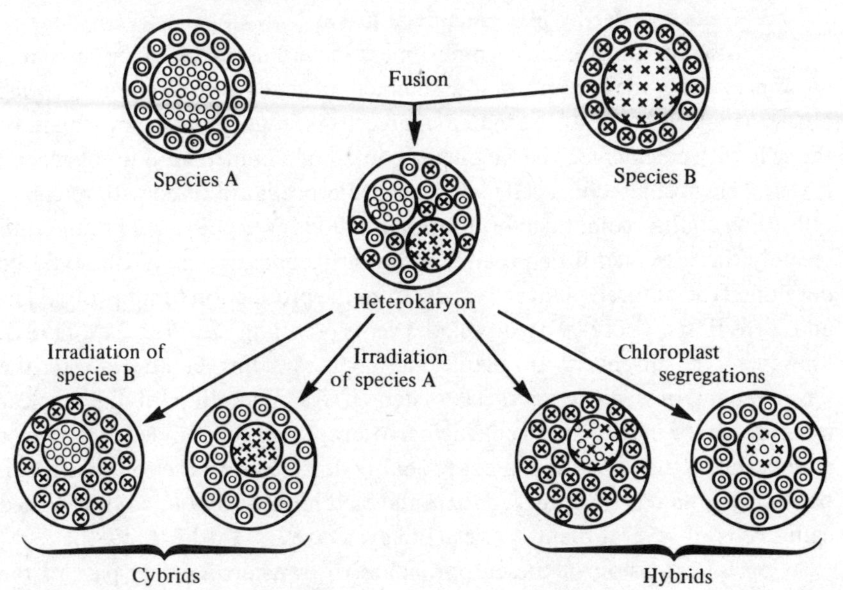

chloroplasts was most probably the result of the strong selections used in the production of these somatic hybrids. It was not possible to determine the amount of chloroplast heterozygosity, for instance by comparing the polypeptide banding patterns produced by Fraction 1 protein (ribulose bisphosphate carboxylase/oxygenase) following gel electrophoresis, because the large-subunit polypeptides (coded by chloroplast DNA) of the Fraction 1 protein are identical in these four *Petunia* species (Gatenby & Cocking, 1977; Kumar, Wilson & Cocking, 1981). No recombination between cp DNAs was detected in any of the three hybrids indicating that cp DNA recombination must be an infrequent event, if it occurs at all. Opportunities for increasing cytoplasmic variability by protoplast fusion may be greater with mitochondria than with chloroplasts. Less detailed analysis has been carried out on mitochondrial DNA. However, Belliard, Vedel & Pelletier (1979) obtained evidence for mitochondrial-based recombination in cytoplasmic hybrids of *Nicotiana tabacum* that were obtained by protoplast fusion. The heterogeneity of most mitochondrial DNAs isolated from Nicotianas is complex, and recent work suggests that those from some *Brassica* species are less complex, making Brassicas preferable for this type of analysis (Lebacq & Vedel, 1981).

The achievements and potentials of the transfer of cytoplasmic male sterility and cytoplasmically based herbicide resistance between sexually incompatible species by the method of protoplast fusion have recently been reviewed (Cocking, 1981). There are major opportunities for both plant breeding and plant cell culture technology in this area.

Gene transfer by plant microcell fusion

Microcell-mediated chromosome transfer has proved to be a promising approach for the chromosome assignment of genes from mammalian cells. Nuclear material of cultured mammalian cells is fragmented by treatment of the cells with colchicine. The fragments consist of a limited amount of genetic material encapsulated in a cell membrane. Microcells are fused with recipient cells using routine somatic hybridisation methods (Wright *et al.*, 1980). Until recently there seemed little possibility that any comparable system could be developed for cultured plant cells apart from the production of subprotoplasts and cytoplasts (Lörz, Paszhowski, Dierks-Ventling & Potrykus, 1981). However, the discovery of plant microplast systems could lead to the development of such a microcell system if fragmentation of the nuclear material can be induced with colchicine. Microplasts, surrounded by an inner membrane of the cell that is most probably derived from the tonoplast, can be readily isolated by rupturing auxin-induced highly vacuolated thin-walled callus cells of several plant species (Bilkey, Davey & Cocking, 1982).

A logical extension of these approaches to transferring only part of the plant genome between diverse species is to utilise isolated chromosomes.

Polyethylene glycol-induced uptake of isolated metaphase chromosome into recipient wheat, parsley and maize protoplasts (Szabados, Hadlaczky & Dudits, 1981) has been obtained.

Gene transfer by fusion between plant and microbial cells

The availability of plant protoplasts and the development of procedures for their fusion, coupled with the finding that fusogens suitable for protoplast fusion are also suitable for animal cell and microbial protoplast fusions (Ferenczy, 1981), has highlighted the possibilities of gene transfer by fusion between plant and microbial protoplasts and between animal cells and plant protoplasts. A significant recent development has been the introduction of Ti plasmids into *Vinca rosea* L. (*Catharanthus roseus* (L.) G. Don) protoplasts by fusion and endocytosis with spheroplasts of *Agrobacterium tumefaciens* (Hasezawa, Nagata & Syono, 1981).

The first stable transformation of higher plants by isolated DNA involved the incubation of *Petunia* suspension cell protoplasts with Ti plasmids in the presence of poly-L-ornithine. Cell colonies were selected for phytohormone independence and opine synthesis (Davey *et al*., 1980) and more recently by DNA–DNA hybridisation (Draper, Davey, Cocking & Cox, 1982). The fusion studies of Hasezawa *et al.* (1981) have shown that because the Ti plasmid is probably introduced into the cytoplasm of the plant protoplasts in an intact state, the frequency of transformation is increased in comparison with that obtained when isolated plasmid is used. In this context it is particularly noteworthy that recently Schaffner (1980) showed that monkey cells were transformed by *Escherichia coli* harbouring a recombinant plasmid of Simian Virus 40. The infection frequency was greatly increased by fusion of lysozyme-treated bacteria with the monkey cells in the presence of polyethylene glycol.

Currently, extensive investigations are being carried out using plasmids containing various transposons in attempts to transform higher plant protoplasts. Parallel assessments will need to be undertaken using *E. coli* spheroplasts harbouring cloned Tn5 and other transposon recombinants at high copy number. These will need to be incubated with plant protoplasts in the presence of suitable fusogens. Selection for enhanced resistance to an appropriate antibiotic can then be undertaken. As previously mentioned, the availability of a regenerating nitrate-reductase-deficient tobacco system will enable comparable evaluations of fusion for the transfer of nitrate reductase genes from fungal protoplasts such as those of *Neurospora*.

General conclusions

As recently discussed by Sandri-Goldin, Goldin, Levine & Glorioso (1981) the limiting factor in transformation may be the extent of stabilisation of the transforming DNA in the recipient cell. Protoplast fusions with bacterial protoplasts present large quantities of plasmid DNA to the recipient plant protoplasts. These workers have suggested that this excess of plasmid or bacterial, DNA enhances the frequency of the transformation event. Whilst lipsomes (Nicholls, 1981) have been effectively utilised for such plasmid-mediated transformations in protoplasts of micro-organisms (Makins & Holt, 1981), protoplast fusions are themselves likely to prove to be a highly efficient method for effecting gene transfer for genetic engineering in higher plants. Moreover, there is no need to isolate and purify the DNA.

References

Belliard, G., Vedel, F. & Pelletier, G. (1979). Mitochondrial recombination in cytoplasmic hybrids of *Nicotiana tabacum* by protoplast fusion. *Nature*, **281**, 401–3.

Bilkey, P. C., Davey, M. R. & Cocking, E. C. (1982). Isolation, origin and properties of enucleate plant microplasts. *Protoplasma*, **110**, 147–51.

Cocking, E. C. (1979). Parasexual reproduction in flowering plants. *New Zealand Journal of Botany*, **17**, 665–71.

Cocking, E. C. (1981). Opportunities from the use of protoplasts. *Philosophical Transactions of the Royal Society of London. B*, **292**, 557–68.

Cocking, E. C., Davey, M. R., Pental, D. & Power, J. B. (1981). Aspects of plant genetic manipulation. *Nature*, **293**, 265–70.

Davey, M. R., Cocking, E. C., Freeman, J., Pearce, N. & Tudor, I. (1980). Transformation of *Petunia* protoplasts by isolated *Agrobacterium* plasmid. *Plant Science Letters*, **18**, 307–13.

Draper, J., Davey, M. R., Cocking, E. C. & Cox, B. J. (1982). Ti plasmid homologous sequences present in tissues from *Agrobacterium* plasmid-transformed *Petunia* protoplasts. *Plant and Cell Physiology*, **23**, 255–62.

Dudits, D., Feyer, O., Hadlaczky, G., Koncz, C., Lazar, G. & Horvath, G. (1980). Intergeneric gene transfer mediated by plant protoplast fusion. *Molecular and General Genetics*, **179**, 283–8.

Evans, P. K. & Cocking, E. C. (1975). In *New Techniques in Biophysics and Cell Biology*, vol. 2, pp. 127–58. London: Wiley.

Ferenczy, L. (1981). Microbial protoplast fusion. In *Genetics as a Tool in Microbiology*, ed. S. W. Glover & D. A. Hopwood, pp. 1–33. 31st Symposium of the Society for General Microbiology. Cambridge: Cambridge University Press.

Flavell, R. B. (1981). Needs and potentials of molecular genetic modification in plants. In *Genetic Engineering for Crop Improvement*, ed. K. O. Rachie & J. Lymen, pp. 208–22. New York: Rockefeller Press.

Gatenby, A. A. & Cocking, E. C. (1977). Polypeptide composition of Fraction 1 protein subunits in the genus *Petunia*. *Plant Science Letters*, **10**, 97–101.

Gleba, Y. Y. & Hoffmann, F. (1979). 'Arabidobrassica' plant genome engineering by protoplast fusion. *Naturwissenschaften*, **66**, 547–54.

Hasezawa, S., Nagata, T. & Syono, K. (1981). Transformation of *Vinca* protoplasts mediated by *Agrobacterium* spheroplasts. *Molecular and General Genetics*, **182**, 206–10.

Krumbiegel, G. & Schieder, O. (1979). Selection of somatic hybrids after fusion of protoplasts from *Datura innoxia* Mill and *Atropa belladonna* L. *Planta*, **145**, 371–9.

Kumar, A., Cocking, E. C., Bovenberg, W. A. & Kool, A. K. (1982). Restriction endonuclease analysis of chloroplast DNA in interspecies somatic hybrids of *Petunia. Theoretical and Applied Genetics*, **62**, 377–83.

Kumar, A., Wilson, D. & Cocking, E. C. (1981). Polypeptide composition of Fraction 1 protein of the somatic hybrid between *Petunia parodii* and *P. parviflora. Biochemical Genetics*, **19**, 255–61.

Lebacq, P. & Vedel, F. (1981). Sal-1 restriction enzyme analysis of chloroplast and mitochondrial DNAs in the genus *Brassica. Plant Science Letters*, **23**, 1–9.

Lörz, H., Paszhowski, J., Dierks-Ventling, C. & Potrykus, I. (1981). Isolation and characterization of cytoplasts and miniprotoplasts derived from protoplasts of cultured cells. *Physiologia Plantarum*, **53**, 385–91.

Makins, J. F. & Holt, G. (1981). Liposome-mediated transformation of *Streptomyces* by chromosomal DNA. *Nature*, **293**, 671–3.

Melchers, G., Sacristan, M. D. & Holder, A. A. (1978). Somatic hybrid plants of potato and tomato regenerated from fused protoplasts. *Carlsberg Research Communications*, **43**, 203–18.

Nicholls, P. (1981). Liposomes – as artificial organelles, topochemical matrices, and therapeutic carrier systems. In *International Review of Cytology*, Supplement 12, pp. 327–88.

Patnaik, G., Cocking, E. C., Hamill, J. & Pental, D. (1982). A simple procedure for the manual isolation and identification of plant heterokaryons. *Plant Science Letters*, **24**, 105–110.

Patnaik, G., Wilson, D. & Cocking, E. C. (1981). Importance of enzyme purification for increased plating efficiency and plant regeneration from single protoplasts of *Petunia parodii. Zeitschrift für Pflanzenphysiologie*, **102**, 199–205.

Pental, D., Cooper-Bland, S., Harding, K., Cocking, E. C. & Müller, A. J. (1982). Cultural studies on nitrate reductase deficient *Nicotiana tabacum* mutant protoplasts. *Zeitschrift für Pflanzenphysiologie*, **105**, 219–27.

Power, J. B., Berry, S. F., Chapman, J. V. & Cocking, E. C. (1980). Somatic hybridization of sexually incompatible *Petunias: Petunia parodii* and *Petunia parviflora. Theoretical and Applied Genetics*, **57**, 1–4.

Power, J. B., Cummins, S. E. & Cocking, E. C. (1970). Fusion of isolated plant protoplasts. *Nature*, **225**, 1016–18.

Power, J. B. & Davey, M. R. (1979). Laboratory Manual: *Plant Protoplasts* (*Isolation, Fusion, Culture, Genetic Transformation*). Nottingham: Department of Botany, University of Nottingham.

Power, J. B., Frearson, E. M., Hayward, C. & Cocking, E. C. (1975). Some consequences of the fusion and selective culture of *Petunia* and *Parthenocissus* protoplasts. *Plant Science Letters*, **5**, 197–207.

Sandri-Goldin, R. M., Goldin, A. L., Levine, M. & Glorioso, J. C. (1981). High-frequency transfer of cloned herpes simplex virus type 1 sequences to mammalian cells. *Molecular and Cell Biology*, **1**, 743–52.

Schaffner, W. (1980). Direct transfer of cloned genes from bacteria to mammalian cells. *Proceedings of the National Academy of Sciences, USA*, **77**, 2163–7.

Szabados, L., Hadlaczky, G. Y. & Dudits, D. (1981). Uptake of isolated plant chromosomes by plant protoplasts. *Planta*, **151**, 141–5.

Vasil, I. K., Ahuja, M. R. & Vasil, V. (1979). Plant tissue cultures in genetics and plant breeding. *Advances in Genetics*, **20**, 127–215.

Vienken, J., Ganser, R., Hampp, R. & Zimmermann, U. (1981). Electric field-induced fusion of isolated vacuoles and protoplasts of different developmental and metabolic provenience. *Physiologia Plantarum*, **53**, 64–70.

Wright, J. A., Lewis, W. H. & Parfett, C. L. J. (1980). Somatic cell genetics: a review of drug resistance, lectin resistance and gene transfer in mammalian cells in culture. *Canadian Journal of Genetics and Cytology*, **22**, 443–96.

S. BRIGHT, V. JARRETT, R. NELSON,
G. CREISSEN, A. KARP, J. FRANKLIN,
P. NORBURY, J. KUEH, S. ROGNES
and B. MIFLIN

Modification of agronomic traits using in vitro technology*

Introduction

Conventional plant breeding has had considerable success during this century in improving the yield and quality traits of a wide range of crop species (Bingham, 1981). This has been particularly so for the major cereals where wheat and barley yields have been consistently improved over the last 40 years. About half of this increase in yield is due to the use of superior varieties, the rest coming from better agronomic practice. In vitro techniques for treating plant material have progressed in the last 20 years to a state where a considerable contribution can be expected in those crops for which conventional plant breeding has been less successful.

To make an impact on any breeding scheme, in vitro methods must satisfy two criteria. Firstly, they should produce a plant which is fit to enter a breeding or screening programme and compete effectively with plants produced by conventional improvement programmes. Secondly, the 'improved' plants should be obtained in an economical and efficient way so that the methods may be applied, in conjunction with conventional breeding and assessment programmes, with the minimum need for highly skilled manpower or expensive equipment.

In this chapter, we discuss the use of plant material of different levels of organisational complexity for the production, screening or selection of desirable variation in agronomic characters. We concentrate on two areas: firstly, the use of protoplasts and complex cultures of potato for the generation of variability and secondly, the use of embryos and seeds of barley to select for particular mutations.

Protoplasts and complex cultures of potato as sources of useful variation

The naked single cell or protoplast has many attractions as a starting-point for in vitro manipulation (Cocking, 1981). Amongst these are the great numbers of protoplasts which can be treated in a small volume

* This paper is dedicated to the memory of Emrys Thomas, who died in May 1981.

(up to 10^7 ml^{-1}) and the ready accessibility of the cell, at this stage, to foreign genetic material, either directly or through protoplast fusion.

A prerequisite for any use of protoplasts in crop improvement is the ability to regenerate plants. This has been achieved at a high efficiency for tobacco (Takebe, Labib & Melchers, 1971). Other crop species which have been successfully regenerated include rape (Kartha *et al.*, 1974; Thomas, Hoffmann, Potrykus & Wenzel, 1976), alfalfa (Kao & Michayluk, 1980; Santos, Ontka, Cocking & Davey, 1980), cassava (Shahin & Shepard, 1980) and potato (Shepard & Totten, 1977). Much work remains to be done in order to be able to extend these successes to a full range of commercial cultivars and to get reproducible and efficient regeneration. For the potato, plants have been regenerated from protoplasts isolated from West German dihaploid lines (Binding *et al.*, 1978), five US cultivars (Shepard, Bidney & Shahin, 1980) and, recently, 13 UK cultivars (Gunn & Shepard, 1981; Thomas, 1981; Table 1). Protoplasts are isolated from plants grown in growth chambers or as axenic shoot cultures. The efficiency of each of the stages of protoplast isolation, the division of protoplast-derived cells, the formation of protoplast-derived cell colonies and shoot regeneration is dependent not only upon the medium but also upon the genotype and physiology of the donor plants. For instance, cv. Maris Bard protoplasts treated by the methods of Shepard gave fewer regenerated plants than cv. Maris Piper (Gunn & Shepard, 1981), whereas with shoot-culture-derived protoplasts of these two varieties the opposite was

Table 1. *Protoplast isolation, culture and plant regeneration from ten potato cultivars*

	Protoplast isolation					
Cultivar	Leaves	Shoot cultures	Divisions	Callus formation	Shoot formation	Plants
Champion	+	0	+	+	+	+
Desirée	−	+	+	+	+	−
King Edward	+	+	+	+	+	+
Majestic	+	+	+	+	+	+
Maris Bard	−	+	+	+	+	+
Maris Piper	−	+	+	+	+	+
Myatts Ashleaf	+	−	+	+	−	−
Pentland Crown	+	+	+	+	−	−
Record	+	+	+	+	+	−
Fortyfold	+	+	+	+	+	+

Key: +, successful; −, unsuccessful; 0, not yet attempted.

true (Thomas, 1981). The process of going from protoplast to regenerated plant first potted out in soil takes a minimum of 16 weeks (Shepard, Bidney & Shahin, 1980) and in our hands the most rapid sequence we have found takes 18 weeks.

In addition to using protoplast culture as a source of useful variation, it is possible to use tissue cultures made from cultured explants such as leaf, stem, rachis, petiole or tuber pieces. Such complex cultures form callus at the explant margins and then, under an appropriate hormonal regime (Roest & Bokelmann, 1976; K. J. Webb, personal communication), shoots are initiated; these can be rooted and grown on to plants. This sequence of events is more rapid than that of the protoplast culture system, giving potted plants in a minimum of seven weeks. It is applicable to a number of potato cultivars, although again there is a pronounced effect of genotype as well as explant source.

The reason for the great interest in the production of plants from protoplasts and complex cultures of potato is a series of reports from the laboratory of J. F. Shepard (Matern, Strobel & Shepard, 1978; Shepard *et al.*, 1980; Secor & Shepard, 1981; Shepard, 1981) that amongst protoplast-derived plants of cv. Russet Burbank there exists marked variation for a number of agronomic characteristics and that these characteristics can be carried through the following tuber generations. In particular, improved yield, photoperiod response, growth habit, maturity date and tuber morphology were observed in lines from an original population of 1700 protoplast-derived plants (Shepard *et al.*, 1980). In the same population disease-resistance characters were also stably altered in a number of lines. Resistance to several races of *Phytophthora infestans* was observed in eight lines (Shepard *et al.*, 1980) and to *Alternaria solani* in four lines out of 500 tested (Matern *et al.*, 1978). In contrast, Wenzel *et al.* (1979) observed very little variation amongst 200 plants derived from dihaploid protoplasts.

Considerable variation was also observed in plants regenerated from complex cultures of cv. Desirée (van Harten, Bouter & Broertjes, 1981). In this case the aim of the experiment was to produce mutations by radiation treatment but it was found that the control group of 18 plants derived from rachis tissue cultures contained nine mutants. Fewer but still significant numbers of mutants were observed among untreated plants regenerated from cultured leaf pieces. These observations pose a number of practical questions about the relevance of this variation in potato breeding schemes and a number of scientific questions about its origins, causes and control.

Consideration of the variation found in regenerated potato plants

Practical questions about variation

(1) Does variation in protoplast-derived plants occur in cultivars other than Russet Burbank?

(2) Is the variation in protoplast-derived plants of the same kind as that in explant-derived plants?

(3) Are there simple and standard methods for the generation of variant plants, which can be applied to a wide range of breeders' material?

(4) Are tissue culture and/or protoplast culture useful techniques to add to the standard methods of the potato (and other crop) breeders?

Scientific questions about variation

(1) What are the origins and causes of variation? Some of the suggestions are that (a) old vegetatively propagated varieties might accumulate genetic changes in leaf cells, (b) leaf cells accumulate genetic changes during differentiation and (c) variation arises during the time that the cells are in culture (Shepard *et al.*, 1980; Thomas *et al.*, 1982).

(2) What is the basis of variation? Amongst possible factors are (a) single gene mutations, (b) chromosome or gene rearrangements, (c) gene amplification or depletion, (d) karyotype changes and (e) extranuclear changes (Larkin & Scowcroft, 1981).

Fig. 1. Plants regenerated from protoplasts of potato cv. Maris Bard. The two centre plants were derived from shoot cultures of a single regenerated shoot.

An evaluation of variation found in regenerated potato plants

We have been regenerating plants from protoplasts and complex cultures of a range of British cultivars in order to answer some of the questions raised. Plants regenerated from protoplasts of cv. Maris Bard (Fig. 1) showed a great range of variability in 10 morphological characters (Thomas *et al.*, 1982). Some examples of differing leaf morphology are shown in Fig. 2. From these it is clear that there is variability between protoplast-derived plants of cultivars other than cv. Russet Burbank. As cv. Maris Bard is a cultivar of recent origin (released in 1974), it seems also that the age of the cultivar is unimportant in the expression of variation. Forty-one shoots were taken in this experiment from 10 separate calluses, each of which was regenerated from a single protoplast. It is noteworthy that there were differences between each of the regenerated shoots and plants. As shoots from the same callus were different, it implies that at least part of the observed variation arose during the culture phase rather than pre-existing within the original leaf. To examine this further it will be necessary to perform an experiment to evaluate how much variability is apparent between protoplast-derived plants regenerated from different calluses (i.e. from different protoplasts), the same callus and the same particular area of callus.

The characteristics of plants regenerated from complex cultures have also been examined. Plants have been obtained from tuber, stem, leaf, rachis or petiole pieces from seven cultivars (V. Jarrett, unpublished data) and have also been regenerated on cut stems of greenhouse-grown plants (Thomas *et*

Fig. 2. Leaf morphology of cv. Maris Bard (A) and some plants regenerated from leaf protoplasts of cv. Maris Bard (B–H).

al., 1982). In general, the plants obtained in this way were more alike than those from protoplasts. Seventeen plants from an *in situ* callus on a stem did not show more variation than greenhouse-grown plants of the same variety (Thomas *et al.*, 1982). Like protoplast-derived plants, some of those regenerated from complex cultures had leaf abnormalities and other morphological changes. Several hundred plants of seven cultivars were regenerated for field testing in a standard potato breeding assessment at Loughgall, Northern Ireland, in 1982. From this we hope to find out if any useful variation was produced.

A particular advantage of the cultivar Desirée in these experiments is its red tuber skin. This cultivar is genetically red in all three meristem layers (L1, L2, L3), which contribute to the tuber skin colour, and so can be used to score mutations and for the production of chimeral progeny in which only one layer is mutated (van Harten *et al.*, 1981). Yellow tuber skin types occur spontaneously in the field at low frequency. In their experiments, van Harten *et al.* (1981) found that yellow-skinned types occurred more frequently in control plants regenerated from leaves (2/18) than from rachis pieces (0/18); interestingly, we have found the reverse (Table 2). Furthermore, we found no variegation, nor fusion of terminal and lateral leaves in any of the 100 leaf-derived plants whereas from the rachis-derived plants the figures were 23% and 32% respectively. Use of the red tuber skin marker allows quantitative examination of the rate of mutation for a character which is controlled by one or a few genes (Howard, 1970). It should facilitate comparison of different explant sources and medium formulations upon the rate of mutation, for which there is little data at present. Unfortunately we

Table 2. *Tuber skin colour*[a] *in plants regenerated from complex cultures of potato cv. Desirée*

	Number of plants with tubers			
Source of culture	Red	Pale pink	Red splashed	Yellow
Tuber	8	0	0	0
Leaf	100	0	0	0
Stem	7	0	0	0
Rachis or petiole	42	15	1	12

[a] Tuber skin colour was scored after harvesting plants grown for four months in controlled environment conditions. The colours did not change appreciably after storage at 4 °C for several months.

have so far been unable to regenerate plants from protoplasts of cv. Desirée (Table 1).

As part of an investigation into the reasons for the variation in protoplast-derived plants of cv. Maris Bard, a karyological analysis is being undertaken. The original shoots were maintained for over 12 months as sterile shoot cultures, from which a piece of stem with attached leaf was transferred at four-to-eight week intervals to fresh medium after the axillary bud had grown into a new shoot (Thomas, 1981). Fresh plants were rooted from these cultures and the root tips analysed. Of the 17 plants analysed so far (A. Karp, personal communication) only one has the correct number of 48 chromosomes as found in cv. Maris Bard plants (both tuber-grown and from shoot cultures). The remainder, with one exception, contain 49–95 chromosomes. This demonstrates that many of the protoplast-derived plants are very unlikely to be useful in breeding schemes and it is probable that the altered morphologies are caused by the aneuploidy; the one plant (P40) with 48 chromosomes is also the one classified as being most like cv. Maris Bard (Thomas *et al.*, 1982). Another plant (P35) that is similar to cv. Maris Bard, differing only in having fused terminal and lateral leaves, contains 46 chromosomes. Obviously more work is needed to analyse tuber roots from the first harvests to check that the shoot culture stocks have the same chromosome numbers as the first regenerated plants scored for morphology. We also need to know how many of the plants from complex cultures are aneuploid (the first seven plants tested from cv. Maris Bard leaf cultures all had 48 chromosomes). This preliminary result indicates considerable chromosomal instability during the production of plants from tetraploid potato protoplasts. One theory to account for this is that tetraploid potato protoplasts may frequently undergo chromosome doubling in the early stages of culture to give cells with 96 chromosomes. We speculate that this chromosome number is unstable and chromosomes are then lost to give a variable number. Although this may explain the nature of the chromosome instability in cv. Maris Bard regenerants more work is needed in other potato varieties before generalisations can be made. Chromosome doubling of haploid protoplasts is quite widespread and dihaploid potato protoplasts were found to produce 97% tetraploid plants after regeneration (Wenzel *et al.*, 1979) with a low frequency of aneuploids. Dihaploid leaf culture also give rise to tetraploids (Jacobsen, 1978). If the majority of variant forms are caused by aneuploidy then it is not surprising that Wenzel *et al.* (1979) saw little variability as the doubled dihaploid chromosome number of 48 is stable.

Changes in chromosome number cannot be the sole explanation of the variation observed in protoplast-derived plants of cv. Russet Burbank as five high-yielding variant lines were found to have a normal chromosome number

(Shepard *et al.*, 1980). Shepard *et al.* report that of 1700 original protoplast-derived plants they picked out 60 lines (4%) which looked similar to cv. Russet Burbank. In this process of selection they would undoubtedly remove most of the aneuploids, which could be in a large majority. The question then arises as to the causes of the interesting variation in yield and disease resistance which they found in these lines and whether these changes are due to more subtle forms of chromosomal change such as rearrangements or whether they are caused by mutations of single genes (Larkin & Scowcroft, 1981). We do not know the answer to this. From the preliminary results with cv. Maris Bard described above, it appears that the frequency of chromosome doubling (and the production of aberrant forms) may be higher in protoplast-derived plants than in those regenerated from complex cultures. It remains to be determined whether the frequency of interesting variation, such as that for disease resistance characteristics, follows this pattern too. If useful variation is forthcoming from complex cultures as has been found in a number of other species (Larkin & Scowcroft, 1981) then these cultures may be useful in potato improvement. The choice between protoplast and complex cultures would then depend on the relative simplicity of the techniques and the frequency at which the plants with particular desired traits occur using the two methods.

Whatever the final outcome of the debate on the relative merits of culture-derived variation for potato improvement it is clear that the fact of variation poses other particular problems. In order to introduce specific genes into protoplasts, such as alien genes for disease resistance, by fusion or direct DNA transformation it is highly desirable that the genetic background of the protoplast should be stable. There is thus every reason to try to understand the causes of variation in order to be able to control or eliminate them when necessary.

Embryos and seeds of barley as systems for selection of biochemical mutants

The preceding section has dealt essentially with the generation of diversity in plants amongst which useful types are identified in field or glasshouse screening. If the protoplast techniques were available it should be possible to select some specific mutant types directly at this low level of organisation. Possible candidates for selection in these cases are resistance to phytotoxins and herbicides. With potato this is very difficult to envisage given such a high background level of variability. In the cereals, protoplast culture is still in its infancy with only one report (Vasil & Vasil, 1980) of plant regeneration. It has been possible to circumvent these problems in cereals by using complex cultures derived from immature embryos to select for toxin-resistant mutants (Gengenbach, Green & Donovan, 1977; Brettell, Thomas

& Ingram, 1980) or amino-acid accumulation (Hibberd, Walter, Green & Gengenbach, 1980). However, in complex cultures there are problems in (a) knowing how many cells or shoots are being screened and (b) regenerating fertile plants (Hibberd *et al.*, 1980). We have developed an alternative system for barley in which we screen mature embryos grown in sterile culture. The advantages of this system are that (a) each embryo gives rise to a normal fertile plant, (b) the embryos germinate uniformly and respond directly to medium constituents because they have very few reserves stored within the scutellum (Bright, Wood & Miflin, 1978), (c) it is possible to use established seed mutagenesis techniques such as sodium azide (Kleinhofs, Sander, Nilan & Konzak, 1974) to induce high numbers of mutations with little chromosomal damage and (d) after growth and self-fertilisation of plants from mutagen-treated seed, recessive mutations can be isolated in the diploid homozygous condition. The principal disadvantage of this system lies in the number that can be screened. We have generally aimed to screen 20000 embryos in any particular selection system. These populations have generally given from two to four interesting mutants. Where appropriate (see below) seeds rather than embryos can be screened, the only disadvantage being that the nutrients supplied to the embryo from the endosperm are not under the strict control of the experimenter.

We have been seeking barley mutants with enhanced production of the amino acids lysine and threonine. If these amino acids were to accumulate in the soluble nitrogen pool of the seed in sufficient quantity they could improve the nutritional quality of the grain. Lysine and threonine are the first two limiting amino acids when young pigs are fed a barley diet (Fuller, Mennie & Crofts, 1979). These two amino acids are synthesised together with methionine from a common, branched pathway whose key enzymes are regulated by feedback inhibition (Bryan, 1980). The first enzymic step in the pathway is regulated primarily by the concentrations of the feedback regulators lysine and S-adenosylmethionine (Rognes, Lea & Miflin, 1980). In order to isolate the mutants, we have used two selection systems. Either the lysine analogue S-(2-aminoethyl)-cysteine (AEC) or a combination of lysine and threonine (LT) was included in the medium at concentrations which inhibit growth. Thus plants that have mutated so that they accumulate lysine should be resistant to AEC and those that have mutated so that they accumulate methionine should be resistant to LT (Bright *et al.* 1978; Bright, Norbury & Miflin, 1979). Because variable amounts of either of these amino acids in the endosperm would also relieve inhibition we found it necessary to carry out the screening with embryos rather than whole seeds.

From our screens we isolated two mutants (R(Rothamsted) 906 and R4402) that are resistant to AEC and three (R2501, R3004 and R3202 (Fig. 3)) that are resistant to lysine plus threonine. Both R906 (Bright *et al.*, 1979) and R4402 fail to take up AEC rather than overproduce lysine. The resistance genes are recessive and allelic (Bright *et al.*, 1982*a*); Bright, Kueh & Rognes, 1983). The mutants R2501 (Bright *et al.*, 1982*b*) and R3004 (Rognes, Bright & Miflin, 1983) have been analysed for free amino acids and for alterations in regulatory properties. Both of these mutants overproduce threonine, this leading to increases in the content of threonine in the soluble fraction of the seed from 10 nmol mg^{-1} N to 150 nmol mg^{-1} N in the case of R2501 and 120 nmol mg^{-1} N in the case of R3004. These increases are sufficient to increase the total seed threonine content by 6–10%, which could lead to a considerable decrease in the requirement for threonine supplementation of a barley diet. Neither the content of lysine nor that of any of the other soluble amino acids was dramatically increased in the soluble fractions of seeds but in young plants of R2501 the content of methionine was doubled and that of alanine increased threefold (Bright, Miflin & Rognes, 1982*b*). When analysed at the biochemical level both mutants were found to have altered

Fig. 3. Barley plants grown from mature embryos after seven days on basal medium (upper row) or (lower row) medium containing lysine plus threonine (8 mM each) and arginine (1 mM). Left, control plants; centre, mutant R3202; right, mutant R3004.

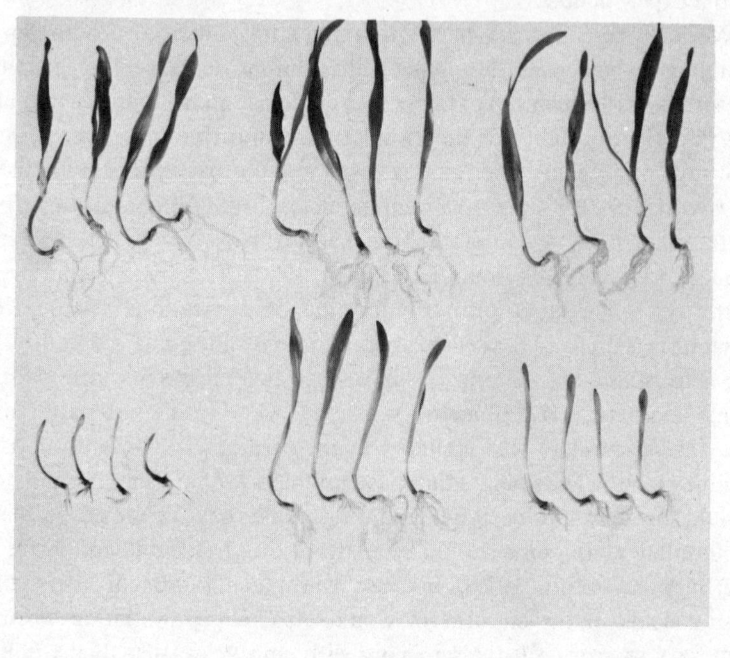

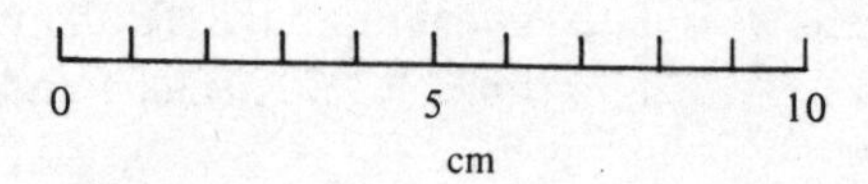

aspartate kinases. This enzyme exists in at least three forms in barley (AKI, AKII, AKIII). AKI is inhibited by threonine, but constitutes less than 10% of the total activity. AKII and AKIII are both inhibited by lysine or lysine plus S-adenosylmethionine. In R2501, AKII is much less sensitive to lysine inhibition but AKIII has unaltered sensitivity whereas in R3004 the situation is reversed. Preliminary genetic analysis indicates that the single dominant genes involved are unlinked in R3004 and R2501.

The reasons for the accumulation of threonine almost certainly depend on the way in which the pathway is regulated. An important factor may be the stimulation of threonine synthesis by S-adenosylmethionine (Thoen, Rognes & Aarnes, 1978). We know that there are two lysine-sensitive enzymes on the pathway to lysine so it is perhaps not surprising that mutational alteration of only one does not allow for lysine overproduction. In tobacco, which contains more threonine-sensitive aspartate kinase, there is probably only one reaction, i.e. that catalysed by dihydrodipicolinate synthase, which is fully lysine sensitive. Consistent with this, a lysine-accumulating, AEC-resistant tobacco mutant has been obtained (Negrutiu, Cattoir-Reynaerts, Verbruggen & Jacobs, 1981). From this we have been encouraged to repeat mutagenesis on one of the threonine-accumulating mutants, in order to attempt to obtain a second mutation that would lead to both lysine and threonine overproduction. We have also used the embryo screening method to select for hydroxyproline-resistant mutants which accumulate proline in the leaf

Fig. 4. Growth of a single, green, chlorate-resistant barley seedling against a background of dead and dying M_2 seedlings after germination for five days plus a further nine days in the presence of 10 mM potassium chlorate.

(Kueh & Bright, 1981). Such mutants may be useful in elucidating the role of proline in stress metabolism.

As a final example of ways of selecting for changes in agronomic characters we suggest that screening of seedlings can be a very simple and powerful method. Seed screening has been used to generate a number of mutants that have altered photorespiratory pathways and thereby provide valuable information about regulation (Somerville & Ogren, 1981) although none so far with enhanced growth. Seedling screening can be done on quite a large scale with barley, where 5 kg of seed contains about 10^5 individuals. Herbicide tolerance can be selected for using cell cultures (Chaleff, 1980) but it is probably much simpler to select resistant plants directly and so avoid the problems of transfer of resistance from cell culture to whole plant (Gressel, Zilkah & Ezra, 1978). An example of such a selection is given in Fig. 4, which shows the isolation of a chlorate-resistant barley plant. In agricultural terms, chlorate-resistant plants may not be much use as they frequently lack nitrate reductase (Oostindier-Braaksma & Feenstra, 1973; S. W. J. Bright & P. B. Norbury, unpublished observations) but the principle is the same for other herbicides.

Conclusions

The entire range of organisational complexity from the isolated protoplast to the whole seed can be used for the selection or identification of variation in desired characters. The disadvantages of tissue cultures in terms of chromosomal instability may be matched by the greater numbers that can be screened and the possibility of generating novel forms of variation. Mature embryos or seeds may be useful systems to use but are limited by numbers. The choice of a system for selecting for variation must depend on a sound knowledge of the biology, including tissue culture responses, and the genetics of the chosen crop plant. In the end simplicity is to be preferred.

We thank Dr K. J. Webb for letting us know of her leaf regeneration medium for potato. Parts of this work were supported by grants from the Potato Marketing Board (to V.J. and G.C.), the Council of Europe (to S.R.), NATO Grant 25580 (to S.R. and S.B.), EEC Grant 473 (to B.M.) and an SRC CASE award (to R.N.).

References

Binding, H., Nehls, R., Schieder, O., Sopory, S. K. & Wenzel, G. (1978). Regeneration of mesophyll protoplasts isolated from dihaploid clones of *Solanum tuberosum*. *Physiologia Plantarum*, **43**, 52–4.

Bingham, J. (1981). The achievements of conventional plant breeding. *Philosophical Transactions of the Royal Society of London, B*, **292**, 441 – 54.

Brettell, R. I. S., Thomas, E. & Ingram, D. S. (1980). Reversion of Texas male-sterile cytoplasm maize in culture to give fertile, T-toxin resistant plants. *Theoretical and Applied Genetics*, **58**, 55–8.

Bright, S. W. J., Kueh, J. S. H., Franklin, J., Rognes, S. E. & Miflin, B. J. (1982*a*). Two genes for threonine accumulation in barley seeds. *Nature*, **299**, 278–9.

Bright, S. W. J., Kueh, J. S. H. & Rognes, S. E. (1983). Lysine transport in two barley mutants with altered uptake of basic amino acids in the root. *Plant Physiology*, submitted.

Bright, S. W. J., Miflin, B. J. & Rognes, S. E. (1982*b*). Threonine accumulation in the seeds of a barley mutant with an altered aspartate kinase. *Biochemical Genetics*, **20**, 229–43.

Bright, S. W. J., Norbury, P. B. & Miflin, B. J. (1979). Isolation of a recessive barley mutant resistant to S-(2-aminoethyl) cysteine. *Theoretical and Applied Genetics*, **55**, 1–4.

Bright, S. W. J., Wood, E. A. & Miflin, B. J. (1978). The effect of aspartate-derived amino acids (lysine, threonine, methionine) on the growth of excised embryos of wheat and barley. *Planta*, **139**, 113–17.

Bryan, J. K. (1980). Synthesis of the aspartate family and branched chain amino acids. In *Biochemistry of Plants*, vol. 5, ed. B. J. Miflin, pp. 403–52. New York: Academic Press.

Chaleff, R. S. (1980). Further characterization of picloram-tolerant mutants of *Nicotiana tabacum*. *Theoretical and Applied Genetics*, **58**, 91–5.

Cocking, E. C. (1981). Opportunities from the use of protoplasts. *Philosophical Transactions of the Royal Society of London, B*, **292**, 557–68.

Fuller, M. F., Mennie, I. & Crofts, R. M. J. (1979). The amino acid supplementation of barley for the growing pig, 2. Optimal additions of lysine and threonine for growth. *British Journal of Nutrition*, **41**, 333–40.

Gengenbach, B. G., Green, C. E. & Donovan, C. M. (1977). Inheritance of selected pathotoxin resistance in maize plants regenerated from cell cultures. *Proceedings of the National Academy of Sciences, USA*, **74**, 5113–7.

Gressel, J., Zilkah, S. & Ezra, G. (1978). Herbicide action, resistance and screening in cultures vs. plants. In *Frontiers of Plant Tissue Culture*, ed. T. A. Thorpe, pp. 427–36. Calgary: University of Calgary.

Gunn, R. E. & Shepard, J. F. (1981). Regeneration of plants from mesophyll protoplast-derived protoplasts of British Potato (*Solanum tuberosum* L.) cultivars. *Plant Science Letters*, **22**, 97–101.

Harten, A. M. van, Bouter, H. & Broertjes, C. (1981). *In vitro* adventitious bud techniques for vegetative propagation and mutation breeding of potato (*Solanum tuberosum* L.). II. Significance for mutation breeding. *Euphytica*, **30**, 1–8.

Hibberd, K. A., Walter, T., Green, C. E. & Gengenbach, B. G. (1980). Selection and characterization of a feedback-insensitive tissue culture of maize. *Planta*, **148**, 183–7.

Howard, H. W. (1970). Genetics of the potato *Solanum tuberosum*. London: Logos Press.

Jacobsen, E. (1978). Doubling dihaploid potato clones via leaf tissue culture. *Zeitschrift für Pflanzenzuchtung*, **80**, 80–2.

Kao, K. N. & Michayluk, M. R. (1980). Plant regeneration from mesophyll protoplasts of alfalfa. *Zeitschrift für Pflanzenphysiologie*, **96**, 135–41.

Kartha, K. K., Michayluk, M. R., Kao, K. N., Gamborg, O. L. & Constabel, F. (1974). Callus formation and plant regeneration from mesophyll protoplasts of rape plants (*Brassica napus* L. cv. Zephyr). *Plant Science Letters*, **3**, 265–71.

Kleinhofs, A., Sander, C., Nilan, R. A. & Konzak, C. F. (1974). Azide mutagenicity-mechanisms and nature of mutants produced. In *Polyploidy and Induced Mutations in Plant Breeding*, STI/PUB/359, pp. 175–99. Vienna: International Atomic Energy Agency.

Kueh, J. S. H. & Bright, S. W. J. (1981). Proline accumulation in a barley mutant resistant to trans-4-hydroxy-L-proline. *Planta*, **153**, 166–71.

Larkin, P. J. & Scowcroft, W. R. (1981). Somaclonal variation – a novel source of variability from cell cultures for plant improvement. *Theoretical and Applied Genetics*, **60**, 197–214.

Matern, U., Strobel, G. & Shepard, J. F. (1978) Reaction to phytotoxins in a potato population derived from mesophyll protoplasts. *Proceedings of the National Academy of Sciences, USA*, **75**, 4935–9.

Negrutiu, I., Cattoir-Reynaerts, A., Verbruggen, I. & Jacobs, M. (1981). Lysine overproduction in an S(2-aminoethyl)-cysteine resistant mutant isolated in protoplast culture of *Nicotiana sylvestris*. *Archives Internationales de Physiologie et de Biochimie*, **89**, 188–9.

Oostindier-Braaksma, F. J. & Feenstra, W. J. (1973). Isolation and characterisation of chlorate-resistant mutants of *Arabidopsis thaliana*. *Mutation Research*, **19**, 175–85.

Roest, S. & Bokelmann, G. S. (1976). Vegetative propagation of *Solanum tuberosum* L. *in vitro*. *Potato Research*, **19**, 173–8.

Rognes, S. E., Bright, S. W. J. & Miflin, B. J. (1983). Feedback insensitive aspartate kinase isoenzyme in barley mutant resistant to lysine plus threonine. *Planta*, in press.

Rognes, S. E., Lea, P. J. & Miflin, B. J. (1980). S-adenosylmethionine – a novel regulator of aspartate kinase. *Nature*, **287**, 357–9.

Santos, A. V. P., dos, Ontka, D. E., Cocking, E. C. & Davey, M. R. (1980). Organogenesis and somatic embryogenesis in tissues derived from leaf protoplasts and leaf explants of *Medicago sativa*. *Zeitschrift für Pflanzenphysiologie*, **99**, 261–70.

Secor, G. A. & Shepard, J. F. (1981). Variability of protoplast-derived potato clones. *Crop Science*, **21**, 102–5.

Shahin, E. A. & Shepard, J. F. (1980). Cassava mesophyll protoplasts: isolation, proliferation, and shoot formation. *Plant Science Letters*, **17**, 459–65.

Shepard, J. F. (1981). Protoplasts as sources of disease resistance in plants. *Annual Reviews of Phytopathology*, **19**, 145–66.

Shepard, J. F., Bidney, D. & Shahin, E. A. (1980). Potato protoplasts in crop improvement. *Science*, **208**, 17–24.

Shepard, J. F. & Totten, R. E. (1977). Mesophyll protoplasts of potato: isolation, proliferation and plant regeneration. *Plant Physiology*, **60**, 313–16.

Somerville, C. R. & Ogren, W. L. (1981). Photorespiration-deficient mutants of *Arabidopsis thaliana* lacking mitochondrial serine transhydroxymethylase. *Plant Physiology*, **67**, 666–71.

Takebe, I., Labib, G. & Melchers, G. (1971). Regeneration of whole plants from isolated protoplasts of tobacco. *Naturwissenschaften*, **58**, 318–20.

Thoen, A., Rognes, S. E. & Aarnes, H. (1978). Biosynthesis of threonine from homoserine in pea seedlings II: threonine synthase. *Plant Science Letters*, **13**, 113–19.

Thomas, E. (1981). Plant regeneration from shoot culture-derived protoplasts of tetraploid potato (*Solanum tuberosum* cv. Maris Bard) *Plant Science Letters*, **23**, 84–8.

Thomas, E., Bright, S. W. J., Franklin, J., Gibson, R. W., Lancaster, V. & Miflin, B. J. (1982). Variation amongst protoplast-derived potato plants (*Solanum tuberosum* cv. Maris Bard). *Theoretical and Applied Genetics*, **62**, 65–8.

Thomas, E., Hoffmann, F., Potrykus, I. & Wenzel, G. (1976). Protoplast regeneration and stem embryogenesis of haploid androgenetic rape. *Molecular and General Genetics*, **145**, 245–7.

Vasil, V. & Vasil, I. K. (1980). Isolation and culture of cereal protoplasts. II. Embryogenesis and plantlet formation from protoplasts of *Pennisetum americanum*. *Theoretical and Applied Genetics*, **56**, 97–9.

Wenzel, G., Schieder, O., Przewozny, T., Sopory, S. K. & Melchers, G. (1979). Comparison of single cell culture derived *Solanum tuberosum* L. plants and a model for their application in breeding programmes. *Theoretical and Applied Genetics*, **55**, 49–55.

PART V

Genetic engineering of higher plants

A. A. GATENBY

The expression of eukaryotic genes in bacteria and its application to plant genes

Introduction

Recombinant DNA technology has been responsible for the rapid progress in recent years in the fine structural molecular analysis of genome organisation and function. The knowledge coming from the application of recombinant DNA techniques to plant chromosomes has been reviewed by Bedbrook & Kolodner (1979), Bedbrook & Gerlach (1980) and Flavell (1980, 1981). Briefly, these achievements have included the isolation and DNA sequencing of several plant genes (Bedbrook, Smith & Ellis, 1980*b*; Gerlach & Dyer, 1980; McIntosh, Poulsen & Bogorad, 1980), an examination of the mitochondrial genome (Lonsdale, Thompson & Hodge, 1981) and of mitochondrial DNA sequences involved in pollen development in maize (Levings *et al.*, 1980; Thompson, Kemble & Flavell, 1980), the development of a rapid assay for classifying cytoplasmic variants in maize (Kemble, 1980; Kemble, Gunn & Flavell, 1980), the investigation of the DNA sequences in heterochromatin (Bedbrook *et al.*, 1980*a*), the construction of chloroplast genome maps (Bedbrook & Bogorad, 1976; Bowman, Koller, Delius & Dyer, 1981) and the identification and physical mapping of nuclear chromosomes by *in situ* hybridisation (Gerlach, Miller & Flavell, 1980; Hutchinson, Chapman & Miller, 1980; Hutchinson, Flavell & Jones, 1981).

In addition to such purely analytical investigations, recombinant DNA technology has as a principal objective the genetic modification of organisms. This genetic modification includes the transfer of foreign or modified genes into eukaryotic or prokaryotic cells, usually with an ultimate desire of obtaining an altered phenotype. The genetic modification of higher plants is outside the scope of this chapter and the reader is referred to the recent review by Cocking, Davey, Pental & Power (1981). A major purpose of the transfer into, and expression of eukaryotic genes in, bacteria is the exploitation of known mechanisms of prokaryotic gene expression in order to increase the synthesis of the eukaryotic polypeptide. Such investigations give insight into the extent of the permissible interchangeability of prokaryotic and eukaryotic gene control signals. The opportunity is provided to maximise the production of polypeptides that may have an economic value, to synthesise protein subunits for structural investigations and to undertake precise changes in the nucleotide composition of genes for the purpose of examining changes in the structures and functions of polypeptides.

The techniques employed to detect and maximise eukaryotic gene expression in bacteria are similar for animal and plant genes. The paucity of publication on the expression of plant or plant virus genes in bacteria (Meagher, Tait, Betlach & Boyer, 1977; Gatenby, Castleton & Saul, 1981; Gatenby & Castleton, 1982) requires extensive use of examples of animal gene expression in bacteria for the illustration of these techniques. Useful reviews on the techniques for maximising gene expression in bacteria have been published by Bernard & Helinski (1979, 1980), Roberts & Lauer (1979) and Timmis (1981).

Features and strategies involved in eukaryotic gene expression in bacteria

The central requirements for maintaining the synthesis of a eukaryotic polypeptide in *Escherichia coli* are a bacterial or viral promoter, for driving transcription of the DNA sequence to high levels, and a ribosome binding site (RBS) for efficient translation at the initiation codon. In addition, the stability of the eukaryotic transcripts and polypeptides needs to be considered, together with any limitation on expression due to a different codon usage between the foreign gene and the *E. coli* tRNA pool.

Fig. 1. A comparison of two types of construction allowing increased levels of eukaryotic gene expression. In the upper example direct expression of the native polypeptide is obtained by judicious positioning of a promoter and RBS with respect to the initiation codon, or a eukaryotic RBS that is recognised by *E. coli* ribosomes is retained but a bacterial or phage promoter is positioned upstream. In the lower example a gene fusion is obtained between a C-terminal eukaryotic sequence that is in the same reading frame as an N-terminal bacterial sequence.

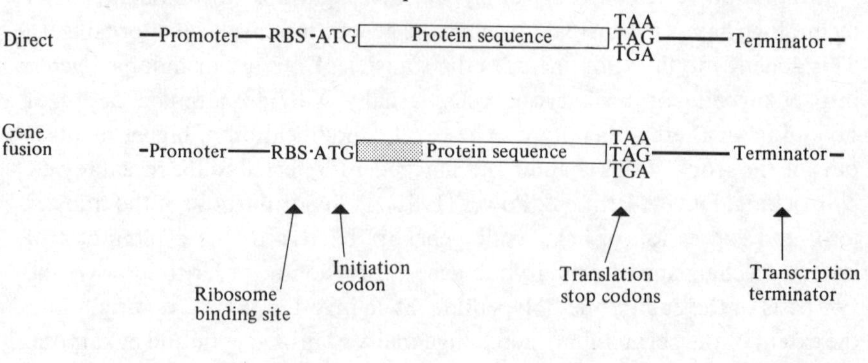

Promoters

Many *E. coli* promoters have been sequenced and they have a general structure of TTGACA and TATAAT centred −35 and −10 nucleotides respectively upstream from the transcription initiation site (Figs. 1, 2 and 3 and reviews by Scherer, Walkinshaw & Arnott, 1978; Rosenberg & Court, 1979). These sequences are thought to be involved in the initial recognition of the promoter region (at the −35 region) by the RNA polymerase prior to the formation of the tightly bound enzyme–DNA initiation complex. Although activator proteins and repressor proteins provide important on/off signals for some operons, many transcription initiation frequencies are determined solely by the interaction of RNA polymerase with the promoter region. The rate of formation of the initiation complex (and the subsequent transcription of the gene) varies with different promoters, and the strength of a promoter is in part determined by its nucleotide composition. In addition the time taken for a free promoter to be regenerated by movement of the polymerase and the time taken for RNA chain elongation may vary with the transcription sequence if pauses or abortive starts occur.

Fig. 2. A comparison of the nucleotide sequences of a prototype *E. coli* promoter and part of the upstream non-transcribed region of chloroplast DNA near the maize ribulose bisphosphate carboxylase large subunit (RuBPCase LS) gene. Homologous regions between the maize sequence and the prototype sequence are underlined. Sequences have been aligned allowing some flexibility in the non-conserved 6–9 bp and 4–7 bp regions, where bp are base pairs. In the prototype sequence highly conserved positions (N > 75%) are indicated by large upper-case letters, other well-conserved positions (N > 50%) are indicated by smaller upper-case letters and weakly conserved positions (n ~ 50%) are indicated by lower-case letters. (Maize sequence from McIntosh *et al.*, 1980; prototype sequence from Rosenberg & Court, 1979.)

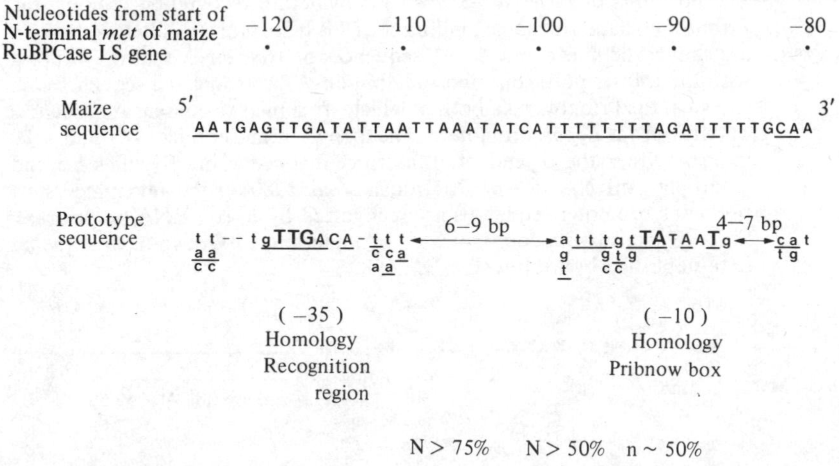

Nucleotides from start of N-terminal *met* of maize RuBPCase LS gene −120 −110 −100 −90 −80

Maize sequence 5′ AATGAGTTGATATTAATTAAATATCATTTTTTTTAGATTTTTGCAA 3′

Prototype sequence t t - - t g TTG A C A - t t t ←6–9 bp→ a t t t g t TA T A A T g ←4–7 bp→ c a t
a a / c c; c c a / a a; g / t; g t g / c c; t g

(−35) Homology Recognition region

(−10) Homology Pribnow box

N > 75% N > 50% n ~ 50%

Four promoters have been used frequently in the in vitro construction of transcriptional units. These are the promoters from the *lac* operon (P_{lac}), the bacteriophage lambda *N* operon (P_L), the *trp* operon (P_{trp}) and the β-lactamase promoter ($P_{\beta\text{-}lact}$) of the *E. coli* plasmid pBR322 (Table 1). However, for the purposes of this discussion, only the first three are considered in any detail since the $P_{\beta\text{-}lact}$ promoter is inadequate in cases where induced expression of a cloned gene is required.

P_{lac}. Transcription from P_{lac} is regulated by the *lac* repressor, the catabolite gene activator protein (CAP) and RNA polymerase. Negative control is exerted by the *i* gene product, the repressor, which binds to the operator in

Table 1. *Main promoters used for increased expression of genes in* E. coli

Promoter	Source	Control of Expression
P_{lac}	*lac* operon	Constitutive owing to titrating out of repressor[a]
P_L	lambda *N* operon	Induced by thermoinactivation of *c*I repressor
P_{trp}	*trp* operon	Induced by 3-β-indolylacrylic acid
$P_{\beta\text{-}lact}$	β-lactamase promoter of pBR322	Constitutive

[a] When P_{lac} is on a multicopy plasmid there is insufficient repressor in the cell to bind to all of the *lac* operators. The operators therefore titrate out the repressor and expression is constitutive.

Fig. 3. Structural features of the RuBPCase LS gene of maize chloroplasts. The positions of putative –35 RNA polymerase recognition regions, the Pribnow box and ribosome binding site (RBS) are indicated. Below the RBS are shown the maize and *E. coli* sequences of 16S rRNA that may duplex with the transcript during ribosome binding. Two staggered sequences are shown at the Pribnow box, both of which are similar to consensus sequences obtained from *E. coli* promoters. The arrows at nucleotides –61 and –80 indicate either the 5′ end of transcripts mapped using S1 nuclease and chloroplast mRNA at –61 (McIntosh *et al.*, 1980) or the anticipated start site if the promoter sequences are recognized by *E. coli* RNA polymerase. Shine–Dalgarno sequence indicated by underlining. (Drawn from sequence data published by McIntosh *et al.*, 1980.)

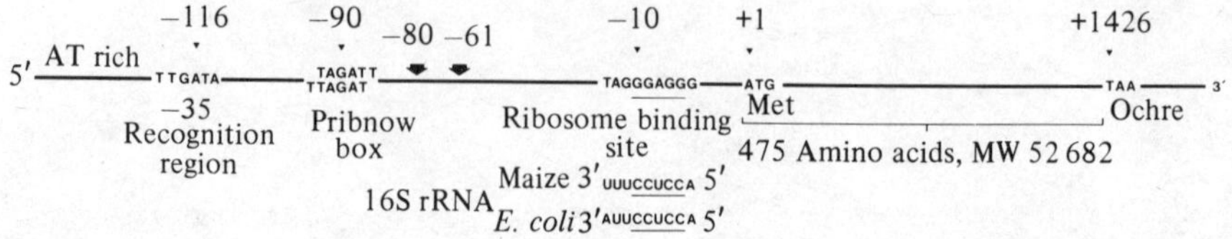

–116 –90 –80 –61 –10 +1 +1426
5′ AT rich TTGATA TAGATT TTAGAT TAGGGAGGG ATG TAA 3′
–35 Recognition region
Pribnow box
Ribosome binding site
Met
Ochre
475 Amino acids, MW 52 682
16S rRNA
Maize 3′ UUUCCUCCA 5′
E. coli 3′ AUUCCUCCA 5′

the absence of a β-galactosidase inducer. Positive control is exerted by cyclic AMP activated CAP. Expression vectors using P_{lac} usually make use of the promoter mutation UV5 which makes the promoter insensitive to catabolite repression and increases its efficiency for expression (see Maquat & Reznikoff, 1978 and Scherer *et al.*, 1978 for further details). When P_{lac} is present on a multicopy plasmid the multiple copies of the *lac* operator compete with the *lac* operator on the bacterial chromosome for the binding of the few *lac* repressor molecules in the cell. This results in constitutive expression of the *lac* operon or any transcriptional units constructed with the *lac* operator. Expression can be enhanced by the addition of the *lac* inducer isopropyl-β-D-thiogalactoside (IPTG) but this does not usually give a substantial increase in yield because the *lac* operators titrate out more repressor molecules than can be synthesised and are therefore usually active without inducers. Strains are available that overproduce *lac* repressor and therefore respond to IPTG induction more noticeably (Mercereau-Puijalon *et al.*, 1978); a useful property when it is desirable to demonstrate that expression of a eukaryotic gene is under *lac* control. P_{lac} is a powerful promoter and in appropriate constructions can lead to expression of *lac* Z protein synthesis to levels constituting 30% of the total soluble protein of the cell (Roberts & Lauer, 1979).

P_L. The early leftward promoter P_L of the bacteriophage lambda *N* operon is also a powerful promoter. In addition to promoting a very high level of expression, P_L-initiated transcription can extend beyond termination signals in the presence of the lambda *N* gene product, which functions as a transcriptive antitermination factor (Gottesman, Adhya & Das, 1980). The RNA polymerase is modified at a lambda *N* utilisation site (*nut*) by the *N* protein into a form that no longer responds to most transcription termination sites (reviewed by Greenblatt, 1981).

The achievement of expression using P_L and *N* depends on the absence of the lambda *c*I repressor and the loss of the additional negative control normally provided by the *cro* gene product (reviewed by Ptashne *et al.*, 1980; Johnson *et al.*, 1981). This is most readily achieved by transferring P_L and part of the *N* operon (including the *nut* L site) from the phage chromosome into a multicopy plasmid (Bernard *et al.*, 1979; Remaut, Stanssens & Fiers, 1981), thus physically separating P_L from its repressors. Control of expression from P_L on the plasmid is, however, still required since unrepressed transcription leads to loss of the plasmid, possibly as a result of interference with plasmid replication. Regulation is accomplished by the use of a temperature-sensitive repressor gene, cI_{857}, which is located either on a chromosome as a defective prophage or on a plasmid that is compatible with the expression plasmid (Bernard *et al.*, 1979). At 32 °C transcription is repressed by the binding of the product of the cI_{857} gene to the leftward operator o_L, leaving P_L inaccessible to RNA polymerase. Raising the temperature to 45 °C denatures the repressor and allows RNA polymerase to bind to the leftward operator so that transcription is initiated from P_L. Thus expression from this promoter is subject to controlled activation of transcription by heat induction.

P_{trp}. Transcription from P_{trp} is regulated by the *trp* repressor and only low levels of *trp* gene products are synthesised when *E. coli* cells are grown in media supplemented with tryptophan. There is sufficient *trp* repressor present in the cell to maintain almost complete repression even when P_{trp} is present on a multicopy plasmid (reviewed by Bernard & Helinski, 1980). Operon induction can be achieved using the tryptophan analogue 3-β-indolylacrylic acid and expression can therefore be easily regulated in any strain. Unfortunately, P_{trp} expression plasmids are unlikely to be as useful as plasmids based on P_{lac} or P_L in the near future because the most promising plasmids available (Tacon, Carey & Emtage, 1980) are not being released by the employers of the cited authors without recipients entering into a contractual agreement.

The ribosome binding site

The initiation of protein synthesis in *E. coli* is partly governed by the availability of a RBS on a transcript. The nature of the RBS is still incompletely defined but sequence analysis has identified a region, located upstream from the translational initiation codon on bacterial and phage transcripts, that is conserved. This region is known as the Shine–Dalgarno (SD) sequence and is complementary to the 3′ end of the 16S ribosomal RNA (Shine & Dalgarno, 1975). Base pairing between the SD sequence and the 3′ end of the 16S ribosomal RNA occurs during the initiation of protein synthesis. The SD sequence has different lengths (3–11 bases) in different mRNAs and it has been suggested from thermodynamic considerations that the efficiency of ribosome binding is directly proportional to the extent of homology (Jay, Khoury, Seth & Jay, 1981). The efficiency of the initiation of translation is also determined by the location of the SD sequence relative to the translational initiation codon (Roberts, Kacich & Ptashne, 1979). Recently, synthetic RBS of high efficiency have been produced for the construction of expression vectors (Jay *et al.*, 1981).

Gene fusions

There are two general approaches to the expression of eukaryotic gene sequences in bacteria, namely the construction of transcriptional and translational gene fusions (Fig. 1). Transcriptional fusions are constructed from a promoter which derives transcription of the eukaryotic sequence, and a RBS and initiation codon which allow direct expression of the native polypeptide. Translational fusions are assembled from a eukaryotic coding sequence that is inserted into an internal position of a bacterial coding sequence so that the eukaryotic sequences are in the same translational phase as the bacterial coding sequences. Translational fusions make use of the bacterial gene promoter, RBS and initiation codon and result in a hybrid polypeptide with an N-terminal bacterial sequence and a C-terminal eukaryotic sequence. It is essential for intervening sequences of eukaryotic genes to be absent. This can be achieved by using either copy DNA (cDNA) clones or chemically synthesised genes (Edge *et al.*, 1981).

Transcriptional fusions. There are two general ways in which transcriptional fusions have been constructed. These are: (a) by positioning a promoter upstream from a gene possessing a RBS that is recognised in *E. coli* and (b) by the formation of a hybrid RBS composed of a prokaryotic SD sequence and a eukaryotic, or a synthetic, initiation codon that is transcribed from an upstream promoter. Both of these types of constructions allow direct expression of the eukaryotic polypeptide (Table 2), although the N-terminal methionine may not be cleaved.

The positioning of a strong promoter upstream from a gene possessing a RBS has been widely used to amplify the expression of many *E. coli* genes (Steffen & Schleif, 1977; Murray & Kelley, 1979; Wilson & Murray, 1979). This approach has also been useful for enhancing the expression of chloroplast genes in *E. coli* (Gatenby *et al.*, 1981; Gatenby & Castleton, 1982). On the evidence of their efficient translation in *E. coli*, chloroplast genes would appear to possess a SD sequence that is recognised by the *E. coli* ribosome (Fig. 3).

The expression of other eukaryotic genes in *E. coli* is not often accomplished

Table 2. *Examples of the expression of eukaryotic genes in* E. coli *by transcriptional fusions*

Gene	Promoter	Amount[a]	Reference
β-globin	P_{lac} + hybrid RBS[b]	15000	Guarente *et al.* (1980)
Growth hormone	P_{lac} + *lac* RBS	186000	Goeddel *et al.* (1979*a*)
Maize RuBPCase	P_L + own RBS	60000	Gatenby & Castleton (1982)

[a] Amount is expressed as molecules per cell.
[b] RBS, ribosome binding site.

as easily as with chloroplast genes. More attention must usually be given to the correct positioning of the SD sequence, the initiation codon and the particular eukaryotic coding sequence. This is illustrated by the work of Goeddel *et al.* (1979*a*) on the DNA sequence coding for human growth hormone (HGH). A cDNA coding for amino acids 24–191 of the hormone was enzymatically prepared and then ligated to a chemically synthesised adaptor fragment containing an ATG initiation codon and coding sequences for amino-acid residues 1–23 of the hormone. The gene was placed under the transcriptional control of P_{lac} using a reconstructed *lac* RBS. This system resulted in the high level of synthesis of 186000 monomers of HGH per cell.

The construction of this system required a detailed knowledge of the DNA sequence of parts of the gene and the correct location of the initiation codon. Other systems have been developed that do not require such detailed information and rely on the random placement of a SD sequence, by the use of nucleases in vitro, at various positions between a promoter and the initiation codon. Jay *et al.* (1981) have used this technique to position a synthetic SD sequence at various positions at the 5′ side of a Simian Virus 40 (SV40) t-antigen gene to obtain expression of the native polypeptide using $P_{\beta\text{-}lact}$. The synthetic SD sequence also included a translational termination codon upstream from the RBS to prevent translational fusions between β-lactamase and the SV40 t-antigen. The construction of a hybrid RBS using P_{lac} has also been described by Backman & Ptashne (1978) and Roberts *et al.* (1979). An important advance in the random placement of a SD sequence near an initiation codon is the development of an in vivo assay for clones with an optimum positioning of these sequences (Guarente, Lauer, Roberts & Ptashne, 1980). In this system the region of a eukaryotic gene encoding the N-terminal portion of the protein is fused to DNA encoding an enzymatically active C-terminal fragment of β-galactosidase. A fragment containing P_{lac} and the *lac* SD sequence is placed at many positions in front of the fused gene using nucleases. Those promoter placements that allow efficient expression of the fused gene are identified by the β-galactosidase activity that they cause to be expressed and can be identified on selective plates. The gene of interest is then reconstituted intact, still under P_{lac} control, by recombination either in vitro or in vivo. About 10000–15000 monomers of native rabbit β-globin were obtained with this method (Guarente *et al.*, 1980).

Translational fusions. Translational fusions are usually easier to construct than transcriptional fusions in that they make use of a pre-existing arrangement of promoter, SD sequence and initiation codon because they consist of a eukaryotic sequence that has been inserted into the coding region of a bacterial gene. The main consideration in this form of construction is the correct arrangement of reading frames between the eukaryotic and prokaryotic sequences. The polypeptide obtained with this method is usually a fusion between the N-terminal bacterial sequence and the C-terminal eukaryotic sequence, except in the unusual examples of translational reinitiation (see below). Problems may therefore arise in reconstituting the mature eukaryotic protein by cleavage of the prokaryotic sequence, although this has been achieved in some cases by using trypsin digestion, for example, to release β-endorphin after protecting lysine residues by reaction with citraconic anhydride (Shine *et al.*, 1980) or cyanogen bromide treatment to cleave the N-terminal methionine of the eukaryotic sequence in a fused polypeptide when there is no internal methionine residue in the eukaryotic polypeptide (Itakura *et al.*, 1977; Goeddel *et al.*, 1979*b*).

Correct phasing of reading frames has been simplified for *Eco*RI, *P*ST I and *Hind* III sites by the construction of vectors in all three reading frames based on β-galactosidase (Charnay *et al.*, 1978), β-lactamase (Talmadge, Stahl & Gilbert, 1980) and *trp* E (Tacon, Carey & Emtage, 1980 and Table 3). Other plasmids exist to allow fusions between β-galactosidase (Fraser & Bruce, 1978; Mercereau-Puijalon *et al.*, 1978), β-lactamase (Remaut *et al.*, 1981), *trp*D (Hallewell & Emtage, 1980) and MS2 polymerase (Remaut *et al.*, 1981) (Table 4). Phasing in any of these plasmids can be adjusted by the use of nucleases or linkers, or by the addition of tails of different lengths to randomise the reading frames (see Charnay *et al.*, 1978; Erwin, Maurer & Donelson, 1980; Talmadge *et al.*, 1980 and Remaut *et al.*, 1981 for examples of these techniques). Examples of these plasmids and the levels of expression obtained using them are shown in Table 5.

Table 3. *Plasmids used for positioning cloned inserts in the correct translational reading frame*[a]

Gene	Site	Vectors	Reference
β-lactamase	*Pst* I	pKT 279, pKT 280, pKT 287	Talmadge & Gilbert (1980)
lac Z	*Eco* RI	pCP ϕ 1, pCP ϕ 2, pCP ϕ 3	Charnay *et al.* (1978)
trp E	*Hind* III	pWT 111, pWT 121, pWT 131	Tacon *et al.* (1980)

[a] See also Guarente *et al.* (1980).

Table 4. *Plasmids used for amplifying gene expression*

Plasmid	Promoter	Size (kb)	Notes	Reference
pHUB 2	P_L	7.6	*Hpa* I, *Eco* RI, *Bam* HI and *Sal* I sites	Bernard *et al.* (1979)
pHUB 4	P_L	6.5	*Hpa* I, *Bam* HI and *Sal* I sites. *N* gene also on plasmid	Bernard *et al.* (1979)
pLc 24	P_L	3.3	Many different cloning sites allowing translational fusions with β-lactamase or MS2 polymerase or transcriptional fusions with P_L	Remaut *et al.* (1981)
pLa 2311	P_L	3.8		
pLa 8	P_L	3.8		
pLa 236	P_L	4.5		
pLc 28	P_L	2.8		
pOMPO	P_{lac}	7.1	*Eco* RI site allowing translational fusion with β-galactosidase	Mercereau–Puijalon *et al.* (1978)
p*trp ED*5-1	P_{trp}	6.7	*Hind* III, *Bam* HI and *Sal* I sites	Hallewell & Emtage (1980)
pLG 200 pLG 300 pLG 400	Plasmids for constructing fusions between a portable P_{lac}, a eukaryotic sequence and β-galactosidase			Guarente *et al.* (1980)

The use of fusions between β-lactamase and a eukaryotic sequence frequently allows the penicillinase signal sequence to transport the fused polypeptide into a periplasmic space. This is often an advantage for the purification of the eukaryotic polypeptide since most *E. coli* proteins are not transported into this space. The plasmids described by Talmadge & Gilbert (1980) not only have the advantage of having *Pst* I sites in all three reading frames but also have the penicillinase signal sequence which is cleaved after transport.

Translational reinitiation. Another type of construction, although originally designed as a translational fusion, turned out to be a transcriptional fusion because of translational reinitiation at the eukaryotic initiation codon. Garapin *et al.* (1981) placed a herpes simplex virus type I thymidine kinase (TK) gene in three different reading frames near the beginning of the *lac Z* gene under control of P_{lac}. The anticipated result was a fused lac Z/TK gene in one of the reading frames; instead a native TK polypeptide was detected in all three reading frames. DNA sequence data indicate that translational reinitiation occurred at the 5′ end of the TK gene after translational stop

Table 5. *Examples of the expression of eukaryotic genes in* E. coli *by translational fusions*

Gene	Promoter	N-terminal polypeptide	Amount[a]	Reference
Insulin	P_{lac}	β-galactosidase	20%	Goeddel *et al.* (1979*b*)
Insulin	$P_{\beta\text{-}lact}$	β-lactamase	6000 mols.	Talmadge *et al.* (1980)
Ovalbumin	P_{lac}	β-galactosidase	30000–90000 mols.	Mercereau–Puijalon *et al.* (1978)
Ovalbumin	P_{lac}	β-galactosidase	1.5%	Fraser & Bruce (1978)
Growth hormone	P_{trp}	*trpD*	3%	Martial, Hallewell, Baxter & Goodman (1979)
Somatostatin	P_{lac}	β-galactosidase	0.03%	Itakura *et al.* (1977)
β-endorphin	P_{lac}	β-galactosidase	80000 mols.	Shine *et al.* (1980)

[a] The amount is expressed either as the number of molecules per cell (mols.) or as a percentage of cell protein.

signals in all three phases in the vicinity of the *Eco* RI site used for the DNA ligation. Similarly Nagata *et al.* (1980) found that human leukocyte interferon could be synthesised in all three reading frames using $P_{\beta\text{-}lact}$ as a non-fused polypeptide. Emtage *et al.* (1980) observed expression of fowl plague virus haemagglutinin as a non-fused polypeptide in all three reading frames using P_{trp} as a promoter. They concluded that translational initiation occurred from the natural haemagglutinin AUG and that the eukaryotic RBS was recognised with an efficiency of 20% compared with the prokaryotic site.

The use of eukaryotic sequences that function as promoters and ribosome binding sites in E. coli

Several examples are now known where expression of a cloned eukaryotic gene is detectable in *E. coli* even when no special measures are taken to increase the rate of transcription or the efficiency of translation. These genes have been isolated from yeasts (Clarke & Carbon, 1978; Dickson & Markin, 1978; Citron, Feiss & Donelson, 1979; Struhl, Stinchcomb & Davis, 1980), *Neurospora* (Vapnek *et al.*, 1977) and chloroplasts (Gatenby *et al.*, 1981). In some cases it seems likely that this expression relies on eukaryotic sequences. For example, the analysis of the DNA sequence of the chloroplast gene for the large subunit (LS) of ribulose bisphosphate carboxylase (RuBPCase) of maize illustrates strong homologies between known *E. coli* promoters and SD sequences (McIntosh, Poulsen & Bogorad, 1980; Gatenby *et al.*, 1981 and Figs. 2 and 3).

The stabilisation of plasmids, transcripts and polypeptides

The high levels of transcription attained in unregulated expression plasmids can lead to eventual loss of the plasmid, possibly as a result of interference with plasmid replication. Hallewell & Emtage (1980) have reported that a P_{trp} expression plasmid p*trp ED*5-1 is maintained stably in *E. coli* during normal growth but is rapidly lost from cells synthesising large amounts of *trp* proteins. Plasmid loss after prolonged induction has also been observed by Remaut *et al.* (1981) with plasmids containing P_L. In both of these examples the plasmids are stable when the powerful promoters are regulated by either the *trp* repressor or the cI_{857} repressor. Induction with 3-β-indolylacryclic acid (P_{trp}) or by raising the temperature to 42 °C (P_L) allows expression, but as the duration of the experiments is usually only a few hours the loss of plasmid is not important. Such difficulties have not been reported for plasmids containing P_{lac} and would not be expected in plasmids containing $P_{\beta\text{-}lact}$ as this promoter is not particularly powerful.

Very little information has been published about methods for stabilising in vivo eukaryotic transcripts or polypeptides. Kusunic & Kushner (1980)

used a polynucleotide phosphorylase (*pnp*) deficient strain of *E. coli* and found enhanced expression of the imidazoleglycerol phosphate dehydratase gene of yeast. This appeared to be a specific stabilisation of the mRNA since the half-life increased from 1·5 to 18·7 min. Cheng *et al.* (1981) demonstrated that overproduction of a protein in *E. coli* can lead to its stabilisation against degradation. They suggested that this may be due to the formation of proteinaceous aggregates which protect the protein from the bacterial proteases. *E. coli* contains eight soluble proteolytic activities (Swamy & Goldberg, 1981) and a highly specific and efficient mechanism for the degradation of proteins with aberrant or truncated structures.

The culture medium can also affect polypeptide stability. Goeddel *et al.* (1979*a*) observed that HGH was more stable in cells grown in a rich medium than in cells grown in minimal medium. Backman and Ptashne (1978) describe variation in the synthesis of λ repressor when cells were grown in different media with the *c*I gene under transcriptional control of P_{lac}. The translational fusion of small eukaryotic polypeptides and bacterial polypeptides has also resulted in the stabilisation of the eukaryotic sequence, presumably by protection against proteases. Itakura *et al.* (1977) stabilised the hormone somatostatin by a fusion which caused replacement of a few of the C-terminal codons of the β-galactosidase gene.

The detection and identification of expression

The expression of eukaryotic polypeptides in *E. coli* can be detected and identified using immunological assays, functional assays for enzyme activities or biological activity, and the complementation of mutants.

Immunological assays. Immunological techniques are widely used to characterise expression of a previously identified DNA fragment or to screen random recombinant clones for the synthesis of a specific antigen.

Methods for screening phage plaques or bacterial colonies include *in situ* immunoassay on agar overlays (Anderson, Shapiro & Skalka, 1979), reaction of radioactive antibodies with antigens (Broome & Gilbert, 1978; Clark, Hitzeman & Carbon, 1979), reaction of radioactive protein A with antigen–antibody complexes (Erlich, Cohen & McDevitt, 1979) and chromogenic detection with horseradish peroxidase–antibody conjugates (Kaplan, Naumovski & Collier, 1981). These methods are very sensitive with the theoretical ability of detecting one antigen molecule per *E. coli* cell using radioiodinated antibody or protein A, or 10–20 molecules per cell using *in situ* immunoassay (Anderson *et al.*, 1979).

For a more detailed examination of individual clones, analysis using minicells (Reeve, 1979; Gatenby *et al.*, 1981), maxicells (Sancar, Hack &

Rupp, 1979; Guarente *et al.*, 1980) or phage infection of UV-irradiated cells (Jaskunas, Lindahl, Nomura & Burgess, 1975; Gatenby *et al.*, 1981) is preferable. These techniques rely on the inhibition of chromosomal-encoded protein synthesis either by UV-induced damage (maxicells and phage infection) or by physical separation on sucrose gradients of small cells lacking chromosomal DNA (minicells). The synthesis of proteins encoded by plasmids or phages is then followed with the use of a radioisotope. Maxicells are intact cells which contain mutations (*rec*A, *uvr*A, *phr*-1) (Sancar *et al.*, 1979) that make the chromosomal DNA particularly sensitive to UV irradiation. Incubation of the culture after UV treatment causes degradation of the chromosomal DNA but plasmic copy number increases due to the inhibition of cell protein synthesis.

The expression of genes cloned into lambda can be examined with either lysogens or strains carrying a plasmid with the *c*I repressor gene (Reeve, 1978). Both repress transcription from the phage promoters and thus allow the investigation of the expression of the cloned gene using its own promoters (Brunel, Davison, Thi & Merchez, 1980). Radioimmunoassays can also be used to quantify and confirm expression of particular clones (Mercereau-Puijalon *et al.*, 1978; Shine *et al.*, 1980; Talmadge *et al.*, 1980).

Enzymatic assays and bioassay. Functional assays have been widely used for the analysis of human leukocyte or fibroblast interferon activity in *E. coli* extracts either as a screening method for the identification of cDNA clones (Nagata *et al.*, 1980) or for the characterisation of purified clones (Derynck *et al.*, 1980; Mory *et al.*, 1981). Human growth hormone has also been bioassayed (Olson *et al.*, 1981). Enzymatic assays have been carried out to characterise clones for herpes simplex thymidine kinase (Garapin *et al.*, 1981), mouse dihydrofolate reductase (Chang *et al.*, 1978), yeast argininosuccinate lyase (Clarke & Carbon, 1978), yeast galactokinase (Citron *et al.*, 1979), yeast imidazoleglycerol-phosphate dehydratase (Struhl & Davis, 1977), yeast β-galactosidase (Dickson & Markin, 1978) and *Neurospora* catabolic dehydroquinase (Vapnek *et al.*, 1977).

Functional expression in vivo. Genes have also been identified by their ability to complement *E. coli* mutants. The source of these genes has usually been lower eukaryotes. The following mutant *E. coli* alleles have been complemented by yeast genes: *arg*H (Clarke & Carbon, 1978), *gal*K (Citron *et al.*, 1979), *trp*AB (Walz, Ratzkin & Carbon, 1978), *his*B (Struhl & Davis, 1977), *leu*B (Ratzkin & Carbon, 1977) and *lac*Z39 (Dickson & Markin, 1978). An *aro*D mutant has been complemented by the *Neurospora* gene (Vapnek *et al.*, 1977) and a *tdk*-1 mutant has been completed by a herpes simplex virus TK gene

(Garapin *et al.*, 1981). All of these complementations utilised genomic fragments of DNA which thus presumably represented genes lacking intervening sequences. An interesting example of a higher eukaryotic gene being selected for in *E. coli* is that of the gene for mouse dihydrofolate reductase; selection was for resistance to the antimetabolite trimethoprim (Chang *et al.*, 1978). Yeast fragments have also been selected that suppress an *E. coli* mutation (Clarke & Carbon, 1978). An important feature of the identification of eukaryotic genes by specific complementation or the acquisition of drug resistance is that selection can be applied to increase the level of expression. Usually the unmodified foreign gene functions poorly in *E. coli*. For example, the expression of the yeast galactokinase gene in *E. coli* produced transformants that had a generation time of 14·3 h and 0·7% of the levels of the enzyme that wildtype cells contained (Citron *et al.*, 1979). Similarly, with yeast β-galactosidase the generation time was more than 24 h (Dickson & Markin, 1978). Although the level of yeast *his*3 gene expression is relatively high (Struhl *et al.*, 1980), Clarke & Carbon (1978) were able to select for variants of yeast argininosuccinate lyase with spontaneously increased levels of expression at a frequency of 10^{-5}; some of these variants produced more enzyme per cell than the cell's own derepressed operon. Brennan & Struhl (1980) showed that by selecting for *his*3 overproducers, *E. coli* could use a variety of mechanisms for increasing the expression of a eukaryotic gene including point mutations, deletions and IS2 insertions. Walz *et al.* (1978) also reported increased expression of yeast genes in *trp*AB mutants by IS2 insertion.

Expression in bacterial hosts other than E. coli

Other bacterial hosts may be suitable for the expression of eukaryotic genes. In particular *Bacillus subtilis* has attracted attention since strains of this bacterium are widely used commercially for producing enzymes and antibiotics, and existing methods for large-scale growth and protein isolation could be useful for producing eukaryotic polypeptides (Hardy, Stahl & Küpper, 1981). Reviews on *B. subtilis* as a host for cloning have been published by Lovett & Keggins (1979) and Dubnau, Gryczan, Contente & Shivakumar (1980). Reeve (1979) has described the use of *B. subtilis* minicells. Plasmids for cloning into gram-positive and gram-negative bacteria have been described by Bernard & Helinski (1980). Hardy *et al.* (1981) have recently demonstrated the production in *B. subtilis* of hepatitis B core antigen and of the major antigen of foot and mouth disease virus; synthesis of the latter reached about 1% of cellular protein.

Plant gene expression in *E. coli*

At present there are only two publications on the expression of plant genes in *E. coli* and both of these concern the chloroplast genes for the LS of RuBPCase (Gatenby *et al.*, 1981; Gatenby & Castleton, 1982). It is likely that plant nuclear gene expression in *E. coli* will soon be demonstrated. The expression of nuclear genes in *E. coli* will no doubt present similar difficulties to those experienced in the expression of mammalian genes described previously. For example the demonstration of intervening sequences in the genes for bean phaseolin (Sun, Slightom & Hall, 1981) and soybean leghaemoglobin (Jensen *et al.*, 1981) would require the use of cDNA clones for expression in bacteria.

Chloroplast gene expression in plasmids

Our decision to examine chloroplast gene expression in *E. coli* was strongly influenced by the known similarities between chloroplasts and prokaryotes. In both there are 70S ribosomes containing 23S, 16S and 5S rRNA (reviewed by Bedbrook & Kolodner, 1979), there is strong sequence homology between maize chloroplast and *E. coli* 16S rRNA genes (Schwarz & Kossel, 1980) and N-formylmethionyl-tRNA is used for the initiation of polypeptide chain synthesis (Schwartz, Meyer, Eisenstadt & Brawerman, 1967). In addition chloroplast genes have been transcribed and translated successfully in vitro using cell-free extracts of *E. coli* (Whitfeld & Bottomley, 1980), indicating that the prokaryotic protein-synthesising components may recognise chloroplast genes in vivo. More recently Zech, Hartley & Bohnert (1981) demonstrated the binding of the *E. coli* DNA-dependent RNA polymerase to specific sites on spinach chloroplast DNA and were able to obtain an overall correlation of polymerase binding sites with a map of chloroplast mRNAs which accumulate in the organelle in vivo.

In order to examine the in vivo expression of chloroplast genes in *E. coli*, my colleagues and I used the maize and wheat genes for the LS of RuBPCase (McIntosh *et al.*, 1980; Bowman, Koller, Delius & Dyer, 1981) cloned into the *Bam* H1 site of plasmid pBR322. Plasmids were constructed with the 4·3 kbp maize fragment and the 9·6 kbp wheat fragment in both of the two possible orientations (Fig. 4). Following transformation into a minicell-producing strain and isolation by sucrose gradient centrifugation, the minicells were incubated with [^{35}S]methionine. The presence of a polypeptide of the same molecular weight as the RuBPCase LS was demonstrated by sodium dodecylsulphate (SDS) polyacrylamide gel electrophoresis (Fig. 5) (Gatenby *et al.*, 1981). This polypeptide could be immunoprecipitated by anti-RuBPCase serum, but not control serum, and was only found in cells containing the plasmid. Furthermore, synthesis was independent of the orientation of the fragment in the plasmid, suggesting that the expression of the gene was arising from chloroplast sequences since expression from external plasmid promoters would be expected to be orientation dependent. Similar examples of expression from within a cloned fragment were described by Henikoff, Tatchell, Hall & Nasmyth (1981), Clarke & Carbon (1978), Struhl *et al.* (1980) and Brunel *et al.* (1980). The level of expression of the maize gene was approximately 100 to 200 subunit monomers per cell.

An examination of the DNA sequence of the maize LS gene reveals

Fig. 4. The two possible orientations of the maize RuBPCase LS gene cloned into the *Bam* HI site of the plasmid pBR322. The orientations can be distinguished using the asymmetric distribution of *Pst* I sites within the plasmid and the insert and the sizes of the *Pst* I fragments produced by cutting the new plasmids pZmB1 (A) and pZmB1 (B) are shown in the gel tracks (*a*) and (*b*), respectively, on the left. The orientation of the relevant insert can be deduced from these fragments. The pBR322 sequence is drawn as a thick line and the maize sequence as a thin line. Ap^R^, ampicillin resistance.

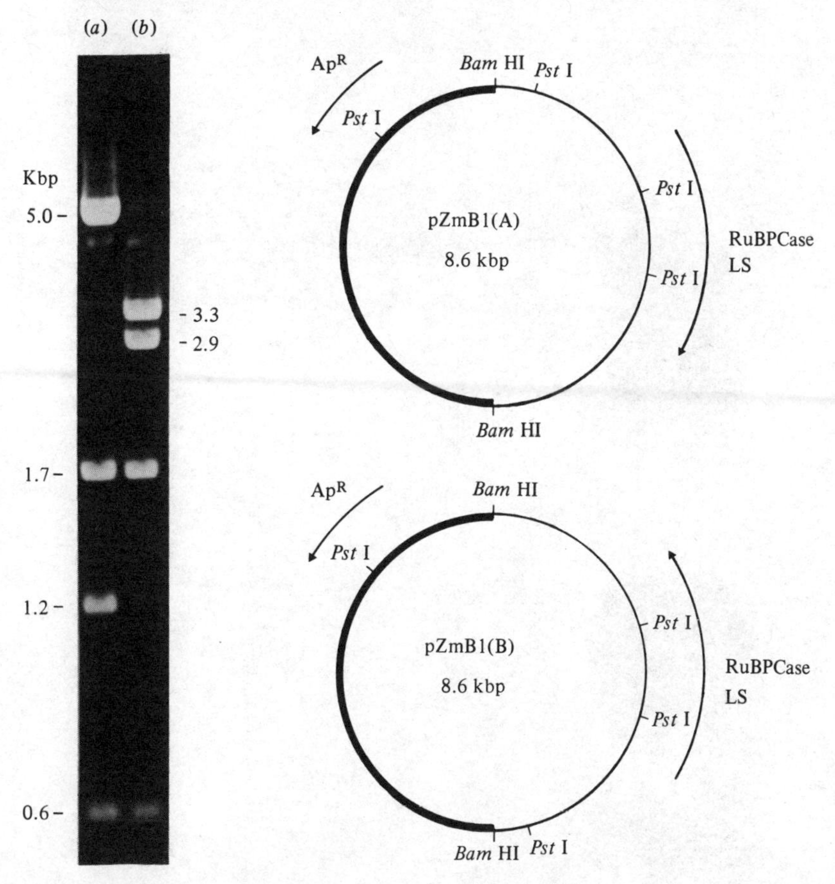

Fig. 5. Expression of the maize RuBPCase LS gene in *E. coli* minicells. ^{35}S-labelled samples were run on a 10% polyacrylamide gel containing SDS and autoradiographed. The samples from left to right are (*a*) total proteins synthesised from the plasmid pZmB1(B), (*b*) proteins absorbed to an anti-RuBPCase immunoabsorbent column, (c) protein incubated with anti-RuBPCase serum and absorbed to protein A Sepharose, (*d*) protein incubated with pre-immune serum and absorbed to protein A Sepharose and (*e*) molecular weight marker tracks. The use of anti-RuBPCase serum absorbed to protein A Sepharose enhances detection of the specific protein (cf. columns *b* and *c*). Reprinted from Gatenby *et al.*, 1981: *Nature*, vol. 291, no. 5811, pp. 117–21.

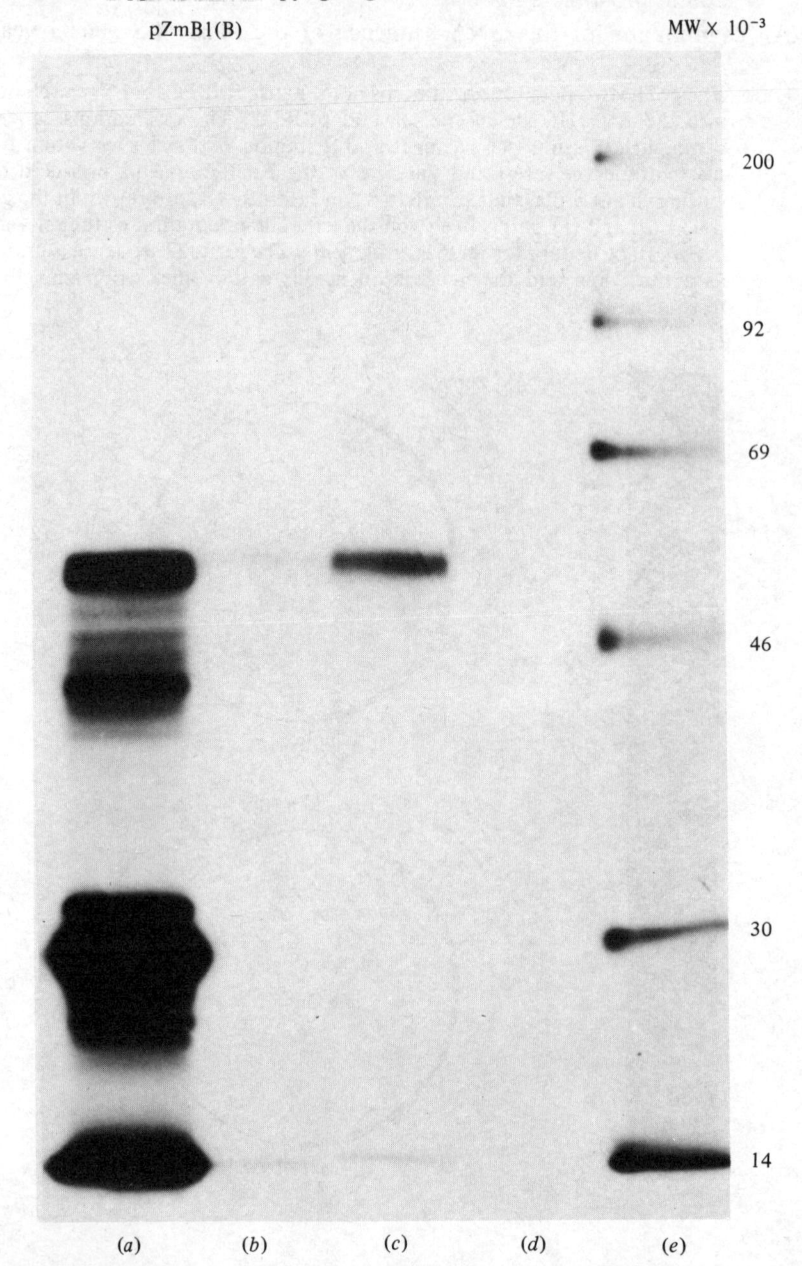

structural features that may explain this expression. McIntosh *et al.* (1980) reported the presence of a sequence that is partially complementary to a SD sequence located five nucleotides upstream from the N-terminal methionine codon on the LS mRNA. Further upstream is a region bearing striking similarities to a prokaryotic promoter sequence with an approximately 80% homology between a bacterial prototype sequence and the maize sequence (Gatenby *et al.*, 1981; Figs. 2 and 3). Whether such sequences function as promoters and RBS in *E. coli* remains to be determined but the sequence homologies are so similar that it will be surprising if they do not. If the putative maize promoter sequences shown in Figs. 2 and 3 are functional then an interesting problem arises. *E. coli* RNA polymerase, on recognising these features, would be expected to initiate transcription at or close to the nucleotide at position −80 upstream from the translational initiation codon owing to the strong steric constraints on the site at which RNA polymerase initiates transcription (Rosenberg & Court, 1979). However, using RNA/DNA hybridisation and S1 mapping techniques McIntosh *et al.* (1980) have placed the 5′ end of the LS mRNA 59 to 63 nucleotides upstream from the start of the initiation codon. This difference of 20 or so nucleotides (equivalent to two turns of the double helix) may indicate that either (a) maize chloroplast RNA polymerase recognises the sequences shown in Figs. 2 and 3 but because of differences in its structure it initiates transcription two turns of the helix further away from the site where *E. coli* RNA polymerase would initiate transcription, or (b) maize chloroplast RNA polymerase does not recognise these promoter-like sequences which are present fortuitously.

Expression of a chloroplast gene cloned in bacteriophage lambda during lytic infection

The sequence 5′-GGAGG-3′ located five nucleotides upstream from the N-terminal methionine codon on the LS mRNA should allow efficient duplexing with the 3′ end of *E. coli* 16s rRNA (3′-AUU<u>CCUCCA</u>-5′) to form a RBS (Fig. 3). From thermodynamic considerations this five-base SD sequence in maize may be more efficient in duplexing than the four-base SD sequence of *lac* (Jay *et al.*, 1981). That this SD sequence in maize is recognised by *E. coli* ribosomes is supported by the finding that the polypeptide synthesised in minicells was of the correct size (assuming that the ochre stop codon of the LS gene is also recognised as a translational stop in vivo). It therefore seemed probable that a simple transcriptional fusion of the LS gene to a strong external promoter would lead to higher levels of expression because of efficient translation at the endogenous RBS.

My colleagues and I placed the maize LS gene into two positions in the bacteriophage lambda chromosome where we could make use of the lambda transcriptional pathways (Fig. 6). Transcriptional fusions were constructed using the lambda *N* operon transcribed from the promoter P_L (λZmB1^l), and the *Q* operon transcribed from the promoter P'_R(λZmB1^r) (Fig. 6). It was anticipated that high levels of expression would be obtained but at different times after infection, since P_L is an early promoter in the lytic cycle and P'_R is a late promoter. This result was obtained and enhanced levels of RuBPCase LS synthesis during lytic infection were observed either as a delayed early protein transcribed from P_L, or as a late protein transcribed from P'_R (Gatenby *et al.*, 1981; Figs. 7 and 8). It was further demonstrated that repression of transcription of the phage chromosome during infection of a homoimmune lysogen resulted in abolition of the enhanced expression of LS, thus indicating that the regulatory mechanism of lambda transcription could be utilised to control expression of a cloned chloroplast gene. Approximately 1% of total protein synthesis in the UV-irradiated infected cells was accounted for by synthesis of LS using P_L as the promoter. Unfortunately, the RuBPCase LS synthesised in *E. coli* during lytic infection did not accumulate to high levels because the advantage that accrued from P_L-initiated transcription of the maize gene was rapidly reduced by repression of early transcription from P_L by the product of the lambda *cro* gene (Fig. 8). The lambda *cro* gene codes for a product that depresses transcription from the early promoters (Ptashne *et al.*, 1980; Johnson *et al.*, 1981).

Fig. 6. Transcription in lambda clones containing the maize RuBPCase LS gene. The orientation of the gene in λZmB1^l and λZmB1^r are shown. The positions of the major phage transcripts initiated from the promoters P_L, P_R and P'_R are marked, together with the *nin* 5 deletion (gap), the *imm*^{21} substitution (dashed line) and lambda genes *A* and *J*. The chromosomes are drawn in a linear form but in vivo the DNA circularises. The arrow-heads show the direction of transcription. Reprinted from Gatenby *et al.*, 1981; *Nature*, vol. 291, no. 5811, pp. 117–21. Copyright © 1981 Macmillan Journals Limited.

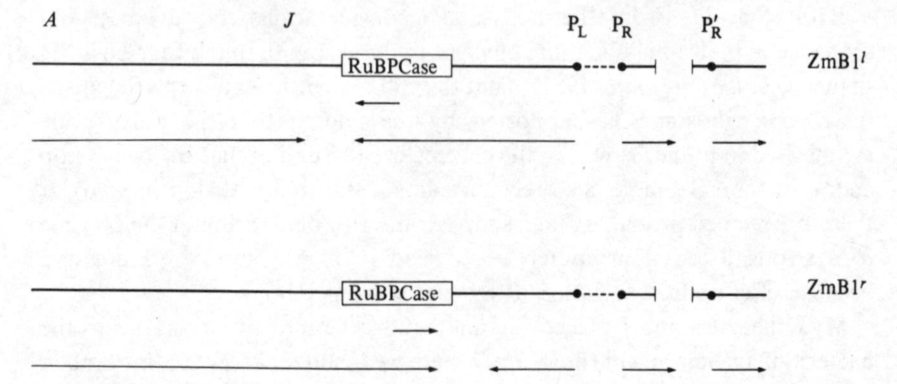

Amplification of chloroplast gene expression by transcriptional fusion with the lambda N *operon on a multicopy plasmid*

In order to overcome the problem of repressed transcription from early promoters in bacteriophage lambda we have now used the thermo-inducible expression plasmid pHUB4, which contains P_L (Bernard *et al.*, 1979), to construct a transcriptional fusion between the lambda *N* operon and the chloroplast gene. Expression of the LS from P_L is subject to control by a temperature-sensitive repressor (cI_{857}) that is located in the chromosome as part of a defective prophage. Thus at 32 °C LS synthesis is repressed, but incubation at 41–45 °C denatures the repressor to allow expression of LS from P_L. The prophage used carries two useful deletions. With the first (ΔH1 (Castellazzi, Brachet & Eisen, 1972)), genes to the right of o_R at least through *chl*A are deleted. This removes *cro* and the right host–prophage junction POB′, thus preventing *cro* product repression of P_L and prophage excision

Fig. 7. Synthesis of ³⁵S-labelled polypeptides during lytic infection of *E. coli* cells with lambda clones showing the expression of the maize RuPBCase LS gene using phage promoters. The autoradiographs show that the maize protein is synthesised as a delayed early protein when transcribed from the promoter P_L. Reprinted from Gatenby *et al.*, 1981: *Nature*, vol. 291, no. 5811, pp. 117–21. Copyright © 1981 Macmillan Journals Limited.

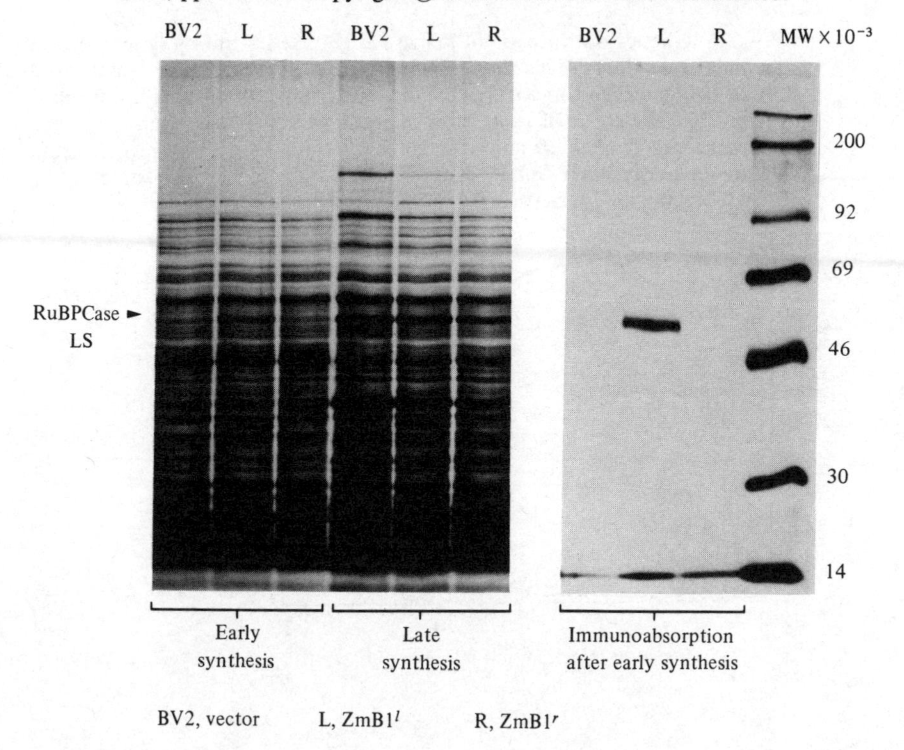

respectively. In the second (ΔBAM (Kourilsky *et al.*, 1978)), between 58·0 and 71·3% λ is removed; a useful feature since it removes the λkil gene responsible for host killing upon induction of a λN^+ prophage (Greer, 1975). Therefore with this prophage, induction of transcription from P_L on the multicopy plasmids is unhindered by the repressor products of *c*I and *cro*.

In one of our plasmids, pPBI3 (Gatenby & Castleton, 1982), which contains a tandem duplication of the LS gene transcribed from P_L, transcription continued after induction for more than 120 min, resulting in the synthesis of 60000 subunit monomers of LS per cell, i.e. about 2% of the total cell protein. This expression system relies on direct translation at the chloroplast SD sequence on P_L-initiated transcripts. An SD sequence has also been identified in the RuBPCase LS gene of spinach (Zurawski, Perrot, Bottomley & Whitfeld, 1981) and upstream from a putative translation start site for an unidentified polypeptide on a maize chloroplast DNA fragment (unpublished data of Z. Schwarz, A Steinmatz and L. Bogorad cited in McIntosh *et al.*, 1980). If, as we anticipate, a SD sequence is a common feature of chloroplast genes then it can be expected that other chloroplast proteins will be synthesised in large quantities by the simple transcriptional fusion to the lambda *N* operon or other bacterial operons. So far the presence of

Fig. 8. Kinetics of synthesis of maize RuBPCase LS during lytic infection with lambda clones. This result confirms that LS is synthesised as a delayed early protein when transcribed from P_L and a late protein when transcribed from P'_R. The effect of repression of transcription by the phase *cro* gene product can be seen 10 min after infection with λZmBl^l. Reprinted from Gatenby *et al.*, 1981: *Nature*, vol. 291, no. 5811, pp. 117–21. Copyright © 1981 Macmillan Journals Limited.

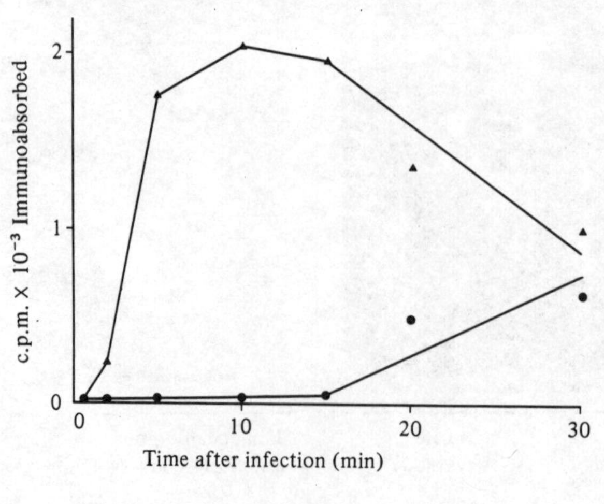

intervening sequences in chloroplast protein genes has not been detected; their absence will facilitate direct translation in *E. coli.*

Concluding remarks

The expression of eukaryotic genes in bacteria is a relatively new field of research and only very recently has the expression of higher plant genes been demonstrated in *E. coli.* An advantage with chloroplast genes is that expression can probably be achieved by simple transcriptional fusions because of the chloroplast SD sequence. The expression of plant nuclear genes in *E. coli* will be a little harder to achieve but with the range of cloning vectors available for high levels of expression it will probably be achieved fairly soon. Whether there is any economic benefit of plant gene expression in bacteria will really depend on the assessment of the potential value of suitable polypeptides and the difficulty or ease with which they can be obtained from traditional sources. What is likely, however, is that the subject may help to improve our understanding of the fundamental problems of plant gene organisation, regulation and structure.

I would like to thank R. B. Flavell for his critical reading of the manuscript and useful suggestions and J. A. Castleton for assistance with some of the experiments described here. Financial support was provided by the National Research Development Corporation.

References

Anderson, D., Shapiro, L. & Skalka, A. M. (1979). *In situ* immunoassays for translation products. In *Methods in Enzymology*, vol. 68, ed. R. Wu, pp. 428–36. London: Academic Press.

Backman, K. & Ptashne, M. (1978). Maximising gene expression on a plasmid using recombination *in vitro. Cell*, **13**, 65–71.

Bedbrook, J. R. & Bogorad, L. (1976). Endonuclease recognition sites mapped on *Zea mays* chloroplast DNA, *Proceedings of the National Academy of Sciences, USA*, **73**, 4309–13.

Bedbrook, J. R. & Gerlach, W. L. (1980). Cloning of repeated sequence DNA from cereal plants. In *Genetic Engineering*, vol. 2, ed. J. K. Setlow & A. Hollaender, pp. 1–19. New York: Plenum Press.

Bedbrook, J. R., Jones, J., O'Dell, M., Thompson, R. D. & Flavell, R. B. (1980*a*). A molecular description of telomeric heterochromatin in *Secale* species. *Cell*, **19**, 545–60.

Bedbrook, J. R. & Kolodner, R. (1979). The structure of chloroplast DNA. *Annual Review of Plant Physiology*, **30**, 593–620.

Bedbrook, J. R., Smith, S. M. & Ellis, R. J. (1980*b*). Molecular cloning and sequencing of cDNA encoding the precursor to the small subunit of chloroplast ribulose-1,5-bisphosphate carboxylase. *Nature*, **287**, 692–7.

Bernard, H. U. & Helinski, D. R. (1979). Use of the λ phage promoter P_L to promote gene expression in hybrid plasmid cloning vehicles. In *Methods in Enzymology*, vol. 68, ed. R. Wu, pp. 482–92. London: Academic Press.

Bernard, H. U. & Helinski, D. R. (1980). Bacterial plasmid cloning vehicles. In *Genetic Engineering*, vol. 2, ed. J. K. Setlow & A. Hollaender, pp. 133–67. New York: Plenum Press.

Bernard, H. U., Remaut, E., Hershfield, M. V., Das, H. K., Helsinki, D. R., Yanofsky, C. & Franklin, N. (1979). Construction of plasmid cloning vehicles that promote gene expression from the bacteriophage lambda P_L promoter. *Gene*, **5**, 59–76.

Bowman, C. M., Koller, B., Delius, H. & Dyer, T. A. (1981). A physical map of wheat chloroplast DNA showing the location of the structural genes for the ribosomal RNAs and the large subunit of ribulose 1,5-bisphosphate carboxylase. *Molecular and General Genetics*, **183**, 93–101.

Brennan, M. B. & Struhl, K. (1980). Mechanism of increasing expression of a yeast gene in *Escherichia coli. Journal of Molecular Biology*, **136**, 333–8.

Broome, S. & Gilbert, W. (1978). Immunological screening methods to detect specific translation products. *Proceedings of the National Academy of Sciences, USA*, **75**, 2746–9.

Brunel, F., Davison, J., Thi, V. & Merchez, M. (1980). Cloning and expression of *Trypanosoma brucei* kinetoplast DNA in *Escherichia coli. Gene*, **12**, 223–34.

Castellazzi, M., Brachet, P. & Eisen, H. (1972). Isolation and characterization of deletions in bacteriophage λ residing as prophage in *E. coli* K12. *Molecular and General Genetics*, **117**, 211–18.

Chang, A. C. Y., Nunberg, J. H., Kaufman, R. J., Erlich, H. A., Schimke, R. T. & Cohen, S. N. (1978). Phenotypic expression in *E. coli* of a DNA sequence coding for mouse dihydrofolate reductase. *Nature*, **275**, 617–24.

Charnay, P., Perricaudet, M., Galibert, F. & Tiollais, P. (1978). Bacteriophage lambda and plasmid vectors, allowing fusion of cloned genes in each of the three translational phases. *Nucleic Acids Research*, **5**, 4479–94.

Cheng, Y.-S. E., Kwoh, D. Y., Kwoh, T. J., Soltvedt, B. C. & Zipser, D. (1981). Stabilisation of a degradable protein by its overexpression in *Escherichia coli. Gene*, **14**, 121–30.

Citron, B. A., Feiss, M. & Donelson, J. E. (1979). Expression of yeast galactokinase gene in *Escherichia coli. Gene*, **6**, 251–64.

Clarke, L. & Carbon, J. (1978). Functional expression of cloned yeast DNA in *Escherichia coli*: specific complementation of arginino-succinate lyase (arg H) mutations. *Journal of Molecular Biology*, **120**, 517–32.

Clarke, L., Hitzeman, R. & Carbon, J. (1979). Selection of specific clones from colony banks by screening with radioactive antibody. In *Methods in Enzymology*, vol. 68, ed. R. Wu, pp. 436–42. London: Academic Press.

Cocking, E. C., Davey, M. R., Pental, D. & Power, J. B. (1981). Aspects of plant genetic manipulation. *Nature*, **293**, 265–70.

Derynck, R., Remaut, E., Saman, E., Stanssens, P., De Clercq, E., Content, J. & Fiers, W. (1980). Expression of human fibroblast interferon gene in *Escherichia coli. Nature*, **287**, 193–7.

Dickson, R. C. & Markin, J. S. (1978). Molecular cloning and expression in *E. coli* of a yeast gene coding for β-galactosidase. *Cell*, **15**, 123–30.

Dubnau, D., Gryczan, T., Contente, S. & Shivakumar, A. G. (1980). Molecular cloning in *Bacillus subtilis*. In *Genetic Engineering*, vol. 2, ed. J. K. Setlow & A. Hollaender, pp. 115–31. New York: Plenum Press.

Edge, M. D., Greene, A. R., Heathcliffe, G. R., Meacock, P. A., Schuch, W., Scanlon, D. B., Atkinson, T. C., Newton, C. R. & Markham, A. F. (1981). Total synthesis of a human leukocyte interferon gene. *Nature*, **292**, 756–62.

Emtage, J. S., Tacon, W. C. A., Catlin, G. H., Jenkins, B., Porter, A. G. &

Carey, N. H. (1980). Influenza antigenic determinants are expressed from haemagglutinin genes cloned in *Escherichia coli*. *Nature*, **283**, 171–4.

Erlich, H. A., Cohen, S. N. & McDevitt, H. O. (1979). Immunological detection and characterisation of products translated from cloned DNA fragments. In *Methods in Enzymology*, vol. 68, ed. R. Wu, pp. 443–53. London: Academic Press.

Erwin, C. R., Maurer, R. A. & Donelson, J. E. (1980). A bacterial cell that synthesizes a protein containing the antigenic determinants of rat prolactin. *Nucleic Acids Research*, **8**, 2537–46.

Flavell, R. B. (1980). The molecular characterisation and organisation of plant chromosomal DNA sequences. *Annual Review of Plant Physiology*, **31**, 569–96.

Flavell, R. B. (1981). The analysis of plant genes and chromosomes by using DNA cloned in bacteria. *Philosophical Transactions of the Royal Society of London B*, **292**, 579–88.

Fraser, T. H. & Bruce, B. J. (1978). Chicken ovalbumin is synthesised and secreted by *Escherichia coli*. *Proceedings of the National Academy of Sciences, USA*, **75**, 5936–40.

Garapin, A. C., Colbère-Garapin, F., Cohen-Solal, M., Horodniceanu, F. & Kourilsky, P. (1981). Expression of herpes simplex virus type I thymidine kinase gene in *Escherichia coli*. *Proceedings of the National Academy of Sciences, USA*, **78**, 815–19.

Gatenby, A. A. & Castleton, J. A. (1982). Amplification of maize ribulose bisphosphate carboxylase large subunit synthesis in *E. coli* by transcriptional fusion with the lambda *N* operon. *Molecular and General Genetics*, **185**, 424–9.

Gatenby, A. A., Castleton, J. A. & Saul, M. W. (1981). Expression in *E. coli* of maize and wheat chloroplast genes for large subunit of ribulose bisphosphate carboxylase. *Nature*, **291**, 117–21.

Gerlach, W. L. & Dyer, T. A. (1980). Sequence organisation of the repeating units in the nucleus of wheat which contains 5S rRNA genes. *Nucleic Acids Research*, **8**, 4851–65.

Gerlach, W. L., Miller, T. E. & Flavell, R. B. (1980). The nucleolus organizers of diploid wheats revealed by *in situ* hybridization. *Theoretical and Applied Genetics*, **58**, 97–100.

Goeddel, D. V., Heyneker, H. L., Hozumi, T., Arentzen, R., Itakura, K., Yansura, D. G., Ross, M. J., Miozzari, G., Crea, R. & Seeburg, P. H. (1979*a*). Direct expression in *Escherichia coli* of a DNA sequence coding for human growth hormone. *Nature*, **281**, 544–8.

Goeddel, D. V., Kleid, D. G., Bolivar, F., Heyneker, H. L., Yansura, D. G., Crea, R., Hirose, T., Kraszewski, A., Itakura, K. & Riggs, A. D. (1979*b*). Expression in *Escherichia coli* of chemically synthesised genes for human insulin. *Proceedings of the National Academy of Sciences, USA*, **76**, 106–10.

Gottesman, M. E., Adhya, S. & Das, A. (1980). Transcription antitermination by bacteriophage lambda *N* gene product. *Journal of Molecular Biology*, **140**, 57–75.

Greenblatt, J. (1981). Regulation of transcription termination by the *N* gene protein of bacteriophage lambda. *Cell*, **24**, 8–9.

Greer, H. (1975). The *kil* gene of bacteriophage lambda. *Virology*, **66**, 589–604.

Guarente, L., Lauer, G., Roberts, T. M. & Ptashne, M. (1980). Improved methods for maximising expression of a cloned gene: a bacterium that synthesises rabbit β-globin. *Cell*, **20**, 543–53.

Hallewell, R. A. & Emtage, S. (1980). Plasmid vectors containing the tryptophan operon promoter suitable for efficient regulated expression of foreign genes. *Gene*, **9**, 27–47.

Hardy, K., Stahl, S. & Küpper, H. (1981). Production in *B. subtilis* of hepatitis B core antigen and of major antigen of foot and mouth disease virus. *Nature*, **293**, 481–3.

Henikoff, S., Tatchell, K., Hall, B. D. & Nasmyth, K. A. (1981). Isolation of a gene from *Drosophila* by complementation in yeast. *Nature*, **289**, 33–7.

Hutchinson, J., Chapman, V. & Miller, T. A. (1980). Chromosome pairing at meiosis in hybrids between *Aegilops* and *Secale* species: a study by *in situ* hybridisation using cloned DNA. *Heredity*, **45**, 245–54.

Hutchinson, J., Flavell, R. B. & Jones, J. (1981). Physical mapping of plant chromosomes by *in situ* hybridisation. In *Genetic Engineering*, vol. 3, ed. J. K. Setlow & A. Hollaender, pp. 207–22. New York: Plenum Press.

Itakura, K., Hirose, T., Crea, R., Riggs, A. D., Heyneker, H. L., Bolivar, F. & Boyer, H. W. (1977). Expression in *Escherichia coli* of a chemically synthesised gene for the hormone somatostatin. *Science*, **198**, 1056–63.

Jaskunas, S. R., Lindahl, L., Nomura, M. & Burgess, R. R. (1975). Identification of two copies of the gene for the elongation factor EF-Tu in *E. coli*. *Nature*, **257**, 458–62.

Jay, G., Khoury, G., Seth, K. & Jay, E. (1981). Construction of a general vector for efficient expression of mammalian proteins in bacteria: Use of a synthetic ribosome binding site. *Proceedings of the National Academy of Sciences, USA*, **78**, 5543–8.

Jensen, E. O., Paludan, K., Hyldig-Nielsen, J. J., Jørgensen, P. & Marcker, K. A. (1981). The structure of a chromosomal leghaemoglobin gene from soybean. *Nature*, **291**, 677–9.

Johnson, A. D., Poteete, A. R., Lauer, G., Sauer, R. T., Ackers, G. K. & Ptashne, M. (1981). λ repressor and *cro* – components of an efficient molecular switch. *Nature*, **294**, 217–23.

Kaplan, D. A., Naumovski, L. & Collier, R. J. (1981). Chromogenic detection of antigen in bacteriophage plaques: a microplaque method applicable to large scale screening. *Gene*, **13**, 211–20.

Kemble, R. J. (1980). A rapid single leaf assay for detecting the presence of 'S' male-sterile cytoplasm in maize. *Theoretical and Applied Genetics*, **57**, 97–100.

Kemble, R. J., Gunn, R. E. & Flavell, R. B. (1980). Classification of normal and male-sterile cytoplasms in maize. II. Electrophoretic analysis of DNA species in mitochondria. *Genetics*, **95**, 451–8.

Kourilsky, P., Perricaudet, M., Gros, D., Garapin, A., Gottesman, M., Fritsch, A. & Tiollais, P. (1978). Description and properties of bacteriophage lambda vectors useful for the cloning of *Eco* RI DNA fragments. *Biochimie*, **60**, 183–7.

Kusunic, D. A. & Kushner, S. R. (1980). Expression of the *HIS3* gene of *Saccharomyces cerevisiae* in polynucleotide phosphorylase-deficient strains of *Escherichia coli* K-12. *Gene*, **12**, 1–10.

Levings, C. S., Kim, B. D., Pring, D. R., Conde, M. F., Mans, R. J., Laughnan, J. R. & Gabay-Laughnan, S. J. (1980). Cytoplasmic reversion of cms-S in maize: association with a transpositional event. *Science*, **209**, 1021–3.

Lonsdale, D. M., Thompson, R. D. & Hodge, T. P. (1981). The integrated forms of the S1 and S2 DNA elements of maize male sterile mitochondrial

DNA are flanked by a large repeated sequence. *Nucleic Acids Research*, **9**, 3657–69.

Lovett, P. S. & Keggins, K. M. (1979). *Bacillus subtilis* as a host for molecular cloning. In *Methods in Enzymology*, vol. 68, ed. R. Wu, pp. 342–57. London: Academic Press.

McIntosh, L., Poulsen, C. & Bogorad, L. (1980). The DNA sequence of a chloroplast gene: the large subunit of ribulose bisphosphate carboxylase of maize. *Nature*, **288**, 556–60.

Maquat, L. E. & Reznikoff, W. S. (1978). *In vitro* analysis of the *Escherichia coli* RNA polymerase interaction with wild-type and mutant lactose promoters. *Journal of Molecular Biology*, **125**, 467–90.

Martial, J. A., Hallewell, R. A., Baxter, J. D. & Goodman, J. M. (1979). Human growth hormone: complementary DNA cloning and expression in bacteria. *Science*, **205**, 602–7.

Meagher, R. B., Tait, R. C., Betlach, M. & Boyer, H. W. (1977). Protein expression in *E. coli* minicells by recombinant plasmids. *Cell*, **10**, 521–36.

Mercereau-Puijalon, O., Royal, A., Cami, B., Garapin, A., Krust, A., Gannon, F. & Kourilsky, P. (1978). Synthesis of an ovalbumin-like protein by *Escherichia coli* K12 harbouring a recombinant plasmid. *Nature*, **275**, 505–10.

Mory, Y., Chernajovsky, Y., Feinstein, S. L., Chen, L., Nir, U., Weissenbach, J., Malpiece, Y., Tiollais, P., Marks, D., Ladner, M., Colby, C. & Revel, M. (1981). Synthesis of human interferon β1 in *Escherichia coli* infected by a lambda phage recombinant containing a human genomic fragment. *European Journal of Biochemistry*, **120**, 197–202.

Murray, N. E. & Kelley, W. S. (1979). Characterization of λ*pol A* transducing phages; effective expression of the *E. coli pol A* gene. *Molecular and General Genetics*, **175**, 77–87.

Nagata, S., Taira, H., Hall, A., Johnsrud, L., Streuli, M., Ecsödi, J., Boll, W., Cantell, K. & Weissmann, C. (1980). Synthesis in *E. coli* of a polypeptide with human leukocyte interferon activity. *Nature*, **284**, 316–20.

Olson, K. C., Fenno, J., Lin, N., Harkins, R. N., Snider, C., Kohr, W. H., Ross, M. J., Fodge, D., Prender, G. & Stebbing, N. (1981). Purified human growth hormone from *E. coli* is biologically active. *Nature*, **293**, 408–11.

Ptashne, M., Jeffrey, A., Johnson, A. D., Maurer, R., Meyer, B. J., Pabo, C. O., Roberts, T. M. & Saucer, R. T. (1980). How the λ repressor and *cro* work. *Cell*, **19**, 1–11.

Ratzkin, B. & Carbon, J. (1977). Functional expression of cloned yeast DNA in *Escherichia coli*. *Proceedings of the National Academy of Sciences, USA*, **74**, 487–91.

Reeve, J. N. (1978). Selective expression of transduced or cloned DNA in minicells containing plasmid pKB280. *Nature*, **276**, 728–9.

Reeve, J. N. (1979). Use of minicells for bacteriophage-directed polypeptide synthesis. In *Methods in Enzymology*, vol. 68, ed. R. Wu, pp. 493–503. London: Academic Press.

Remaut, E., Stanssens, P. & Fiers, W. (1981). Plasmid vectors for high-efficiency expression controlled by the P_L promoter of coliphage lambda. *Gene*, **15**, 81–93.

Roberts, T. M., Kacich, R. & Ptashne, M. (1979). A general method for maximising the expression of a cloned gene. *Proceedings of the National Academy of Sciences, USA*, **76**, 760–4.

Roberts, T. M. & Lauer, G. D. (1979). Maximising gene expression on a plasmid using recombination *in vitro*. In *Methods in Enzymology*, vol. 68, ed. R. Wu, pp. 473–82. London: Academic Press.

Rosenberg, M. & Court, D. (1979). Regulatory sequences involved in the promotion and termination of RNA transcription. *Annual Review of Genetics*, **13**, 319–53.

Sancar, A., Hack, A. M. & Rupp, W. D. (1979). Simple method for identification of plasmid-coded proteins. *Journal of Bacteriology*, **137**, 692–3.

Scherer, G. E. F., Walkinshaw, M. D. & Arnott, S. (1978). A computer-aided oligonucleotide analysis provides a model sequence for RNA polymerase promoter recognition in *E. coli. Nucleic Acids Research*, **5**, 3759–73.

Schwartz, J. H., Meyer, R., Eisenstadt, J. M. & Brawerman, G. (1967). Involvement of *N*-formylmethionine in initiation of protein synthesis in cell-free extracts of *Euglena gracilis. Journal of Molecular Biology*, **25**, 571–4.

Schwarz, Z. & Kossel, H. (1980). The primary structure of 16S rDNA from *Zea mays* chloroplast is homologous to *E. coli* 16S rRNA. *Nature*, **283**, 739–42.

Shine, J. & Dalgarno, L. (1975). Determinant of cistron specificity in bacterial ribosomes. *Nature*, **254**, 34–8.

Shine, J., Fettes, I., Lan, N. C. Y., Roberts, J. L. & Baxter, J. D. (1980). Expression of a cloned β-endorphin gene sequence by *Escherichia coli. Nature*, **285**, 456–61.

Steffen, D. & Schleif, R. (1977). Overproducing *araC* protein with lambda-arabinose transducing phage. *Molecular and General Genetics*, **157**, 333–9.

Struhl, K. & Davis, R. W. (1977). Production of a functional eukaryotic enzyme in *Escherichia coli*: cloning and expression of the yeast structural gene for imidazoleglycerolphosphate dehydratase (*his*3). *Proceedings of the National Academy of Sciences, USA*, **74**, 5255–9.

Struhl, K., Stinchcomb, D. T. & Davis, R. W. (1980). A physiological study of functional expression in *Escherichia coli* of the cloned yeast gene for imidazoleglycerolphosphate dehydratase gene. *Journal of Molecular Biology*, **136**, 291–307.

Sun, S. M., Slightom, J. L. & Hall, T. C. (1981). Intervening sequences in a plant gene – comparison of the partial sequence of cDNA and genomic DNA of French bean phaseolin. *Nature*, **289**, 37–41.

Swamy, K. H. S. & Goldberg, A. L. (1981). *E. coli* contains eight soluble proteolytic activities, one being ATP dependent. *Nature*, **292**, 652–4.

Tacon, W., Carey, N. & Emtage, S. (1980). The construction and characterisation of plasmid vectors suitable for the expression of all DNA phases under the control of the *E. coli* tryptophan promoter. *Molecular and General Genetics*, **177**, 427–38.

Talmadge, K. & Gilbert, W. (1980). Construction of plasmid vectors with unique *Pst*I cloning sites in a signal-sequence coding region. *Gene*, **12**, 235–41.

Talmadge, K., Stahl, S. & Gilbert, W. (1980). Eukaryotic signal sequence transports insulin antigen in *Escherichia coli. Proceedings of the National Academy of Sciences, USA*, **77**, 3369–73.

Thompson, R. D., Kemble, R. & Flavell, R. B. (1980). Variations in mitochondrial DNA organisation between normal and male sterile cytoplasms of maize. *Nucleic Acids Research*, **8**, 1999–2008.

Timmis, K. N. (1981). Methods for the manipulation of DNA *in vitro*. In *Genetics as a Tool in Microbiology*, ed. S. W. Glover & D. A. Hopwood, pp. 49–109. Cambridge: Cambridge University Press.

Vapnek, D., Hautala, J. A., Jacobsen, J. W., Giles, N. H. & Kushner, S. R.

(1977). Expression in *Escherichia coli* K-12 of the structural gene for catabolic dehydroquinase of *Neurospora crassa. Proceedings of the National Academy of Sciences, USA*, **74**, 3508–12.

Walz, A., Ratzkin, B. & Carbon, J. (1978). Control of expression of a cloned yeast (*Saccharomyces cerevisiae*) gene (*trp5*) by a bacterial insertion element (IS2). *Proceedings of the National Academy of Sciences, USA*, **75**, 6172–6.

Whitfeld, P. R. & Bottomley, W. (1980). Mapping of the gene for the large subunit of ribulose bisphosphate carboxylase on spinach chloroplast DNA. *Biochemistry International*, **1**, 172–8.

Wilson, G. G. & Murray, N. E. (1979). Molecular cloning of the DNA ligase gene from bacteriophage T4. I. Characterisation of the recombinants. *Journal of Molecular Biology*, **132**, 471–91.

Zech, M., Hartley, M. R. & Bohnert, H. J. (1981). Binding sites of *E. coli* DNA-dependent RNA polymerase on spinach chloroplast DNA. *Current Genetics*, **4**, 37–46.

Zurawski, G., Perrot, B., Bottomley, W. & Whitfeld, P. R. (1981). The structure of the gene for the large subunit of ribulose 1,5-bisphosphate carboxylase from spinach chloroplast DNA. *Nucleic Acids Research*, **9**, 3251–70.

Note added in proof

It has recently been demonstrated that a *c* DNA clone for a plant nuclear encoded protein has been expressed in *E. coli.*

Edens, L., Heslinga, L., Klok, R., Ledeboer, A. M., Maat, J., Toonen, M. Y., Visser, C. & Verrips, C. T. (1982). Cloning of *c* DNA encoding the sweet-tasting plant protein thaumatin and its expression in *Escherichia coli. Gene*, **18**, 1–12.

R. HULL

The current status* of plant viruses as potential DNA/RNA vector systems

Introduction

Although there have been numerous reports of the introduction of 'foreign' (either bacterial or plant) DNA into plant cells (for reviews see Kleinhofs & Behki, 1977; Lurquin & Kado, 1979) in no case has it been proved that this DNA has been replicated and expressed. However, viruses and some other pathogens can be considered to be forms of 'foreign' nucleic acid which do replicate and express in susceptible plant cells. As they cannot possibly code for all the products needed for their full replication cycle they must depend upon and possibly adapt host functions. Their nucleic acids must have the appropriate recognition sequences for the enzymes responsible for nucleic acid replication and transcription. These nucleic acids possibly also code for products which modify host functions in such a way as to facilitate their own replication and expression. Thus these sequences and coding regions have the potential of being used to replicate and express 'foreign' genes within plant cells. Since plant viruses usually replicate to give a very high copy number of nucleic acid molecules (up to or more than 10^6 molecules per infected cell; e.g. tobacco mosaic virus replicates to give 10^7 particles per cell) they could potentially provide a very efficient means of obtaining good expression of a 'foreign' genc. However there are two apparent disadvantages in the use of viruses as vectors; firstly, they are pathogenic and secondly, in no case have they been shown to integrate with the host chromosomes. These points will be discussed later.

Three types of plant viruses, those with single-stranded RNA, those with single-stranded DNA and those with double-stranded DNA, are being considered as possible vectors.

Single-stranded RNA viruses

More than 90% of plant viruses have single-stranded RNA as their genetic material; the RNA is positive stranded (i.e. is mRNA) and is infectious. At present no RNA virus vector system has been developed though

* As of January 1982 at the time of the Leicester meeting.

some work has been started on one system (Koziel & Siegel, 1981). An outline of the basic means by which the viral RNA could be used as a vector is shown in Fig. 1. The single-stranded RNA would be converted to double-stranded copy DNA (cDNA) using the enzymes reverse transcriptase and DNA polymerase and the DNA would then be cloned into a prokaryotic plasmid or cosmid, i.e. a combined plasmid/lambda vector, system. The desired gene would be inserted into the cDNA moiety of the recombinant plasmid. To reintroduce the viral vector and 'foreign' gene into the plant host either of two routes could be tried. It has recently been shown that a cDNA clone of poliovirus (+-strand RNA virus) is infectious (Racaniello & Baltimore, 1981); it is unknown if such a clone of a plant RNA virus is also infectious. If it is not, the viral vector and 'foreign' gene could be transcribed to RNA which would then be used to infect plant cells.

The system being developed by Koziel & Siegel (1981) uses tobacco rattle virus (TRV) RNA. The RNA needed for TRV infection is divided into two pieces each of which is encapsidated in a separate rod-shaped particle (see Lane, 1979 for review). The RNA of the longer particle replicates in plants but does not code for coat protein and hence produces only unencapsidated RNA; the coat protein information is on the shorter piece of RNA, which can only be replicated in the presence of the longer piece. Koziel & Siegel (1981) are cloning the shorter piece of RNA and will use this as the vector. They will coinfect with the two pieces of RNA, using the functions encoded on the longer RNA to replicate and express the shorter RNA vector and its gene(s). If they interfere with coat protein expression they will still have

Fig. 1. Scheme for using viral RNA as a vector. See text for details. Hatched area, foreign gene; cDNA, copy DNA.

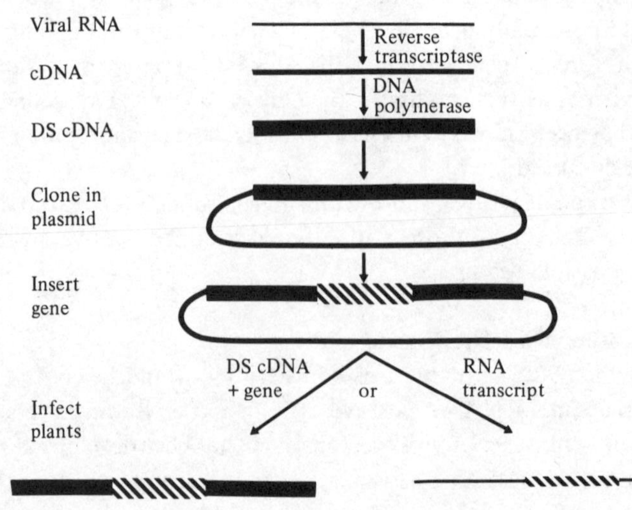

systemic spread of the viral RNA albeit somewhat slowly. If the coat protein is expressed there should be no steric hindrance of the encapsidation of the modified short RNA vector since the particles are rod shaped.

Single-stranded DNA viruses

There is one group of single-stranded DNA plant viruses, the group of geminiviruses. The properties of these viruses have recently been reviewed by Goodman (1981 *a*, *b*). There are more than a dozen viruses in this group, each with a different host range. In some cases the host range is narrow (e.g. Abutilon mosaic virus), in others it is broad (e.g. curley top virus). Some will infect legumes (e.g. bean gold mosaic virus (BGMV)) and others will infect monocotyledons (e.g. maize streak virus). BGMV is the best characterised of the geminiviruses. The encapsidated DNA is mainly circular single-stranded DNA though there are some linear molecules; both the circular and linear forms are infectious. Restriction endonuclease mapping suggests that there are two species of DNA each of which contains approximately 2150 nucleotides. Geminiviruses are thought to replicate in the nucleus and a relaxed circular double-stranded form of BGMV has been isolated; it is also infectious.

Nothing is known about the functions of geminivirus DNA except that presumably it codes for the viral coat protein. Most of the work examining the potential of geminiviruses as vectors is likely to follow the lines discussed below for double-stranded DNA plant viruses.

Double-stranded DNA viruses

The only group of plant viruses which has double-stranded DNA as the genetic material is the caulimovirus group. This group comprises 12 viruses (Table 1) each with a relatively narrow host range; none has yet been found which will infect legumes or monocotyledons. The caulimovirus group has been reviewed by Shepherd (1976, 1977, 1979, 1981).

Properties of the viruses

The type member, cauliflower mosaic virus (CaMV) is the most studied of this group. The virus spreads systemically through many of its *Brassica* hosts. For the Cabb B-JI isolate of CaMV it has been estimated that there are about 50000 particles per cell (R. Hull, unpublished observation). The particles of all caulimoviruses are found in proteinaceous inclusion bodies (Fig. 2) in the cytoplasm of most of the mesophyll cells of infected plants. With CaMV very few particles are observed outside the inclusion bodies (Fujisawa, Rubio-Huertos, Matsui & Yamaguchi, 1967) though particles of carnation etched ring virus (CERV) (Lawson & Hearon, 1974) and petunia

vein-clearing virus (Lesemann & Casper, 1973) are often found free in the cytoplasm. The inclusion bodies of different isolates of CaMV vary in their morphology (Shalla, Shepherd & Petersen, 1980; Shepherd, Richins & Shalla, 1980) and there are inclusion bodies in which no virus particles are observed. Similarly there are morphological differences between the inclusion bodies of different caulimoviruses (Fig. 2). Partially purified preparations of CaMV inclusion bodies comprise a major protein, the molecular weight of which has been variously estimated to be between 55000 and 65000 (Al Ani *et al.*, 1980; Odell & Howell, 1980; Shepherd *et al.*, 1980; Covey & Hull, 1981). As will be shown later its true molecular weight is about 62000.

The function of inclusion bodies is at present unknown. It was thought (Favali, Bassi & Conti, 1973; Shepherd, 1979) that they were sites of viral replication but there is recent evidence (see below) that CaMV replicates in the nucleus. It is possible that they are sites of virus assembly; they are often found close to the nucleus (Shepherd, 1976).

The inclusion bodies have to be disrupted to purify the particles of caulimoviruses. This is best effected by treatment of sap extracts with Triton X-100 and urea (Hull, Shepherd & Harvey, 1976); the inclusion bodies of different caulimoviruses require different amounts of these chemicals for their disruption (Hull & Donson, 1982).

Table 1. *The caulimovirus group*[a]

Virus	Hosts
Members	
Cauliflower mosaic	Cruciferae (Solanaceae)[b]
Carnation etched ring	Caryophyllaceae
Dahlia mosaic	Compositae (Amaranthaceae, Solanaceae, Chenopodiaceae)
Figwort mosaic	Scrophulariaceae (Solanaceae)
Mirabilis mosaic	Nyctaginaceae
Strawberry vein banding	Rosaceae
Possible members	
Blueberry red ringspot	Ericaceae
Cassava vein mosaic	Euphorbiaceae
Cestrum	Solanaceae
Petunia vein clearing	Solanaceae
Plantago virus 4	Plantaginaceae
Thistle mottle	Compositae

[a] Data from Shepherd (1981) and R. Hull, unpublished observations.
[b] Names in parentheses are families of experimental hosts.

Properties of the DNA

It is somewhat difficult to extract the DNA from caulimovirus particles; one has to digest the coat protein with proteases (Shepherd, Bruening & Wakeman, 1970). The DNA isolated from virus particles is mainly double-stranded open circular molecules; there is a certain proportion of linear molecules. The sequence of the DNAs of two isolates of CaMV have recently been published, that of Cabb-S isolate comprising 8024 base pairs (Franck *et al.*, 1980) and that of CM1841 isolate comprising 8031 base pairs (Gardner *et al.*, 1981). Thus the molecular weight is approximately $5{\cdot}2 \times 10^6$.

The DNA isolated from most caulimoviruses has two unusual features. The first is that CaMV DNA has discontinuities at specific sites (Volovitch, Dumas, Drugeon & Yot, 1976; Hull & Howell, 1978; Volovitch, Drugeon,

Fig. 2. Electron micrographs of caulimovirus inclusion bodies. (*a*), (*b*) Figwort mosaic virus; (*c*) *Cestrum* virus; (*d*) Thistle mottle virus. *Cestrum* virus infected tissue kindly provided by Dr G. Martelli. Sectioning and electron microscopy by A. Plaskitt.

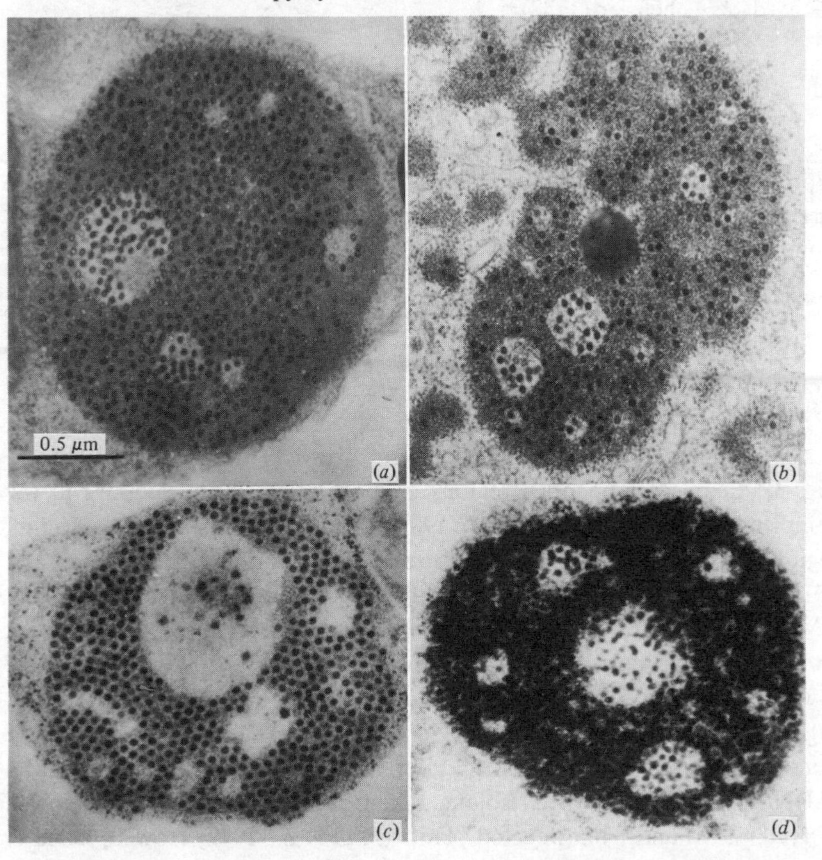

& Yot, 1978). The majority of isolates have one discontinuity in one strand and two in the other; isolate CM4-184, which has a deletion in the region of one discontinuity, has only one in each strand. These discontinuities were originally thought to be nicks or gaps as the DNA is susceptible to the single-strand nuclease S1 at these sites. However, the sequencing work of Franck *et al.* (1980) showed that the discontinuities are overlaps and that the DNA is apparently triple-stranded in these regions; the overlaps are up to 18 residues long. The 5′ ends at the discontinuities are fixed, the 3′ ends can be variable. The DNA of CERV also has three discontinuities, one in one strand and two in the other, but that of figwort mosaic virus has four discontinuities, one in one strand and three in the other (Hull & Donson, 1982). Possible functions of the discontinuities will be discussed later.

The second unusual feature of caulimovirus DNA is that a large proportion of the DNA molecules have a twisted structure when viewed in the electron microscope. Although they resemble supercoils the presence of the discontinuities precludes this. They are discussed by Hull (1981) and by Hohn, Richards & Lebeurier (1982).

Caulimovirus DNA as a vector

There are two main approaches to examining the potential of CaMV as a vector. One is to study at the molecular level the replication and expression of the virus and to identify regions of the DNA which contain essential functions. The other is to ascertain the effects of deletion of DNA from, and/or insertion of DNA into, the CaMV genome on infectivity.

Expression of CaMV

Only one strand of CaMV DNA, that with the single gap (the α strand) is transcribed (Howell & Hull, 1978; Hull, Covey, Stanley & Davies, 1979). Analysis of the sequence of the Cabb-S strain of CaMV revealed six possible open reading regions (I–VI) in the α strand (Franck *et al.*, 1980 and Fig. 3); there were no open regions of more than 370 nucleotides in the complementary strand. Hohn *et al.* (1982) have suggested two further reading regions (VII and VIII), also in the α strand, region VII abutting region I and VIII covering part of region IV in a different frame. The sequence of isolate CM 1841 (Gardner *et al.*, 1981) confirms these reading regions. It can be seen from Fig. 3 that most of the six reading regions suggested by Franck *et al.* (1980) overlap or nearly abut one another. There are small gaps between regions VII and I and between regions V and VI. There is also a larger non-coding region (the intergenic region) around the α strand discontinuity.

RNA transcripts have been isolated both from infected protoplasts (Howell & Hull, 1978) and from infected plants (Hull *et al.* 1979; Howell, Odell &

Dudley, 1980; Odell & Howell, 1980; Covey & Hull, 1981; Covey, Lomonossoff & Hull, 1981). Two species of poly(A)+RNA, about 8 kb and 1.9 kb long, are the most abundant CaMV specific RNAs 20 days after inoculation with the virus (Al Ani *et al.*, 1980; Odell & Howell, 1980; Covey & Hull, 1981; Odell, Dudley & Howell, 1981). The 1.9 kb RNA is the mRNA for a protein of molecular weight 62000 (Odell & Howell, 1980; Covey & Hull, 1981) which Covey & Hull (1981) showed was very similar to the major protein of the inclusion bodies. Mapping of this RNA demonstrated that the protein was derived from open reading region VI. Covey *et al.* (1981) showed that this mRNA has an eleven-base 5′ leader before the coding region and an untranslated 3′ sequence of about 280 bases. About 20 nucleotides upstream from the 3′ end is the sequence AATAAA which is suggested as being a strong signal for polyadenylation (Proudfoot & Brownlee, 1976). There are no strong transcription promoters in the 30 nucleotides upstream of the 5′ end of this mRNA; however, Howell (1981) suggests from ultraviolet mapping data that this RNA does have an independent promoter.

The 8-kb CaMV specific RNA is a more than full-length transcript of the α strand. Its 3′ end maps coterminally with the 3′ end of the inclusion body protein mRNA (Covey *et al.*, 1981) and thus has the strong polyadenylation signal. Its 5′ end maps about 200 bases upstream from the 3′ end (Covey *et*

Fig. 3. In vitro mutagenesis of CaMV DNA. The circle represents the CaMV genome. G1 is the discontinuity in the α strand. The six open reading regions I–VI suggested by Franck *et al.* (1980) are indicated by thicker lines. Arrows indicate sites of insertion; ⊢⊣ regions deleted. Mutations inside the circle result in loss of infectivity; those outside the circle retain infectivity. (Data from Gronenborn *et al.*, 1981; Howarth, Gardner, Messing & Shepherd, 1981; Howell *et al.*, 1981; Shepherd *et al.*, 1981.)

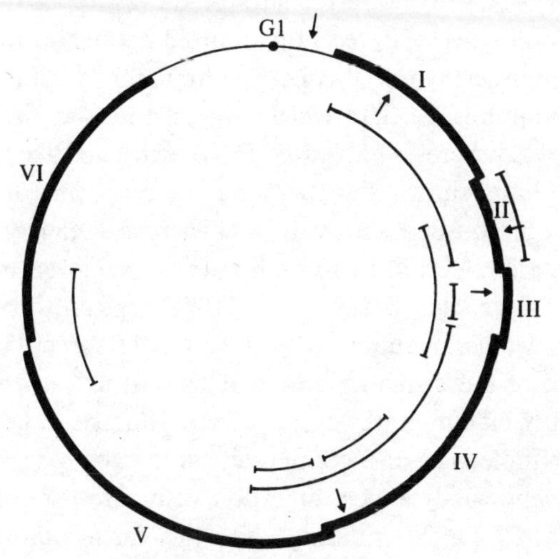

al., 1981) close to the sequence TATAA which is a strong promoter for transcription in eukaryotic cells (see Breathnach & Chambon, 1981 for review). These mapping data raise two interesting points. The first is that the transcript has to pass the polyadenylation signal before continuing around the rest of the DNA. It is unknown whether there are signals for the termination of transcription other than the polyadenylation signal sequence but it is possible that at least one is upstream of the 5′ end of the 8-kb transcript. The second point is that the transcript apparently has to pass the discontinuity in the α strand (about 600 bases from the 5′ end) before continuing around the DNA. The recent findings (see below) which suggest that CaMV DNA exists as covalently closed molecules within the cell would offer the explanation that these molecules are the templates for transcription.

Thus, there is one promoter site (for the 8-kb strand) which is known with any certainty; whether the other major transcript (the 1.9-kb transcript) has an independent promoter or is derived from the 8-kb transcript awaits further study. Following the use of an *Escherichia coli* plasmid system for mapping prokaryotic promoters McKnight & Meagher (1981) suggested that there is a promoter for the open reading of region IV. However, the efficiency of a prokaryotic system for mapping eukaryotic promoters is uncertain. Thus, the strategy by which the information in the other open reading regions is converted to proteins is unknown and it is even uncertain whether all the other open reading regions are used.

Replication of CaMV

It was thought that the inclusion bodies, into which there is considerable incorporation of tritrated thymidine (Kamei, Rubio-Huertos & Matsui, 1969; Favali *et al.*, 1973), were the sites of CaMV replication and expression. However, Guilfoyle (1980) recently reported that transcription of CaMV DNA occurs in nuclei and Shepherd, Gronenborn, Gardner & Daubert (1981) quote unpublished data which suggest that the viral DNA replication also takes place there. Analysis of DNA extracted from nuclei isolated from infected plants shows that there are several forms of CaMV specific DNA (R. Hull, unpublished observations). There is a relatively small proportion of the twisted circles and full-length linear molecules, the forms which are found in virus particles. Much of the DNA appears to be in the form of smaller linear molecules of about 6 kbp and a small fragment of about 600 bp. The significance of these small pieces is at present unknown. There is also a molecular form which migrates during electrophoresis in gels faster than the normal linear molecules and comigrates with known supercoiled CaMV DNA from *Xenopus* oocytes. On digestion with a restriction endonuclease which cuts CaMV DNA at a single site this form migrates as a full-length linear molecule. This suggests that it might be a covalently closed molecule; covalently closed supercoiled forms of CaMV DNA are also alluded to by Hohn *et al.* (1982).

Thus there appear to be two forms of full-length circular CaMV DNA in plants, the relaxed molecule with discontinuities found mainly in virus particles and the covalently closed supercoiled molecules, which are most likely restricted to the nucleus. This raises the question of why are there two forms? There seem to be three possible answers, which are somewhat interrelated. Firstly, the two forms could be a means of switching from transcription to encapsidation. If encapsidation is needed for cell-to-cell spread of the virus there would be selection pressure on features enhancing it. Secondly, the interaction of proteins with either single- or triple-stranded DNA is likely to differ from their interaction with double-stranded DNA. Thus the discontinuities might prime or sterically control encapsidation itself. Thirdly, the discontinuities might in some way facilitate the movement of DNA from the nucleus to the cytoplasm, where it would be encapsidated.

Deletions and insertions in the CaMV genome

In order to perform deletions and insertions in the CaMV genome it is preferable to clone the viral DNA in a prokaryotic plasmid or cosmid system. There are various unique sites for restriction endonucleases in CaMV DNA and the DNA of various isolates has been cloned in various vectors (see e.g. Meagher, Shepherd & Boyer, 1977; Szeto, Hamer, Carlson & Thomas, 1977; Hohn, Hohn, Lesot & Lebeurier, 1980; Howell, Walker & Dudley, 1980; Hohn *et al.* 1982). The DNA is isolated as covalently closed circles and hence lacks the discontinuities found in the viral DNA. The cloned DNA is only infectious to plants if the virus moiety is excised from the recombinant plasmid (Howell *et al.*, 1980; Lebeurier, Hirth, Hohn & Hohn, 1980). The DNA is infectious either as religated circular molecules or as linear molecules and the DNA extracted from viruses obtained from plants infected with this cloned DNA has all the unusual properties associated with CaMV DNA. Thus plants can religate linear molecules but, as shown by Hohn *et al.* (1982), they are more efficient at religating molecules with longer, cohesive, single-stranded ends. The resulting viral DNA has also regained its discontinuities.

The effect of mutagenising CaMV DNA by either the addition of inserts or the deletion of nucleotides is usually the loss of infectivity (Fig. 3). However, at two sites, one in the large intergenic region and the other in open reading region II, inserts have been made which do not result in loss of infectivity (Gronenborn, Gardner, Schaefer & Shepherd, 1981; Howell, Walker & Walden, 1981; Shepherd *et al.*, 1981). The isolate CM4-184 has much of open reading region II deleted. The insert in the intergenic region was an *Eco*RI linker ligated into an *Alu*I site (Howell *et al.*, 1981). Shepherd *et al.* (1981) and Gronenborn *et al.* (1981) inserted bacterial DNA fragments into the *Xho*I site in open reading region II. CaMV DNA with inserted fragments of 60 and 250 bp was infectious and the fragments were retained. When larger fragments (500 or 1200 bp) were inserted infected plants contained viral DNA from which most, if not all, of the bacterial DNA had been deleted. Thus, there appears to be a limit to the amount of 'foreign' DNA which full-length CaMV DNA can accommodate. This is likely to reflect limits on encapsidation and, as noted above, encapsidation may be needed for cell-to-cell spread. This point can only be answered using protoplasts.

Cell systems for CaMV

Infection of turnip leaf protoplasts with CaMV has been shown using production of radioactive virus and transcripts (Howell & Hull, 1978) and production of inclusion bodies and virus (Furusawa *et al.*, 1980). However, it is generally accepted that infection of leaf protoplasts is not a reliable and efficient system. One possible reason for the unreliability is that the majority of the protoplasts might be in the wrong metabolic state. If the viral DNA replicates in the nucleus (as seems likely) and if it requires host DNA polymerase activity to do so (as is quite probable) virus replication should be more efficient in cells close to active division.

Prospects for the future

Of the three types of plant virus vector systems, the development of that based on caulimoviruses is the most advanced. However, at this stage there are various problems which, though they initially relate to the caulimovirus system at present, will also have to be circumvented with the other systems. One of the major problems is that the viruses are pathogens and thus cause undesirable effects on plants. Nothing is known about the molecular biology of symptom production and it is not known whether the gene products which incite symptoms are essential for the replication of the viral genome. Similarly, nothing is known about host range determination and whether the limitations of host range can be overcome. As noted above no plant viruses integrate into the host genome. Hull (1978, 1980) pointed out that this is not necessarily a severe disadvantage in plants which can be vegetatively propagated. However, it would be helpful if a virus-based system could be made to integrate. One further problem is that the spread of plant viruses is restricted in meristematic tissues and thus a cytoplasmic virus based vector might be lost on growing up transformed protoplasts. This would be overcome if an integrating system was developed or, if the problem of symptom expression is overcome, by inoculating the vector + 'foreign' DNA into a seedling and allowing it to spread systemically through the developing plant.

The studies on developing a vector system based on caulimoviruses will yield considerable information on DNA replication and expression in plant cells. However, in view of some of the problems discussed above it seems likely that the next generation of vector systems will comprise features from the viral DNA together with some from other DNAs. One could suggest combining the replication and expression features from CaMV with the integration and host range features from the T-DNA of *Agrobacterium* Ti plasmid (the T-DNA is that moiety of the DNA of the plasmid which integrates into the host cell DNA.)

References

Al Ani, R., Pfeiffer, P., Whitechurch, O., Lesot, A., Lebeurier, G. & Hirth, L. (1980). A virus specified protein produced upon infection by cauliflower mosaic virus. *Annals de Virologie (Institut Pasteur)*, **131E**, 33–53.

Breathnach, R. & Chambon, P. (1981). Organization and expression of eukaryotic split genes coding for proteins. *Annual Review of Biochemistry*, **50**, 349–83.

Covey, S. N. & Hull, R. (1981). Transcription of cauliflower mosaic virus DNA. Detection of transcripts, properties and location of the gene encoding the virus inclusion body. *Virology*, **111**, 463–74.

Covey, S. N., Lomonossoff, G. P. & Hull, R. (1981). Characterisation of cauliflower mosaic virus DNA sequences which encode major polyadenylated transcripts. *Nucleic Acids Research*, **9**, 6735–47.

Favali, M. A., Bassi, M. & Conti, G. G. (1973). A quantitative autoradiographic study of intracellular sites for replication of cauliflower mosaic virus. *Virology*, **53**, 115–19.

Franck, A., Jonard, G., Richards, K., Hirth, L. & Guilley, H. (1980). Nucleotide sequence of cauliflower mosaic virus DNA. *Cell*, **21**, 285–94.

Fujisawa, I. M., Rubio-Huertos, M., Matsui, C. & Yamaguchi, A. (1967). Intracellular appearance of cauliflower mosaic virus particles. *Phytopathology*, **57**, 1130–2.

Furusawa, K., Yamaoka, N., Okuno, T., Yamamoto, M., Kohno, M. & Kunoh, H. (1980). Infection of turnip protoplasts with cauliflower mosaic virus. *Journal of General Virology*, **48**, 431–6.

Gardner, R. C., Howarth, A. J., Brown-Luedi, M., Shepherd, R. J. & Messing, J. (1981). The complete nucleotide sequence of an infectious clone of cauliflower mosaic virus by M13mp7 shotgun sequencing. *Nucleic Acids Research*, **9**, 2871–88.

Goodman, R. M. (1981*a*). Geminiviruses. In *Handbook of Plant Virus Infections and Comparative Diagnosis*, ed. E. Kurstak, pp. 883–910. Amsterdam: Elsevier North-Holland Biomedical Press.

Goodman, R. M. (1981*b*). Geminiviruses. *Journal of General Virology*, **54**, 9–21.

Gronenborn, B., Gardner, R. C., Schaefer, S. & Shepherd, R. J. (1981). Propagation of foreign DNA in plants using cauliflower mosaic virus as vector. *Nature*, **294**, 773–6.

Guilfoyle, T. J. (1980). Transcription of the cauliflower mosaic virus genome in isolated nuclei from turnip *Brassica rapa* cultivar Just Right leaves. *Virology*, **107**, 71–80.

Hohn, T., Hohn, B., Lesot, A. & Lebeurier, G. (1980). Restriction map of native and cloned cauliflower mosaic virus DNA. *Gene*, **11**, 21–31.

Hohn, T., Richards, K. & Lebeurier, G. (1982). Cauliflower mosaic virus on its way to becoming a useful plant vector. *Current Topics in Microbiology & Immunology*, **96**, 193–236.

Howarth, A. J., Gardner, R. C., Messing, J. & Shepherd, R. J. (1981). Nucleotide sequence of naturally occurring deletion mutants of cauliflower mosaic virus. *Virology*, **112**, 678–85.

Howell, S. H. (1981). Ultraviolet mapping of RNA transcripts encoded by the cauliflower mosaic virus genome. *Virology*, **112**, 488–95.

Howell, S. H. & Hull, R. (1978). Replication of cauliflower mosaic virus and transcription of its genome in turnip leaf protoplasts. *Virology*, **86**, 468–81.

Howell, S. H., Odell, J. T. & Dudley, R. K. (1980). Expression of the cauliflower mosaic virus genome in turnips (*Brassica rapa*). In *Genome Organization and Expression in Plants*, ed. C. J. Leaver, pp. 529–36. New York: Plenum Press.

Howell, S. H., Walker, L. L. & Dudley, R. K. (1980). Cloned cauliflower mosaic virus DNA infects turnips *Brassica rapa*. *Science*, **208**, 1265–7.

Howell, S. H., Walker, L. L. & Walden, R. M. (1981). Rescue of *in vitro* generated mutants of cloned cauliflower mosaic virus genome in infected plants. *Nature*, **293**, 483–6.

Hull, R. (1978). The possible use of plant viral DNAs in genetic manipulation in plants. *Trends in Biochemical Sciences*, **3**, 254–6.

Hull, R. (1980). Genetic engineering in plants: – the possible use of cauliflower mosaic virus DNA as a vector. In *Plant Cell Cultures: Results and Perspectives*, ed. F. Sala, B. Parisi, R. Cella & O. Ciferri, pp. 219–24. Amsterdam: Elsevier North-Holland Biomedical Press.

Hull, R. (1981). Cauliflower mosaic virus DNA as a possible gene vector for higher plants. In *Genetic Engineering in the Plant Sciences*, ed. N. J. Panopoulos, pp. 99–109. Praeger Scientific.

Hull, R., Covey, S. N., Stanley, J. & Davies, J. W. (1979). The polarity of the cauliflower mosaic virus genome. *Nucleic Acids Research*, **7**, 669–77.

Hull, R. & Donson, J. (1982). Physical mapping of the DNAs of carnation etched ring and figwort mosaic viruses. *Journal of General Virology*, **60**, 125–34.

Hull, R. & Howell, S. H. (1978). Structure of the cauliflower mosaic virus genome. II. Variation in DNA structure and sequence between isolates. *Virology*, **86**, 482–93.

Hull, R., Shepherd, R. J. & Harvey, R. D. (1976). Cauliflower mosaic virus: an improved purification procedure and some properties of the virus particles. *Journal of General Virology*, **31**, 93–100.

Kamei, T., Rubio-Huertos, M. & Matsui, C. (1969). Thymidine-H^3 uptake by X-bodies associated with cauliflower mosaic virus infection. *Virology*, **37**, 506–8.

Kleinhofs, A. & Behki, R. (1977). Prospects for plant genome modification by nonconventional methods. *Annual Review of Genetics*, **11**, 79–101.

Koziel, M. & Siegel, A. (1981). Cloning of a DNA copy of tobacco rattle virus RNA-2. *Proceedings of the 5th International Congress of Virology*, p. 257. Strasbourg: International Union of Molecular Biologists.

Lane, L. C. (1979). The nucleic acids of multipartite, defective and satellite plant viruses. In *Nucleic Acids in Plants*, vol. 2, ed. T. C. Hall & J. W. Davies, pp. 65–110. Florida: CRC Press Inc.

Lawson, R. H. & Hearon, S. S. (1974). Ultrastructure of carnation etched ring-virus infected *Saponaria vaccaria* and *Dianthus caryophyllus*. *Journal of Ultrastructural Research*, **48**, 201–15.

Lebeurier, G., Hirth, L., Hohn, T. & Hohn, B. (1980). Infectivities of native and cloned DNA of cauliflower mosaic virus. *Gene*, **12**, 139–46.

Lesemann, D. & Casper, R. (1973). Electron microscopy of petunia vein-clearing virus, an isometric plant virus associated with specific inclusions in petunia cells. *Phytopathology*, **63**, 1118–24.

Lurquin, P. F. & Kado, C. I. (1979). Recent advances in the insertion of DNA into higher plant cells. *Plant, Cell and Environment*, **2**, 199–203.

McKnight, T. D. & Meagher, R. B. (1981). Isolation and mapping of small cauliflower mosaic virus DNA fragments active as promoters in *E. coli*. *Journal of Virology*, **37**, 673–82.

Meagher, R. B., Shepherd, R. J. & Boyer, H. W. (1977). The structure of cauliflower mosaic virus. I. Restriction endonuclease map of cauliflower mosaic virus DNA. *Virology*, **80**, 362–75.

Odell, J. T., Dudley, K. & Howell, S. H. (1981). Structure of the 19S RNA transcript encoded by the cauliflower mosaic virus genome. *Virology*, **111**, 377–85.

Odell, J. T. & Howell, S. H. (1980). The identification, mapping and characterization of messenger RNA for P66, a cauliflower mosaic virus coded protein. *Virology*, **102**, 349–59.

Proudfoot, N. J. & Brownlee, G. G. (1976). 3′ non-coding region in eukaryotic messenger RNA. *Nature*, **263**, 211–14.

Racaniello, V. R. & Baltimore, D. (1981). Cloned poliovirus complementary DNA is infectious in mammalian cells. *Science*, **214**, 916–19.

Shalla, T. A., Shepherd, R. J. & Petersen, L. J. (1980). Comparative cytology of 9 isolates of cauliflower mosaic virus. *Virology*, **102**, 381–8.

Shepherd, R. J. (1976). DNA viruses of higher plants. *Advances in Virus Research*, **20**, 305–39.

Shepherd, R. J. (1977). Cauliflower mosaic virus (DNA virus of higher plants). In *The Atlas of Insect and Plant Viruses*, ed. K. Maramorosch, pp. 159–66. New York: Academic Press.

Shepherd, R. J. (1979). DNA plant viruses. *Annual Review of Plant Physiology*, **30**, 405–23.

Shepherd, R. J. (1981). Caulimoviruses. In *Handbook of Plant Virus Infections and Comparative Diagnosis*, ed. E. Kurstak, pp. 847–78. Amsterdam: Elsevier North-Holland Biomedical Press.

Shepherd, R. J., Bruening, G. E. & Wakeman, R. J. (1970). Double-stranded DNA from cauliflower mosaic virus. *Virology*, **41**, 339–47.

Shepherd, R. J., Gronenborn, B., Gardner, R. & Daubert, S. D. (1981). Molecular cloning of foreign DNA in plants using cauliflower mosaic virus as recombinant vector. In *Genetic Engineering in the Plant Sciences*, ed. N. J. Panapoulos, pp. 255–7. Praeger Scientific.

Shepherd, R. J., Richins, R. & Shalla, T. A. (1980). Isolation and properties of the inclusion bodies of cauliflower mosaic virus. *Virology*, **102**, 389–400.

Szeto, W. W., Hamer, D. H., Carlson, P. S. & Thomas, C. A. (1977). Cloning of cauliflower mosaic virus DNA in *Escherichia coli*. *Science*, **196**, 210–12.

Volovitch, M., Drugeon, G. & Yot, P. (1978). Studies on the single-stranded discontinuities of the cauliflower mosaic virus genome. *Nucleic Acids Research*, **5**, 2913–25.

Volovitch, M., Dumas, J. P., Drugeon, G. & Yot, P. (1976). Single-stranded interruptions in cauliflower mosaic virus. In *Nucleic Acids and Protein Synthesis in Plants*, ed. L. Bogorad & J. H. Veil, pp. 635–41. Colloquin No. 261. Paris: Centre National de la Recherche Scientifique.

J. SCHRÖDER, H. DE GREVE, J. P. HERNALSTEENS, J. LEEMANS, M. VAN MONTAGU, L. OTTEN, G. SCHRÖDER, L. WILLMITZER and J. SCHELL

Ti plasmid-mediated gene transfer to higher plant cells

Introduction

Crown gall is a neoplastic disease of many dicotyledonous plants which is caused by the soil bacterium *Agrobacterium tumefaciens*. The tumour-inducing capacity of the bacterium resides in large extrachromosomal plasmids which are therefore called Ti plasmids (Van Larebeke *et al.*, 1974). Although the plasmids are necessary for the induction of tumorous growth, they are not necessary for the maintenance of neoplastic growth after establishment of the growth pattern. Thus it is possible to grow the tumour cells in sterile culture without contamination from bacteria, and this has been very important in the molecular analysis of the disease.

In contrast to normal cells, the tumour cells grow in culture in the absence

Fig. 1. Chemical structure of some opines found in crown gall tumours.

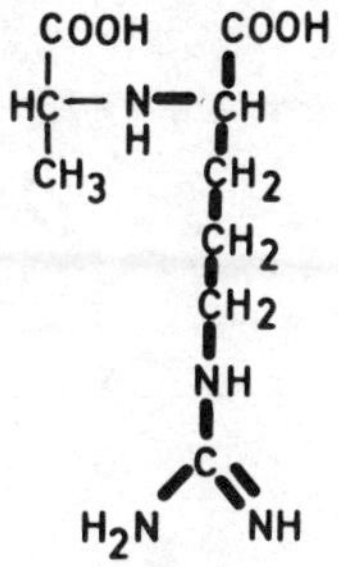

Octopine

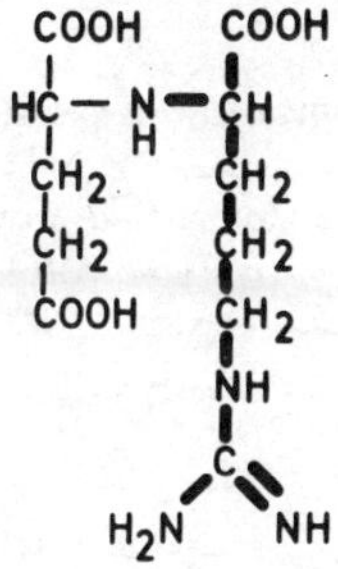

Nopaline

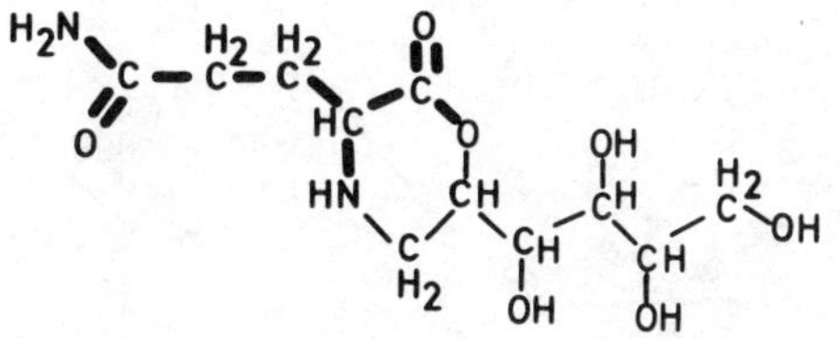

Agropine

of auxins and cytokinins, and this hormone-independent growth defines tumours in plant cells (Braun, 1956). Also, these cells synthesise a number of low molecular weight compounds, called opines, which are not found in normal plant tissues. The opine produced defines crown gall cells as octopine-, nopaline-or agropine-type tumours (Fig. 1 and Guyon, Chilton, Petit & Tempe, 1980). Opines can be utilised selectively by the Agrobacteria as sources for carbon, nitrogen and energy, and therefore the interaction between these bacteria and plants can be considered as a special parasitic relationship which benefits the bacteria (Schell *et al.*, 1979).

A large number of detailed investigations have shown that sterile tumour cells contain a part of the Ti plasmid which is covalently integrated into the plant chromosomes (Chilton *et al.*, 1977; Schell *et al.*, 1979; Lemmers *et al.*, 1980; Thomashow *et al.*, 1980*a*; Thomashow *et al.*, 1980*b*; Yadav *et al.*, 1980; Zambryski *et al.*, 1980). This part of the Ti plasmid is called the T-region in the bacteria and T-DNA in the plant cells. It is found exclusively in the nucleus (Chilton *et al.*, 1980; Willmitzer *et al.*, 1980), it represents about 10% of the Ti plasmid and it is directly responsible for opine synthesis and the hormone-independent growth of the cells. Clearly, the Agrobacteria perform a genetic manipulation of plant cells and Fig. 2 summarises some of the salient features of the biological system.

Thus, nature has provided us in the Ti plasmids with a purified set of genes for the investigation of two important aspects. These are: (1) the control of growth and differentiation in plants by hormones and (2) the feasibility of

Fig. 2. Genetic colonisation of plants by *Agrobacterium tumefaciens*.

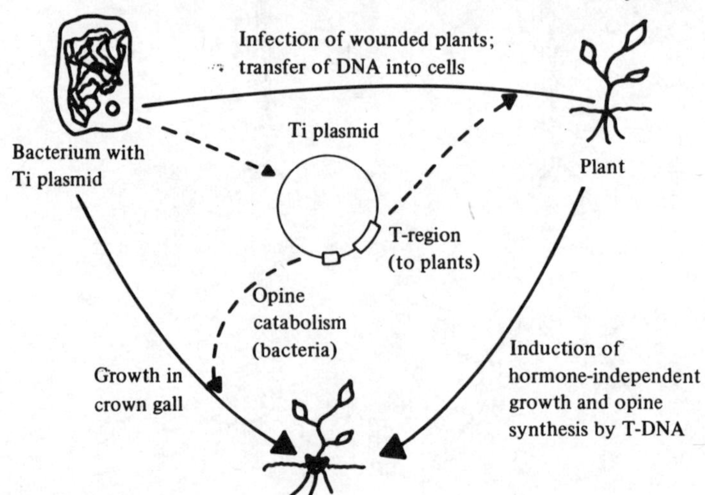

manipulating plants by direct transfer of genes. We describe some recent advances in understanding the functions of the T-DNA and in the use of Ti plasmids as vectors for the controlled genetic engineering of plants.

Expression of the genes encoded on T-DNA

If the T-DNA contains structural genes whose protein products are responsible for the tumorous properties of infected plant cells, both nuclei and polyribosomes would be expected to contain RNA sequences which are complementary to the T-DNA (Fig. 3). The presence of such transcripts has been shown in several reports (Drummond, Gordon, Nester & Chilton, 1977; Gurley *et al.*, 1979; Gelvin, Gordon, Nester & Aronson, 1981; Willmitzer *et al.*, 1981*a*), but in most cases the number, sizes and coding regions were not described. These aspects were recently investigated in some cell suspension cultures from tobacco which harbour the T-DNA from pTiA6, a Ti plasmid which induces octopine synthesis. The cells contain only the T_L-fragment (i.e. the core DNA of Ti plasmids which is to a large extent homologous in both octopine and nopaline types) of the T-region (Fig. 4 and Thomashow *et al.*, 1980*a*; De Beuckeleer *et al.*, 1981). Tumour-specific RNAs were detected and mapped by hybridisation of ^{32}P-labelled Ti plasmid fragments to polyadenylated RNA which had been separated on agarose gels and then transferred to diazobenzyloxymethyl-cellulose (DBM) paper (Willmitzer *et al.*, 1981*a*). The results, summarised in Fig. 5, show that the cells contained distinct transcripts which differed markedly both in their relative abundance and in their sizes. They all bound to oligo(dT)-cellulose, indicating that they were polyadenylated, and thus that the T-DNA, which had been transferred from a prokaryotic organism, must have provided specific poly(A) addition sites. The direction of transcription was determined for most of the transcripts, and the locations of the approximate 5′ and 3′ ends were mapped on the T_L-DNA (Fig. 5).

Fig. 3. Model for the expression of genes on the T-DNA into properties of the cells.

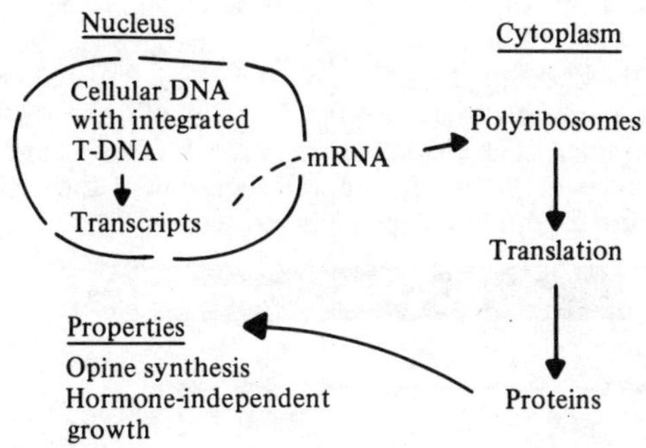

All the RNAs mapped within the T-DNA sequence. This, and the observation that transcription is inhibited by low concentrations of α-amanitin (Willmitzer, Schmalenbach & Schell, 1981*b*), seems to suggest that each transcript is determined by a specific promoter site on the T_L-DNA which is recognised by plant polymerase II. The data presented here do not rule out the possibility that some T-DNA promoters serve for the transcription of more than one species of RNA. Considering the groupwise orientation of several transcripts, the simplest model would assume one promoter site per group of transcripts. If so, one would expect that inactivation of the 5′-proximal gene of a group would also lead to the disappearance of the transcripts from the 5′-distal genes. However, analysis of some cell linings containing the T-DNA of Ti plasmid mutants indicates that groupwise inactivation of genes does not occur (Leemans *et al.*, 1982). The results available so far are consistent with the assumption that each gene of the T-DNA has its own signals for transcription in the eukaryotic plant cells.

Fig. 4. Restriction map of the T_L-region in octopine-coding plasmid pTiAch5 which is identical to that of plasmid pTiAy (Engler, Van Montague, Zaenen & Schell, 1977; De Vos, De Beuckeleer, Van Montagu & Schell, 1981) and the T_L-DNA in tobacco cell line A6-S1 (De Beuckeleer *et al.*, 1981). The dotted lines in the T-DNA indicate that the precise borders of the T-DNA have not yet been determined. The numbers in bold type indicate restriction fragment numbers while those in brackets refer to the size of the fragments in base pairs $\times 10^{-3}$. Fragment *Eco*R1 32 (see last part of this chapter) is the small fragment to the left of *Eco*R1. 7.

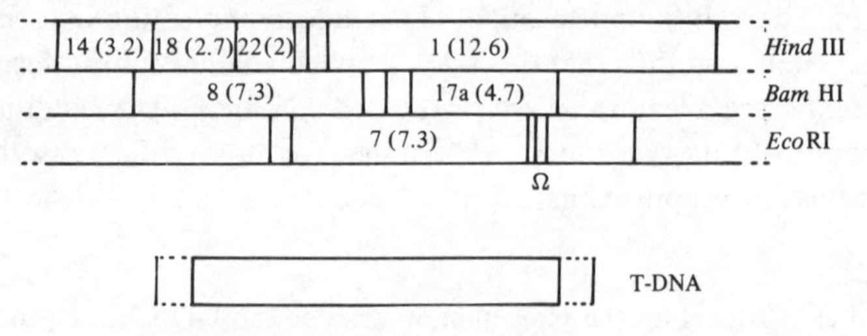

Fig. 5. Transcripts from the T_L-DNA in tumour line A6-S1 (Willmitzer, Simons & Schell, 1982). The numbers refer to the size of the RNAs in bases $\times 10^{-3}$. The thickness of the lines indicates the relative abundance of the RNAs, and the arrowheads describe the direction of transcription where this has been determined. The shaded areas in the T-DNA define the parts which are common to octopine and nopaline plasmids; they contain the genes responsible for hormone-independent growth.

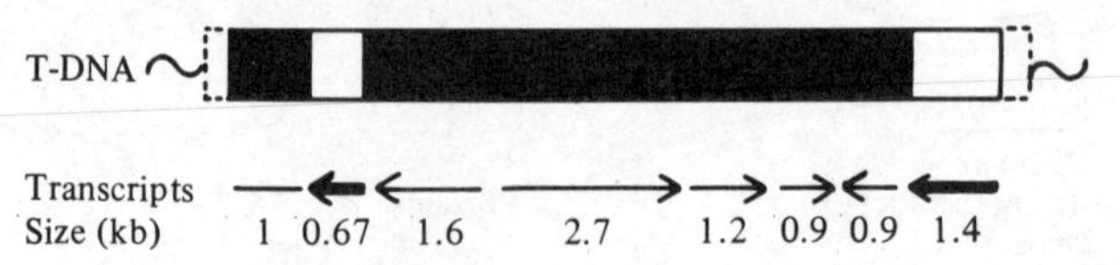

In order to understand the mechanisms of T-DNA actions it is important to know whether these RNAs are translated into proteins and to analyse the functions of these proteins. The hybridisation studies of Willmitzer *et al.* (1981*a*) and a previous report on T-DNA-derived translatable RNAs (McPherson, Nester & Gordon, 1980) had indicated that the concentrations of these RNAs in transformed plant cells are very low. Therefore it was necessary to develop a hybridisation selection procedure that was sufficiently sensitive and specific to detect mRNAs which represent about 0.0001 % of the total mRNA activity in the plant cells. This procedure was used to enrich for T-DNA-derived mRNAs by hybridising them to Ti plasmid fragments that were covalently bound to microcrystalline cellulose; the hybridised RNAs were eluted and translated in vitro in a cell-free system prepared from wheat germ.

The results obtained with this approach are summarised in Fig. 6 (Schröder & Schröder, 1982). The octopine-type tumour cells contained at least three T-DNA-derived mRNAs which were translated in vitro into distinct proteins, and the coding regions correlated with those of three transcripts. The protein encoded at the right end of the T-DNA (molecular mass 39000) was of specific interest since previous genetic analysis had indicated that this part is responsible for octopine synthesis (Koekman, Ooms, Klapwijk & Schilperoort, 1979; De Greve *et al.*, 1981; Garfinkel *et al.*, 1981), and the size of the protein synthesised in vitro was identical with that of the octopine-

Fig. 6. Protein-coding regions on the T_L-DNA of octopine-coding plasmid pTiAch5. Coding regions were identified either by translation of hybridisation-selected mRNA from plants (Schröder *et al.*, 1981; Schröder & Schröder, 1982) or by expression of cloned Ach5 restriction fragments in *Escherichia coli* minicells (Schröder *et al.*, 1981*a*; and G. Schröder, W. Klipp, A. Hillerbrand, R. Ehring, C. Koncz & J. Schröder, unpublished observations). The protein with molecular mass 39000 represents the octopine-synthesising enzyme. The numbers refer to the sizes of the proteins in Kilodaltons. 'Fusion' denotes the fusion protein obtained in minicells from the combined expression of a bacterial gene (located on the pACYC184 plasmid used as the vector in these studies) and part of the structural gene coding for the octopine-synthesising enzyme.

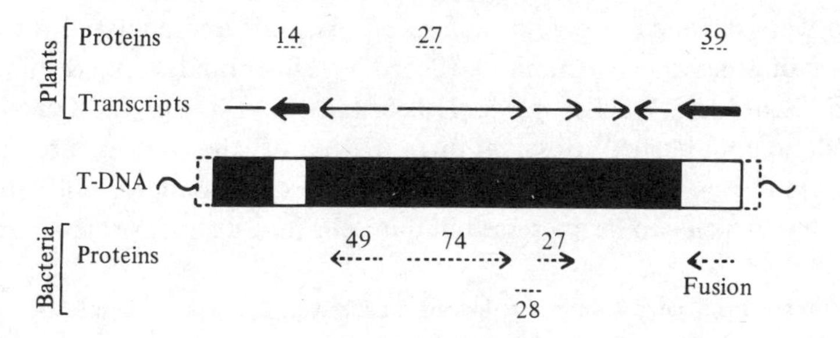

synthesising enzyme in octopine-producing tumours. Immunological studies showed that this protein was recognised by antiserum against the tumour-specific enzyme (Schröder *et al.*, 1981 *a*). These results were confirmed recently (Murai & Kemp, 1982) and demonstrate that the structural gene for this enzyme is on the Ti plasmid. So far, this is the only protein product of the T-DNA with known enzymatic properties; the possible functions of the two smaller T-DNA-derived proteins are not known. The region coding for the octopine-synthesising enzyme has recently been sequenced (H. De Greve, P. Dhaese, J. Seurinck, M. Van Montagu & J. Schell, unpublished observations). According to these data, the sequences for initiation and termination of transcription reveal signals of eukaryotic rather than prokaryotic type, suggesting that this gene is designed for expression in eukaryotic cells.

However, this is not necessarily true for all of the genes, since weak transcription from the T-region was also detected in Agrobacteria (Gelvin *et al.*, 1981). It was therefore of interest whether the T-DNA-derived mRNAs isolated from plant cells shared properties with typical prokaryotic or typical eukaryotic mRNAs. The fact that translation of all three mRNAs was inhibited by the cap* analogue pm^{7}G (Fig. 7) (Schröder & Schröder, 1982) suggests but does not prove that they contain a cap structure at the 5′ end. This would be typical for eukaryotic mRNA, since caps have not been described in prokaryotic RNA. All three mRNAs were found in polyadenylated as well as in non-polyadenylated RNA fractions; this cannot be used to support such a tentative classification, however, because both types of RNA have been described in eukaryotic as well as in prokaryotic cells.

The mRNAs for the three proteins produced by in vitro translation each represent about 0.0001% of the total mRNA activity in the polyribosomal RNA, and at present this appears to be the detection limit for translatable RNA. The other transcripts detected by hybridisation experiments are present at even lower concentrations, and, assuming that they possess mRNA activity, this is likely to be the reason why the corresponding proteins have not been identified by in vitro translation so far.

A different approach has therefore been developed to search for coding regions on the T-DNA and their protein products. Fragments from the T-region were recloned into *Escherichia coli* plasmids and analysed for the expression of genes in *E. coli* minicells (Schröder, Hillebrand, Klipp & Pühler, 1981; G. Schröder, W. Klipp, A. Hillebrand, R. Ehring, C. Koncz & J. Schröder, unpublished observations). Some of the results are also summarised in Fig. 6. There are at least four different coding regions within the T_L-DNA which can be expressed in minicells into distinct proteins from

* The inverted methylated guanosine residue on the 5′ end of many eukaryotic mRNAs.

promoters which are active in the prokaryotic cells. Three of the four regions expressed in *E. coli* correlate with three regions transcribed in plant cells. The plant transcripts are larger than the proteins in *E. coli*, and the regions expressed in minicells appear to lie within the regions transcribed in plant cells. Also, for three of the four proteins expressed in *E. coli* it is known that the direction of transcription is the same in these bacteria and in plant cells (Fig. 6). Although other explanations are not ruled out, it seems possible that plant cells and *E. coli* express at least partly the same coding regions.

A fifth coding region was obtained as a result of fusion between part of a bacterial gene from the vector plasmid and part of the structural gene for

Fig. 7. Inhibition of translation of tumour-specific mRNA by the cap analogue pm^{7}G (Schröder & Schröder, 1982). Affixes (–) and (+) indicate incubations in absence and presence of the drug. Lanes: wheat germ extract incubated (a) without added RNA, (b) with RNA selected by *Hind*III 1 and (c) with RNA selected by *Bam*HI 8 (see Fig. 4 for position of the restriction fragments used to select tumour-specific mRNAs). M, marker proteins. Numbers are the sizes of proteins in Kilodaltons.

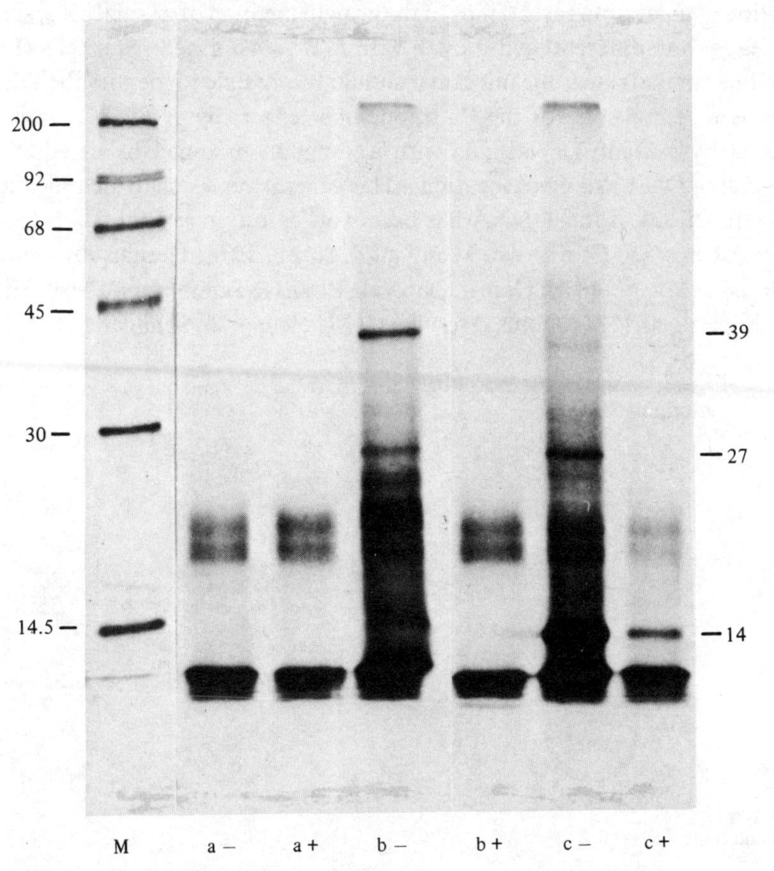

the octopine-synthesising enzyme. The gene product of this hybrid in minicells is a fusion protein of the amino-terminal part of the bacterial protein and the carboxy-terminal part of the plant tumour-specific protein, as shown by specific immunoprecipitation with antisera against both the plant and the bacterial protein (Schröder *et al.*, 1981*b*). This demonstrates by independent evidence that the structural gene for this protein is on the Ti-plasmid.

In summary, these results clearly show that the T-DNA contains genes which are expressed in plant cells, and thus, that Agrobacteria perform a true genetic manipulation of plants. The finding that the T-region also contains coding regions which can be expressed in a prokaryotic background is somewhat unexpected, and more experiments are necessary to investigate the possibility that these regions may be expressed in Agrobacteria in the early stages of infection of plants.

As mentioned above, the exact functions of most T-DNA gene products are not known. However, the action of the genes leading to hormone-independent growth may be formally compared with the effects of a balanced addition of auxins plus cytokinins to the growth medium of normal cells, since this leads to undifferentiated growth with non-transformed plant cells (Fig. 8). If this comparison is meaningful it should be possible to identify the DNA sequences responsible for the 'hormone-like' effects by analysing tumours induced by mutant Ti plasmids with a T-region modified by deletion or insertion of DNA in defined locations. This general approach to identification of the functions of the T-DNA has been worked out in several laboratories (Hernalsteens, De Greve, Van Montagu & Schell, 1978; Hernalsteens *et al.*, 1980; Dhaese *et al.*, 1979; Ooms, Klapwijk, Poulis & Schilperoort, 1980; Hille & Schilperoort, 1981; Ooms, Hooykaas, Moolenaar & Schilperoort, 1981)

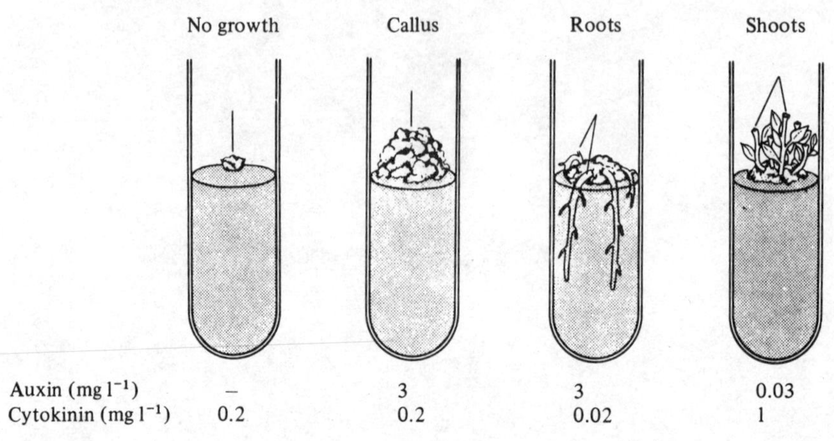

Fig. 8. Control of morphogenesis in tobacco explants by auxins and cytokinins.

and has led to a rough localisation of the regions responsible for octopine and nopaline synthesis in the plant cells (Koekman *et al.*, 1979; Holsters *et al.*, 1980; De Greve *et al.*, 1981; Garfinkel *et al.*, 1981).

Now that the transcripts of the T-DNA have been identified and mapped (see Fig. 5), it is possible to assign the mutational inactivation of functions to specific genes. With respect to the 'hormone-like' effects of the T-DNA discussed above it is of considerable interest that inactivation of two specific genes led to tumours producing shoots, and this resembles the effect of supplying normal cells with excess cytokinins but no auxins (Fig. 8). These two genes, when active, thus appear to produce in undifferentiated tumours containing the complete T-DNA an 'auxin-like' effect by inhibiting shoot formation (Fig. 9). Correspondingly, inactivation of a third gene led to tumours producing roots, and this resembles the effect of supplying normal cells with excess auxins but very little cytokinins (Fig. 8). Apparently this gene codes for a product inhibiting root formation, a 'cytokinin-like' effect (Fig. 9). Thus it is possible to correlate 'auxin-like' and 'cytokinin-like' effects with defined genes on the T-DNA, and the action of both sets of genes together leads to unorganised growth of cells by suppressing both root and shoot formation (Leemans *et al.*, 1981, 1982; Willmitzer *et al.*, 1982).

The general approach to defining functions of the T-DNA which is briefly outlined here has been extended to other genes as well, and the results will not only assist in the eventual analysis of the exact functions of the gene products, but are also important for the development of new Ti plasmid-derived gene vectors which allow subsequent regeneration of plants from single cells.

Introduction of new genes into plants via Ti plasmids

Ti plasmids are natural gene vectors which contain all the functions necessary for the transfer, stable incorporation and expression of genetic information in plants. One disadvantage of the plasmids is the fact that some

Fig. 9. Genes in the T-region of octopine Ti plasmids controlling the formation of shoots and roots (Leemans *et al.*, 1981, 1982). Arrow-heads show the direction of transcription.

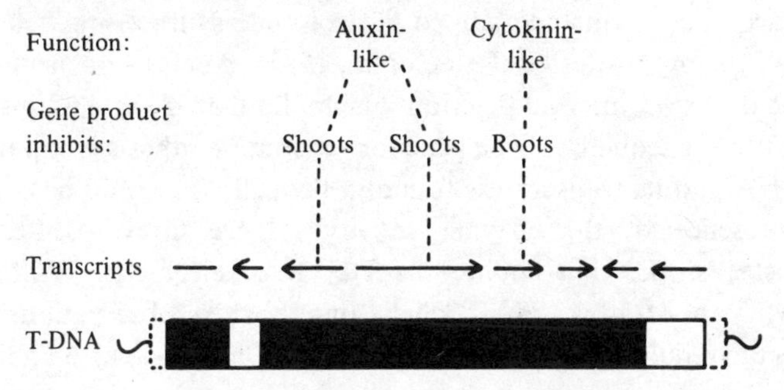

of the functions encoded on the T-DNA force the cells into hormone-independent growth (Fig. 9). Since differentiation into organs is largely controlled by hormones like auxins and cytokinins, the activity of these genes of the T-DNA interferes with the normal morphogenetic potential of the cells, and there is abundant evidence that entirely normal plants are very difficult to regenerate from cells containing the complete T-DNA (Braun, 1959; Braun & Wood, 1976; Turgeon, Wood & Braun, 1976; Wood, Binns & Braun, 1978; Yang *et al.*, 1980; Wullems, Molendijk, Ooms & Schilperoort, 1981 *a*, *b*). Recent results indicate, however that DNA-transfer and tumorous growth are different functions which may be separated (Leemans *et al.*, 1982). The method of choice is therefore to alter the Ti plasmids genetically in such a way that the tumour genes in the T-region are inactivated or eliminated altogether without interfering with the other genes responsible for the transfer of the T-region to the plant cells.

Such a Ti plasmid has been obtained by Tn7 mutagenesis (De Greve *et al.*, 1981). It contains the large transposon Tn7 (obtained from bacterial strains resistant to either spectinomycin/streptomycin or trimethoprim) inserted into the fragment *Eco*RI 32 (see Fig. 4) of pTi B6S3, an octopine-type plasmid, and the insertion of the transposon inactivates the 'auxin-like' functions of the T-DNA. Tumours induced on tobacco with this plasmid are morphogenetically very active and produce numerous shoots.

The vast majority of these shoots can, upon separation from the tumour tissue, develop roots and grow into normal plants. These shoots do not contain any measurable T-DNA-linked activity or function and are, therefore, presumably derived either from untransformed cells present in the initial, uncloned tumour tissue or from cells which have lost the T-DNA by deletion. By extensive screening of several hundred of these shoots, some were found to contain and express the gene for octopine synthesis. These plants have been called rGV-1 to rGV-5 (H. De Greve, J. Leemans, J. P. Hernalsteens, L. Thia-Toong, M. De Beuckeleer, L. Willmitzer, L. Otten, M. Van Montagu & J. Schell, unpublished observations) and one of them, rGV-1, has been studied in great detail. It was found to regenerate into a morphologically normal tobacco plant which contained the octopine-synthesising activity in all cells of all the organs tested (Otten *et al.*, 1981). Analysis of the T-DNA showed that it has an internal deletion which eliminated the Tn7 insertion and adjacent DNA sequences. The gene for octopine synthesis is still present on the T-DNA and its transcript is found in the cells, as would be expected from the presence of the enzyme activity (H. De Greve, J. Leemans, J. P. Hernalsteens, L. Thia-Toong. M. De Beuckeleer, L. Willmitzer, L. Otten, M. Van Montagu & J. Schell, unpublished observations). The experiments proved that the stable incorporation of a well-defined additional

Table 1. *Transmission of the gene for octopine synthesis in tobacco plant rGV-1*[a]

Crosses	No. of progeny tested	Octopine synthesis		
		+		−
rGV-1♂ × rGV-1♀	145	110 (76%)		35 (24%)
rGV-1♂ × Wildtype♀	248	124 (50%)		124 (50%)
rGV-1♀ × Wildtype♂	187	81 (43%)		106 (57%)
Haploid plantlets from anther cultures	102	47 (46%)		55 (54%)
rGV-1♂ × rGV-1♀	200	++[b]	+[b]	−
		42 (21%)	95 (48%)	63 (31 %)

[a] The plant is fertile and the gene is transmitted to progeny as a dominant Mendelian gene which is present once per haploid genome.

[b] ++ and + represent, respectively, enhanced and normal levels of enzyme activity as estimated by means of a semi-quantitative assay on progeny of this parental combination.

gene into higher plant cells is possible without affecting the morphogenetic potential of the cells.

The plant flowered and set seed quite normally, and it was investigated whether the progeny also contained the additional gene activity. The results, summarised in Table 1, show that the gene was transmitted as a stable, dominant Mendelian gene which is present once per haploid genome. Such experiments have been continued for several generations with the same result, indicating that the foreign gene has become a stable part of the plant genome.

The results summarised here clearly indicate that a stable introduction of foreign genes via Ti plasmids into plant cells is possible without affecting the morphogenetic potential of the cells. This is an important prerequisite for successful genetic engineering of plants. Construction of Ti plasmid derivatives in which the genes for tumorous growth are replaced by genes encoding for other selectable markers is under way in order that genetically altered plant cells can be produced.

References

Braun, A. C. (1956). The activation of two growth substance systems accompanying the conversion of normal to tumor cells in crown gall. *Cancer Research*, **16**, 53–6.

Braun, A. C. (1959). A demonstration of the recovery of the crown gall tumor cell with the use of complex tumors of single cell origin. *Proceedings of the National Academy of Sciences, USA*, **45**, 932–8.

Braun, A. C. & Wood, H. N. (1976). Suppression of the neoplastic state with the acquisition of specialized functions in cells, tissues and organs of crown gall teratomas of tobacco. *Proceedings of the National Academy of Sciences, USA*, **73**, 496–500.

Chilton, M.-D., Drummond, H. J., Merlo, D. J., Sciaky, D., Montoya, A. L., Gordon, M. P. & Nester, E. W. (1977). Stable incorporation of plasmid DNA into higher plant cells: the molecular basis of crown gall tumorigenesis. *Cell*, **11**, 263–71.

Chilton, M.-D., Saiki, R. K., Yadav, N., Gordon, M. P. & Quetier, F. (1980). T-DNA from *Agrobacterium* Ti plasmid is in the nuclear DNA fraction of crown gall tumor cells. *Proceedings of the National Academy of Sciences, USA*, **77**, 4060–4.

De Beuckeleer, M., Lemmers, M., De Vos, G., Willmitzer, L., Van Montagu, M. & Schell, J. (1981). Further insight on the transferred-DNA of octopine crown gall. *Molecular and General Genetics*, **193**, 283–8.

De Greve, H., Decraemer, H., Seurinck, J., Van Montagu, M. & Schell, J. (1981). The functional organization of the octopine *Agrobacterium tumefaciens* plasmid pTiB6S3. *Plasmid*, **6**, 235–48.

De Vos, G., De Beuckeleer, M., Van Montagu, M. & Schell, J. (1981). Restriction nuclease mapping of the octopine tumor-inducing plasmid pTiAch5 of *Agrobacterium tumefaciens. Plasmid*, **6**, 249–53.

Dhaese, P., De Greve, H., Decraemer, H., Schell, J. & Van Montagu, M. (1979). Rapid mapping of transposon insertion and deletion mutations in the large Ti plasmids of *Agrobacterium tumefaciens. Nucleic Acids Research*, **7**, 1837–49.

Drummond, M. H., Gordon, M. P., Nester, E. W. & Chilton, M.-D. (1977). Foreign DNA of bacterial plasmid origin is transcribed in crown gall tumours. *Nature*, **269**, 535–6.

Engler, G., Van Montagu, M., Zaenen, I. & Schell, J. (1977). Homology between Ti plasmids of *Agrobacterium tumefaciens*: hybridization studies using electron microscopy. *Biochemical Society Transactions*, **5**, 930–1.

Garfinkel, D. J., Simpson, R. B., Ream, L. W., White, F. F., Gordon, M. P. & Nester, E. W. (1981). Genetic analysis of crown gall: fine structure of the T-DNA by site-directed mutagenesis. *Cell*, **27**, 143–53.

Gelvin, S. B., Gordon, M. P., Nester, E. W. & Aronson, A. I. (1981). Transcription of the *Agrobacterium* Ti plasmid in the bacterium and in crown gall tumors. *Plasmid*, **6**, 17–29.

Gurley, W. B., Kemp, J. D., Alber, M. J., Sutton, D. W. & Gallis, J. (1979). Transcription of Ti-plasmid derived sequences in three octopine-type crown gall tumour lines. *Proceedings of the National Academy of Sciences, USA*, **76**, 2828–32.

Guyon, P., Chilton, M.-D., Petit, A. & Tempe, J. (1980). Agropine in 'null-type' crown gall tumors: evidence for the generality of the opine concept. *Proceedings of the National Academy of Sciences, USA*, **77**, 2693–7.

Hernalsteens, J. P., De Greve, H., Van Montagu, M. & Schell, J. (1978). Mutagenesis by insertion of the drug resistance transposon Tn7 applied to the Ti plasmid of *Agrobacterium tumefaciens. Plasmid*, **1**, 218–25.

Hernalsteens, J. P., Van Vliet, F., De Beuckeleer, M., Depicker, A., Engler, G., Holsters, M., Van Montagu, M. & Schell, J. (1980). The *Agrobacterium tumefaciens* Ti plasmid as a host vector system for introducing foreign DNA in plant cells. *Nature*, **287**, 654–6.

Hille, J. & Schilperoort, R. (1981). The use of transposons to introduce well-defined deletions in plasmids: possibilities for in vivo cloning. *Plasmid*, **6**, 151–4.

Holsters, M., Silva, B., Van Vliet, F., Genetello, D., De Block, M., Dhaese, P., Depicker, A., Inze, D., Engler, G., Villarroel, R., Van Montagu, M. & Schell, J. (1980). The functional organization of the nopaline *A. tumefaciens* plasmid pTiC58. *Plasmid*, **3**, 212–30.

Koekman, B. P., Ooms, G., Klapwijk, P. M. & Schilperoort, R. (1979). Genetic map of an octopine Ti plasmid. *Plasmid*, **2**, 347–57.

Leemans, J., Deblaere, R., Willmitzer, L., De Greve, H., Hernalsteens, J. P., Van Montagu, M. & Schell, J. (1982). Genetic identification of functions of I_L-DNA transcripts in octopine crown galls. *The EMBO Journal*, **1**, 147–52.

Leemans, J., Shaw, Ch., Deblaere, R., De Greve, H., Hernalsteens, J. P., Maes, M., Van Montagu, M. & Schell, J. (1981). Site-specific mutagenesis of *Agrobacterium* Ti plasmids and transfer of genes to plant cells. *Journal of Molecular and Applied Genetics*, **1**, 149–64.

Lemmers, M., De Beuckeleer, M., Holsters, M., Zambryski, P., Depicker, A., Hernalsteens, J. P., Van Montagu, M. & Schell, J. (1980). Internal organization, boundaries and integration of Ti plasmid DNA in nopaline crown gall tumors. *Journal of Molecular Biology*, **144**, 355–76.

McPherson, J. C., Nester, E. W. & Gordon, M. P. (1980). Proteins encoded by *Agrobacterium tumefaciens* Ti plasmid DNA (T-DNA) in crown gall tumors. *Proceedings of the National Academy of Sciences, USA*, **77**, 2666–70.

Murai, N. & Kemp, J. D. (1982). Octopine synthase mRNA isolated from sunflower crown gall callus is homologous to the Ti plasmid of *Agrobacterium tumefaciens. Proceedings of the National Academy of Sciences, USA*, **79**, 86–90.

Ooms, G., Hooykaas, P. J. J., Moolenaar, G. & Schilperoort, R. (1981). Crown call plant tumors of abnormal morphology, induced by *Agrobacterium tumefaciens* carrying mutated octopine Ti plasmids; analysis of T-DNA functions. *Gene*, **14**, 33–50.

Ooms, G., Klapwijk, P. M., Poulis, J. A. & Schilperoort, R. (1980). Characterization of Tn904 insertions in octopine Ti-plasmid mutants of *Agrobacterium tumefaciens. Journal of Bacteriology*, **144**, 82–91.

Otten, L., De Greve, H., Hernalsteens, J. P., Van Montagu, M., Schieder, O., Straub, J. & Schell, J. (1981). Mendelian transmission of genes introduced into plants by the Ti-plasmids of *Agrobacterium tumefaciens. Molecular and General Genetics*, **183**, 209–13.

Schell, J., Van Montagu, M., De Beuckeleer, M., De Block, M., Depicker, A., De Wilde, M., Engler, G., Genetello, C., Hernalsteens, J. P., Holsters, M., Seurinck, J., Silva, B., Van Vliet, F. & Villarroel, R. (1979). Interactions and DNA transfer between *Agrobacterium tumefaciens*, the Ti plasmid and the plant host. *Proceedings of the Royal Society of London, B*, **204**, 251–66.

Schröder, G. & Schröder, J. (1982). Hybridization selection and translation of T-DNA encoded mRNAs from octopine tumors. *Molecular and General Genetics*, **185**, 51–5.

Schröder, J., Hillebrand, A., Klipp, W. & Pühler, A. (1981 *a*). Expression of plant tumor-specific proteins in minicells of *E. coli*: a fusion protein of lysopine dehydrogenase with chloramphenicol acetyltransferase. *Nucleic Acids Research*, **9**, 5187–202.

Schröder, J., Schröder, G., Huisman, H., Schilperoort, R. & Schell, J. (1981 *b*). The mRNA for lysopine dehydrogenase in plant tumor cells is complementary to a Ti plasmid fragment. *FEBS Letters*, **129**, 166–8.

Thomashow, M. F., Nutter, R., Montoya, A. L., Gordon, M. P. & Nester,

E. W. (1980*a*). Integration and organization of Ti plasmid sequences in crown gall tumors. *Cell*, **19**, 729–39.

Thomashow, M. F., Nutter, R., Postle, K., Chilton, M.-D., Blattner, F. R., Powell, A., Gordon, M. P. & Nester, E. W. (1980*b*). Recombination between higher plant DNA and the Ti plasmid of *Agrobacterium tumefaciens*. *Proceedings of the National Academy of Sciences, USA*, **77**, 6448–52.

Turgeon, R., Wood, H. N. & Braun,A. C. (1976). Studies on the recovery of crown gall tumor cells. *Proceedings of the National Academy of Sciences, USA*, **73**, 3562–4.

Van Larebeke, N., Engler, G., Holsters, M., Van den Elsacker, S., Zaenen, I., Schilperoort, R. & Schell, J. (1974). Large plasmid in *Agrobacterium tumefaciens* essential for crown gall inducing ability. *Nature*, **252**, 169–70.

Willmitzer, L., Beuckeleer, M., Lemmers, M., Van Montagu, M. & Schell, J. (1980). The Ti plasmid derived T-DNA is present in the nucleus and absent from plastids of plant crown gall cells. *Nature*, **287**, 359–61.

Willmitzer, L., Otten, L., Simons, G., Schmalenbach, W., Schröder, J., Schröder, G., Van Montagu, M., De Vos, G. & Schell, J. (1981*a*). Nuclear and polysomal transcripts of T-DNA in octopine crown gall suspension and callus cultures. *Molecular and General Genetics*, **182**, 255–62.

Willmitzer, L., Schmalenbach, W. & Schell, J. (1981*b*). Transcription of T-DNA in octopine and nopaline crown gall tumors is inhibited by low concentrations of α-amanitin. *Nucleic Acids Researrch*, **9**, 4801–12.

Willmitzer, L., Simons, G. & Schell, J. (1982). The T_L-DNA in octopine crown gall tumors codes for seven well-defined polyadenylated transcripts. *The EMBO Journal*, **1**, 139–46.

Wood, H. N., Binns, A. N. & Braun, A. C. (1978). Differential expression of oncogenicity and nopaline synthesis in intact leaves derived from crown gall teratomas of tobacco. *Differentiation*, **11**, 175–80.

Wullems, G. J., Molendijk, L., Ooms, G. & Schilperoort, R. (1981*a*). Differential expression of crown gall tumor markers in transformants obtained after in vitro *Agrobacterium tumefaciens*-induced transformation of cell wall regenerating protoplasts derived from *Nicotiana tabacum*. *Proceedings of the National Academy of Sciences, USA*, **78**, 4344–8.

Wullems, G. J., Molendijk, L., Ooms, G. & Schilperoort, R. (1981*b*). Retention of tumor markers in F1 progeny plants from in vitro induced octopine and nopaline tumor tissues. *Cell*, **24**, 719–27.

Yadav, N. S., Postle, K., Saiki, R. K., Thomashow, M. F. & Chilton, M.-D. (1980). T-DNA of crown gall teratoma is covalently joined to host plant DNA. *Nature*, **287**, 458–61.

Yang, F., Montoya, A. L., Merlo, D. J., Drummond, M. H., Chilton, M.-D., Nester, E. W. & Gordon, M. P. (1980). Foreign DNA sequences in crown gall teratomas and their fate during loss of the tumorous traits. *Molecular and General Genetics*, **177**, 707–14.

Zambryski, P., Holsters, M., Kruger, K., Depicker, A., Schell, J., Van Montagu, M. & Goodman, H. M., (1980). Tumor DNA structure in plant cells transformed by *A. tumefaciens*. *Science*, **209**, 1385–91.

INDEX

abscisic acid 93
Acer pseudoplatanus, cell phenolic accumulation 13
aeration and secondary metabolites 99, 100
agar and cell immobilisation 41, 54
agarose and cell immobilisation 40
agitation and secondary metabolites 99, 100
Agrobacterium tumefaciens 313, *see also* crown gall, Ti plasmid
  genetic manipulation of plants 320
agrochemicals, plant resources 3, 5, 8
agropine 313
airlift loop vessel 21, 22, 24
airlift reactors and *Digitalis* sp. cell culture 69
ajmalicine 10
  yield and hormones 16
alginate beads and cell immobilisation 40
alkaloid production in vitro 43
amino acids, biosynthetic pathways in barley 259
anthocyanin, cell culture production and light 81
anthraquinones 10, 11, 16
apple trees
  phloroglucinol and tissue culture 145, 146
  self-rooted 150, 151
aspartate kinase, forms in barley 260
atropine 7
auxin, *see* growth regulators

Bacillus subtilis, eukaryotic gene expression 283
bacteria, eukaryotic gene expression 269–91, *see also Escherichia coli*
bacteriophage lambda
  chloroplast gene cloning 287
  chloroplast gene synthesis kinetics 290, 291
  N operon and chloroplast gene expression 289–91
  ribulose bisphosphate carboxylase gene cloning 288
barley, *see* seedling screening
batch culture 27
  costs 30

6-benzylaminopurine, *see also* growth regulators
  diosgenin production 91
  leaf senescence 93
  sapogenin production 92
Beta vulgaris, axillary shoot proliferation 130
biomass productivity
  batch culture 27
  continuous culture 27
  and vessel size 27
biotransformation
  cardiac glycosides 67–73
  cell culture 19
  immobilised cells, *Digitalis* sp. 72, 73
  β-methyldigitoxin, scaling-up 70, 71

caffeine, sources 7
calcium alginate
  cell immobilisation 54, 55
  Datura sp. cell nutrient uptake 55–7
callus systems 28, 29
  cryopreservation 208
  environment 46
  germplasm storage 207, 208
  microscopy 132
  mixaploid 131
  oil palm 147
  polyploidy 126
  secondary metabolite production 39, 75
  stable regenerating system 130–3
  tissue culture 119
  tree propagation 146–8
capsaicin production and precursor feeding 52, 98
Capsicum frutescens, flatbed and column culture 59
carbon sources, *see also individual sugars*
  depletion effects 96, 97
  and secondary metabolite yield 96–9
carboxymethylcellulose, cell immobilisation 41
cassava, *see Manihot esculenta*
Catharanthus roseus
  biomass yield 24, 27
  carbon source in culture and yield 15, 16

Catharanthus (*cont.*)
cell culture 12
culture media effects 96
fluidised beds 28, 29
gas transfer coefficient and yield 24
hormonal effects on alkaloids 16
serpentine and cell biomass 13
yield and tissue origin 14
cell culture, plant, *see also* media, tissue culture, yield
chemical environment 45, 46
commercial applications of mass 8–33
conditioning period 15
development 9. 10
environmental effects 81–5
factors affecting product yield 13–19, 75–9
instability 188
internal cultural environment 85–100
pretreatment effects 100, 101
product synthesis route 4
product yield and whole plant 10
products 9–13
screening 18
selection 17, 18
substances yielded 10, 11
transfer rates 190
cell fusion 242–5
cell shrinkage
low-temperature preservation 170–2
and survival 169–72
cell wall 20
cells, animal
Chinese hamster ovary, genetic transfer 241
optimum cooling rate variation 173
chemical gradiants, immobilised cells 45
chemicals, *see also* fine chemicals, pharmaceuticals
plant resources 3
chitosan and cell immobilisation 40
chromosome elimination and cell fusion 243
chromosomes, transfer of isolated 246, 247
Chrysanthemum morifolium, callus storage and survival 207
Citrus spp., virus elimination and juvenility 152
codeine 5, 7
collagen and cell immobilisation 41
column culture system 53–9
diagram 53
immobilisation procedure 53
light effects 57. 58
nutrient uptake in agar 55
nutrient uptake in calcium alginate 55–7
precursor feeding 57–9
continuous culture 27
costs and vessel size 30, 31
crown gall 313
plasmids 313, *see also* Ti plasmid
tumour types 314
cryobiology 163
cryopreservation 163, *see also* low-temperature preservation
callus 208
cell suspension cultures 201–3
cost 228
definition 163
germplasm storage 190, 191, 224–6
plant species stored 194–7
pollen 199, 200
potato 230
shoot-tips 211, 212
short- and long-term 213
structural damage 205, 206
cryoprotectants 191, 208
cell suspension 202, 203
dimethylsulphoxide 175, 202, 203
effects on cells 174, 175
ethanediol 176
glycerol 174, 175, 202, 203
pollen storage 199
serum 176
uses 175, 176

Datura innoxia
ornithine and nutrient uptake 56, 59
nutrient uptake in column culture 55, 56
nutrient uptake on flatbed 48, 49
deacetyllanatoside C structure 68
DEAE gels and cell immobilisation 41
dehydrofolate reductase, genetic transfer 241
2,4-dichlorophenoxyacetic acid (2,4-D)
alkaloid biosynthesis 87
anthraquinone production 87
callus culture of trees 146
effects and light reversal 82
yield and cell culture 16, 17
differential thermal analysis 175, 177, 178
differentiation
control by growth regulators 320
and production of immobilised cells 44
and yield of secondary metabolites 76
Digitalis lanata
biotransformtion of cell culture products 67–72
immobilised cells 72, 73
product analysis 67
digitoxin
source and uses 7
structure and biotransformation 68
digoxin
source and uses 5, 7
structure and biotransformation 68
dimethylsulphoxide 199
concentration and saline freezing 170
Dioscorea rotundata (white yam) 188
growth requirements 233
propagation methods 233

diosgenin 5, 7, 9, 10, 39
precursor feeding 98
synthesis and growth regulators 90, 91
DNA recombinant technology 269
objectives 269, 270
DNA vectors, plant viruses 299, 309
downstream processing 29
draught-tube airlift vessel 21
foaming 26
impellar systems 24

ECTEOLA-cellulose and cell immobilisation 41
Elaeis guineensis (oil palm)
micropropagation 147, 152
elicitors 98, 99
embryo, animal
warming rate of cryopreserved and survival 179
embryo, plant
cryopreservation 209, 210
medium modification and storage 209
survival rate of cryopreserved 210
embryogenesis in tissue culture 133, 134, 153
β-endorphin gene fusion 279
enzyme immunoassay (ELISA) 18
erythrocytes
cooling rate and survival 173
post-thaw treatment 180
Escherichia coli, see also plasmids
chloroplast gene expression 284, 285
eukaryotic gene expression structures 270, 275, 279
eukaryotic promoters 280
gene expression assessment 282, 283
gene fusions 274, 275
lambda phage expression of maize genes 289, 290
P_L source and expression 272–4
P_{lac} transcription 272, 273
P_{trp} source and expression 272, 274
plant gene expression 284–91
plasmid stabilisation 280, 281
promoter and gene expression 272
promoter structures and sequences 271, 272, 278
protein synthesis initiation 274
ribosome binding site 274
ribulose bisphosphate carboxylase expression 284–7
Ti plasmid expression 317, 319
ethylene 93

fine chemicals, value 8
flatbed culture apparatus 47–52
diagram 47
drip areas, alkaloid production 51, 52
nutrient uptake 48, 49
oxygen uptake by cells 50, 51
precursor feeding 52
scaling-up 52
suspension culture comparison 48–50
fluidised beds 28, 29
foaming, tobacco cell culture 25
Fraction 1 protein 246
Freesia sp.
callus section 132
tissue culture 129
frost hardiness 165

gas transfer coefficient (K_La) and biomass 24
gelatin and cell immobilisation 40
gene fusions
assay methods 281, 282
bacteria 274, 275, 283
expression of eukaryotic in *E. coli* 270, 275, 279
expression in vivo 282, 283
insert positioning 277
transcriptional 275, 276
translational 276, 277, 279
translation reinitiation 279, 280
gene manipulation 33
gene transfer
plant–microbe fusion 247
plant microcell fusion 246, 247
recipient stabilisation 247, 248
genetic resources, *see also* germplasm conservation and storage
conservation in vitro 219–35
reasons for preserving 219
genetic transformation 241–8
genomes, cytoplasmic
segregation on protoplast fusion 245, 246
genotype
alteration in culture 76, 77, 148, 149
assessment in tissue culture 149
growth regulator effects 125, 149
stability in culture 189, 222, 223
germplasm conservation
in vitro approaches 221–33
international distribution 227
location 220
germplasm storage 187–213, *see also* callus systems, suspension culture
anthers and pollen 198–200
clonal maintenance 221
costs 228
crops 220, 221
cryopreservation 190, 191, 224–6
culture sources 223. 224
embryos 209, 210
genetic stability 222, 223
growth limitation 190, 211, 225
in vitro systems 187
long-term 221
multiplication rates 226

germplasm storage (*cont.*)
non-cellular 189
pests and disease 226, 227
plant protoplasts 192–8
plant regeneration 224
plants stored 228–33
sample selection 189
shoot-tip cultures 211, 212, 223
uses 187, 188
gibberellic acid 93
ginseng 9, 10
glucose as carbon source 15
glycerol, phase diagram 175
growth hormone, gene fusion 279
growth limitation
plant species stored 194–7
storage time 190, 225
growth regulators, *see also individual substances*
germplasm storage 225
morphogenesis control 320
mutation 125
and secondary metabolism 86–94
tissue cultures 113, 116
tree explant culture 142
gymnosperms
rooting of shoots 144
shoot proliferation 143, 144

Haplopappus gracilis, light effects on cell culture 81
harmin 9
harvest, immobilised cells 46, 47
hormones, *see* growth regulators
hyoscyamine 7

ice nucleation and cell freezing 167, 168, 225
immobilisation methods 39–62, *see also* column culture, flatbed culture
adsorption 41, 42
agar 54
calcium alginate 54, 55
column culture system 53–5
inert substrata 40
reasons for use 42–8
substrata used 40–2
immobilised beds 28, 29
immobilised cells
cell–cell contact 44
chemical gradient 45, 46
Digitalis sp. biotransformation 72, 73
environment manipulation 46
light effects 45
phosphate limitation 60, 61
product harvest 46, 47
slow growth 43, 44
in vitro cloning, *see* micropropagation, tissue culture

indole-3-acetic acid (IAA)
nicotine synthesis 17
thebaine synthesis 86
insulin gene fusion 279
International Board for Plant Genetic Resources (IBPGR) 219
sponsorship 220
Ipomoea batatas (sweet potato)
germplasm storage 232
minimal growth storage 232, 233
virus elimination 232
isocapric acid, *Capsicum* sp. cell nutrient uptake 59

jasmine
cost 8
source and uses 5

Kalanchöe blossfeldiana, tissue culture 118
kinetin and nicotine production 89

lactiferous cells and opiate production 11, 12
lanatosides structure 68
light
alkaloid production inhibition 83
'Cool White' fluorescent 81, 82
daylength in tissue culture 113, 114
enzyme effects 81
plant development 81
polyphenol biosynthesis 82
secondary metabolite stimulation 82, 83
and *Solanum* sp. cell culture nutrient uptake 57, 58
Liquidambar styraciflua micropropagation 143, 153
low-temperature preservation 163–82, *see also* cryopreservation
assessment of viability 181, 182
cell dehydration 171
cell shrinkage and survival 169–72
cell survival, factors affecting 172–80
cold shock response 164
cold tolerance and frost hardiness 165
cooling rate 164, 169–73, 191, 204
cryoprotectants 173–6
effects 163–72
ice nucleation and solutes 167, 168
intracellular ice damage 172
manipulation of metabolism 165, 166
optimum rate and cell type 173
post-thaw treatment 180, 191, 192, 207, 212
recovery of *Chlorella* sp. and pretreatment 165, 166
solute concentration 169, 170
supercooling and survival 166–8
two-step cooling 173, 174

low-temperature preservation (*cont.*)
  vitrification 176–8
  warming injury effects 178–80, 191, 205, 207
lymphocytes, cryopreserved survival assessment 181, 182

Manihot esculenta (cassava)
  minimal growth storage conditions 231, 232
  propagation, non-tissue culture 231
  shoot-tip culture 231
marijuana 7
mass growth 20–30
  aeration and mixing 21–5
  biomass and culture systems 27, 28
  factors affecting 20–6
  foaming 25, 26
  metabolic rate 20
  microbes 20, 21
  production systems 28
  vessels 21, 22
media
  carbon sources 15, 96–8
  components and secondary metabolites 86–99
  low-temperature growth 166
  nutrients and secondary metabolites 94–6
  pH 100
  polypeptide stability 281
  production 17
  tissue culture 112, 113
  tree explant culture 142
medicinals, *see* pharmaceuticals
membranes and cell immobilisation 41
meristem tip culture 117
metal hydroxide precipitates and cell immobilisation 41
β-methyldigoxin
  product and tablet production 72
  structure 69
β-methyldigitoxin
  biotransformation reaction 69
  cell culture biotransformation 67–72
  pH and formation 71
microbeads and cell immobilisation 41
microcuttings 234
micropropagation, *see also* tissue culture
  horticultural and agricultural crops 111–34
  procedures 111
  uses 111
mixing and cell damage 22, 23
morphogenesis, *see* differentiation
mutation in tissue culture 124–7
  avoidance 125, 126
  growth regulators 125

naphthaleneacetic acid (NAA)
  anthraquinone production 16, 87
  callus culture of trees 146
  diosgenin production 90, 91
  embryoid production 147
  nicotine production 87, 88
  pretreatment and cell yield of nicotine 100, 101
  sapogenin production 92
Narcissus sp., tissue culture 119
Nicotiana rustica, callus section 132
Nicotiana tabacum
  callus storage 207
  cytoplasmic hybrids 246
  light effects and nicotine production 83
  morphogenesis control 320
  nicotine yield and callus culture 14
  octopine synthesis transfer 322, 323
  shaking rate effects and nicotine 85
  suspension culture methods 79, 80
nicotine 10
  biosynthesis and growth regulators 87, 88
  determination 80
  phosphate effects 94–6
  sucrose depletion 97
  temperature production optimum 84, 85
nitrate and nicotine production 94
nopaline
  structure 313
  synthesis location 321
nutrient uptake in column culture 55–9
nutritional stress 61

octopine
  DNA and protein coding 317, 318
  gene transmission to tobacco 322, 323
  Mendelian inheritance of transferred gene 323
  structure 313
  synthesis location 321
organ culture 3
ornithine, *Datura* sp. cell nutrient uptake 56, 59, 61
Ornithogalum thyrsoides, adventitious shoot culture 129, 130
ovalbumin gene fusion 279
oxygen, cell uptake on flatbed 50, 51

palms, callus culture 147, 148
Papaver somniferum, lactiferous cells and alkaloid synthesis 11, 12
pathogens in germplasm storage 226, 227
perfumes, product value 6
Petroselinum hortense, enzyme activity and light 81
Petunia sp.
  cytoplasmic segregation in protoplasmic function 245, 246
  somatic hybridisation 242

Petunia hybrida 118
pH of media and secondary metabolites 100
pharmaceuticals 6–8
  plant product proportion 6
  plant resources 3, 5, 39
  ten most prescribed plant 7
phenylammonia lyase, eliciting 99
phloroglucinol and apple regeneration in vitro 145, 146
phosphate
  and indole compounds in culture 96
  nicotine production 94–6
phosphodiesterase in cell culture 33
phytoalexins 98
pilocarpine source and use 7
plant breeding 251
plants
  commercial chemical resource 4–8, 39
  DNA introduction 299
  genetic transformation by microbial plasmids 247
  germplasm storage, species and methods 194–7
  resource extent 6
plasmids
  gene expression amplification 278
  lac z 277
  β-lactamase 276, 277, 279
  positioning cloned inserts 277
  stabilisation 280, 281
  Ti, *see* Ti plasmid
  Tn 7 transposon 322, 323
  trp E 277
  vector construction 277
poisons 7, 8
pollen
  embryos 199
  storage methods 198, 199
  survival rates 198, 199
polyacrylamide and cell immobilisation 40
polyploidy
  callus culture 126
  tissue culture in kale 127
polystyrene and cell immobilisation 40
polyvinylpyrrolidone in cryopreservation 167, 168
potato, *see Solanum tuberosum*
precursor feeding 52, 57–9, 98, 99
prefreezing 204
Pribnow box 272
primary metabolites, accumulation curves 78
product accumulation 28
production systems 28, 29
  Digitalis sp. cell culture 70–2
proline and stress metabolism 261
promoter genes 272
protoplasts
  enzymatically isolated 242
  fusion, chloroplast segregation 245, 246
  germplasm storage reports 192–8
  plant regeneration and potato cultivars 252, 253
  potato 251–5
  regeneration of cryopreserved 193
  variability of derived plants 253
purpureaglycoside A structure 68
pyrethrin source and uses 5, 8

quinidine, source and uses 7
quinine, source and uses 5

radioimmunoassay (RIA) 18
repressor gene, *c* I 273
reserpine 7
rheology, cell culture systems 23
ribulose bisphosphate carboxylase
  bacteriophage infection of *E. coli* and expression 289, 290
  expression in *E. coli* 285–7, 289, 290
  gene cloning in bacteriophage 287, 288
  gene orientation 284, 285
  kinetics of synthesis and lambda phage 290
  structural features, maize chloroplasts 272
RNA vectors, plant viruses 299–309
rooting, tree micropropagation 144
rosmarinic acid 98
rutacultin 19

Saintpaulia ionantha, tissue culture method 118
sapogenins, growth regulators and production 92
scaling-up, *Digitalis* sp. cell cultures 70
Schistosoma mansoni
  cooling rate and morphology 170–2
  survival and warming temperature 180
scopolamine, source and uses 5, 7
secondary metabolism
  altered 75
  characteristic curves 78
  culture vessel effects 85
  gene expression 77
  genotype alteration in culture 76, 77
  light effects in cell culture 82, 83
  temperture effects 84, 85
secondary metabolites 15
  cell culture methods 61, 62
  cultural factors affecting 75–102
  development uncoupling 11
  differentiation 42
  extranuclear control 45
  growth rate 43, 44, 79
  precursor feeding 98, 99
  product improvement 77
seed
  recalcitrant 188, 221
  screening 262

seedling screening in barley 259–62
Sequoia sempervirens, micropropagation 143, 144
serpentine 10
  and cell biomass 12, 13
  commercial production costs 31, 32
  growth regulator effects 87
  price 31
  source 10
  yield and hormones 16
sesquiterpene lactones 19
shaking rate 85
shoot-tip culture, potato 229
shoot-tip storage
  callus formation 212
  cryopreservation 211, 212
  genotype storage 223, 224
Solanum spp., minimal growth storage 230
Solanum aviculare
  steroid estimation 80
  steroid GLC separation 89–91
  steroids and growth regulators 89–92
  suspension culture 80
Solanum nigrum
  alkaloid production and flatbed drip areas 51, 52
  cell culture in suspension and on flatbed 48–51
  nutrient uptake and light 57, 58
Solanum tuberosum (potato)
  chromosomal instability 257
  cryopreservation 230
  cultivars 252
  Desirée tube skin colour 256, 257
  genetic stability 229
  germplasm distribution 231
  germplasm storage 229–31
  Maris Bard, micropropagation 257; variation 254, 255
  micropropagation methods 229
  protoplast isolation and plant regeneration 251–3
  Russet Burbank variation sources 257, 258
  shoot-tip culture 229
  tissue culture and rooting 120, 121
  variation in protoplast-derived plants 254–8
somatic hybridisation
  apparatus for heterokaryon isolation 244
  genetic transformation 241–8
  plants hybridised 243
  selection system 243
somatostatin gene fusion 279
Sparaxis tricolour shoot cultures and media 116
'stabilate' 163
stability in cell culture 188
sterilisation 111, 112
  tree explants 141
steroids, production and auxin effects on *Solanum* cells 89, 90
sucrose depletion and nicotine production 96, 97
supercooling 166, 167
suspension culture 200–7
  alkaloid production 48–51
  cryoprotectants and storage 202, 203
  freezing and thawing rates 204, 205
  plant cells stored 201
  pretreatment and storage 201, 202
  secondary metabolism 200
  species stored 205, 206
  storage methods 200–7
sweet potato, *see Ipomoea batatas*

temperature
  nicotine production in cell culture 84, 85
  tissue culture 114
thaumatin source and uses 5
thebaine 10
Theobroma cacao see storage 188
Ti plasmid 313–23
  Agrobacterium sp. 309
  DNA mechanism of action 317
  E. coli expressions 317, 319
  enzyme product 318
  gene expression 315–21
  gene product functions 320, 321
  hormone-independent growth 322
  new gene introduction 322, 323
  octopine, plant functions controlled 320, 321
  octopine synthesis, transfer to tobacco 322, 323
  opine synthesis 315
  plant cell introduction 247
  plant growth control 314
  protein coding, octopine-coding plasmid 317
  restriction map and opine synthesis 315, 316
  sterile tumour cells 314
  T_L DNA transcripts mapping 315, 316, 321
tissue culture 3, *see also* media, micropropagation
  abnormalities 124
  advantages and disadvantages 121, 122, 128
  adventitious shoots 117–21
  axillary shoots 115–17, 126
  bacterial contamination 123, 124
  bulbous species 118, 119
  callus shoots 119, 120, 130–2, 146–8
  daylength 113, 114
  developments 128–34
  disease-free plants 127
  equipment 122

tissue culture (*cont.*)
establishment of plantlets 120, 121
explant origin 112
genetic changes 124–7
growth regulators 113
inhibitory compounds 123, 128
light role 113
mass embryogenesis 133, 134
mechanisation 128, 129
organ tissue shoots 118, 119
plant breeding 127
proliferation methods 114–21
range of uses 122, 127, 128
scheme 115
species propagated 118
systems and preservation methods 192–8
trees 128, 139–54
tobacco, market value and volume 6
totipotency 117
trees, in vitro propagation 139–54
adult, shoot proliferation 143
angiosperm, shoot proliferation 143
applications 149–52
callus culture 146–8
disease elimination 152
embryoids 153
establishment 148
explant choice 141
forest trees 151
fruit tree root stocks 149, 150
genetic variability of products 148, 149
growth regulators and callus culture 146
gymnosperm, shoot proliferation 143, 144
juvenile, shoot proliferation 142, 143
juvenility 140
life-cycle 128, 139
media 142
nucellar tissue 147
oil palms 147, 152
phloroglucinol and rejuvenation 145, 146
propagation methods 138, 140
rejuvenation 144, 145
rooting 144
self-rooted trees 150, 151
shoot culture methods 140–6
sterilisation 141
trophophase–idiophase development 78

ubiquinone 10

variation
causes 254
evaluation in potato varieties 255
genetic bases 254
potato tuber skin colour 256
protoplast-derived potato plants 253–8
vector, virus 299–309, *see also* virus, plant
vincristine, source and uses 5
virus, plant
caulimovirus group members 302
caulimovirus vector; deletions and insertions 307, 308; DNA forms 307; mutagenesis effects 307; mutagenesis in vitro 305; promoter site 306; replication 306, 307; RNA mapping 305; turnip cell infection 308; virus expression and transcription 304–6
double-stranded DNA 301; DNA properties 303, 304; hosts 302; inclusion bodies 302, 303; properties 301
germplasm storage 227
nucleic acid vectors 299–309
prospects 308, 309
RNA vector scheme 300, 301
single-stranded DNA types 301
single-stranded RNA types 299–301
tissue culture and meristems 117
tissue heat treatment 227
tobacco rattle 300
tree tissue culture 152
visnagin 9
vitrification 123, 150, 176–8
barometric pressure 177
phenomenon 176, 177

water, solutes and freezing point 167

X-ray irradiation and cell fusion 243, 244

yam, *see Dioscorea rotundata*
yeast, cooling rate and survival 173
yield in cell culture
biochemical manipulation 18
biomass and oxygen supply 24
carbon source 15, 96
costs 31, 32
culture manipulation 17, 18
culture medium 14–17, 86–100
factors affecting 13–19, 75–102
mass growth 27
novel product synthesis 19, 75
scope for improved 32
tissue origin 14, 76
whole plant comparison 10

Zea mays
cryopreservation of cells and cryoprotectants 202, 203
gene transfer to bacteria 271, 284–7
large subunit (LS) RuBP gene DNA sequence 285–7